ADVANCES IN
BIOSENSORS

Volume 4 • 1999

BIOSENSORS: A
CHINESE PERSPECTIVE

ADVANCES IN BIOSENSORS

BIOSENSORS: A CHINESE PERSPECTIVE

生物傳感器

Editors: ANTHONY P.F. TURNER
Cranfield Biotechnology Centre
Cranfield University
Bedfordshire, England

REINHARD RENNEBERG
Department of Chemistry
The Hong Kong University of Science and Technology

VOLUME 4 • 1999

100 Prospect Street
Stamford, Connecticut 06904

ISBN: 1-7623-0073-6

ISSN: 1061-8945

Printed and bound in the United Kingdom
Transferred to Digital Printing, 2011

CONTENTS

LIST OF CONTRIBUTORS

Xiao Caide	Department of Biological Sciences and Biotechnology Tsinghua University Beijing, China
Chiyui Chan	Department of Chemistry The Hong Kong University of Science and Technology Hong Kong
Hing Leung Chan	Department of Biology and Chemistry City University of Hong Kong Hong Kong
Shong Cheng	Department of Chemistry The Hong Kong University of Science and Technology Hong Kong
Albert Chu	EY Laboratories Hong Kong
Xia Chu	College of Chemistry and Chemical Engineering Hunan University Changsha, China
Jiaqi Deng	Department of Chemistry Fudan University Shanghai, China
Qing Deng	Changchun Institute of Applied Sciences Chinese Academy of Sciences Changchun, China

Feng Derong Biology Institute
Shandong Academy of Sciences
Jinan, China

Shaojun Dong Changchun Institute of Applied Sciences
Chinese Academy of Sciences
Changchun, China

Yuzhi Fang Department of Chemistry
East China Normal University
Shanghai, China

Manliang Feng Department of Chemistry
Shaanxi Normal University
Shaanxi, China

Bixia Ge Shanghai Research Institute of
Industrial Microbiology
Shanghai, China

Jan F.C. Glatz Department of Physiology
Cardiovascular Research Institute
Maastricht, The Netherlands

Bernd Gründig SensLab Ltd.
Leipzig, Germany

Pingang He Department of Chemistry
East China Normal University
Shanghai, China

Jun Hu Shanghai Research Institute of Industrial
Microbiology
Shanghai, China

Litong Jin Department of Chemistry
East China Normal University
Shanghai, China

Xie Jun Department of Life Science and Biomedical
Engineering
Zhejiang University
Zhejiang, China

Wilhelmina A. Kaptein	Department of Chemistry University of Groningen Groningen, The Netherlands
Gotthard Kunze	Institute of Plant Genetics and Crop Plant Research Gatersleben, Germany
Wai-Kuen Kwong	Department of Chemistry The Hong Kong University of Science and Technology Hong Kong
Gao Xiang Li	Institute of Microbiology Chinese Academy of Sciences Beijing, China
Junfeng Liang	Department of Biological Sciences and Biotechnology Tsinghua University Beijing, China
Jian Guo Liu	Institute of Microbiology Chinese Academy of Sciences Beijing, China
Calum J. McNeil	Department of Chemistry University of Newcastle-upon-Tyne Newcastle, England
Wang Ping	Department of Life Science and Biomedical Engineering Zhejiang University Zhejiang, China
Zhang Qintao	Department of Life Science and Biomedical Engineering Zhejiang University Zhejiang, China
Li-Qiang Ren	Department of Biology and Chemistry City University of Hong Kong Hong Kong

Reinhard Renneberg
Department of Chemistry
The Hong Kong University of Science and Technology
Hong Kong

Klaus Riedel
Dr. Bruno Lange Gmbh Berlin
Dusseldorf, Germany

Judith Rishpon
Department of Chemistry
University of Tel Aviv
Tel Aviv, Israel

Li Rong
Department of Life Science and Biomedical Engineering
Zhejiang University
Zhejiang, China

John Sanderson
Cardiology Unit
Prince of Wales Hospital
Hong Kong

Guo-Li Shen
College of Chemistry and Chemical Engineering
Hunan University
Changsha, China

Sen-fang Sui
Department of Biological Sciences and Biotechnology
Tsinghua University
Beijing, China

Pui Yan Tsoi
Department of Biology and Chemistry
City University of Hong Kong
Hong Kong

Wenzhang Xie
Department of Biological Sciences and Biotechnology
Tsinghua University
Beijing, China

Mengsu Yang
Department of Biology and Chemistry
City University of Hong Kong
Hong Kong

Jiannong Ye
Department of Chemistry
East China Normal University
Shanghai, China

Tan Yi
Department of Life Science and Biomedical Engineering
Zhejiang University
Zhejiang, China

Wang Yongqiang
Department of Life Science and Biomedical Engineering
Zhejiang University
Zhejiang, China

Ru-Qin Yu
College of Chemistry and Chemical Engineering
Hunan University
Changsha, China

Guo-Xiong Zhang
Shanghai Institute of Metallurgy
Shanghai, China

Xian-En Zhang
Wuhan Institute of Virology
Chinese Academy of Sciences
Wuhan, China

Zhujun Zhang
Department of Chemistry
Shaanxi Normal University
Shaanxi, China

Zong-Rang Zhang
Shanghai Teachers University
Shanghai, China

Yue Zhou
Department of Biological Sciences and Biotechnology
Tsinghua University
Beijing, China

Jian-Zhong Zhu
Shanghai Institute of Metallurgy
Shanghai, China

ACKNOWLEDGMENTS

In exploring developments in China, the editors had the initial problem of making proper contacts. Fortunately, Academician Professor Erkang Wang and Professor Shaojun Dong from Changchun and Professor Xian-En Zhang from Wuhan helped us greatly early on. Professor Wang was a Visiting Professor in Hong Kong at the right time to help considerably with this book. We had the chance to benefit not only from his sharp mind but also his wisdom. The choices we made were by nature subjective and we apologize if some important groups are missing.

RR thanks the Head of his Department and Chairman of the HK Chapter of the American Chemical Society, Professor Nai-Teng Yu, for bringing him to Hong Kong, and even more importantly, for his continuous mental and material support of all his activities in China including this book.

The Chinese title of this book has quite a history: Mrs. Min-min Chang, Director of the world's most modern library at HKUST, organized an exhibition of Chinese calligraphy for the library. RR was amazed by this art and asked Min-min's help to find an artist who could write "biosensors" in Chinese characters. He did not realize, however, that there is plenty of room for variation hidden in the Chinese characters. This means there can be different translations of "biosensors" in Chinese. To make a final decision, the precious time of Mrs. Chang's husband, Professor H.K. Chang, President of City University of Hong Kong, who is himself a biomedical engineer, was used to discuss various meanings. When RR finally received the artist's calligraphy draft he showed this to his Chinese colleagues who praised the artist over and over. When we asked Min-Min, however, how to pay the artist, we were told that this calligraphy was the gift of Professor Y.S. Wong, now City University's Vice-President.

The Editors wish to thank Camilla Erskine (now back in Cambridge) and Virginia Unkefer of the HK University of Science and Technology (Office of Contract & Grant Administration) for first-round English corrections and help, and Elaine Sarney for her endless patience in handling manuscripts at the Cranfield end.

Financial support from the Hong Kong Croucher Foundation for a meeting of some of the contributors to this book is also gratefully acknowledged.

The making of this book has also had a wonderful "side effect". We learned a lot more about the different biosensor groups and met many of their representatives. Now RR has more visits to the Chinese mainland on his agenda. We wish to thank all the authors for their cooperation, friendship, patience, and support! Finally, RR would like to thank his wife Ilka for her understanding and help in the most difficult, challenging and rewarding time in his life, and APFT would like to ask his wife and family's forgiveness for being a workaholic.

APFT
RR

PREFACE

Biosensors have captured the imagination of the world's scientific and commercial communities by combining the multidisciplinary skills of biologists, physicists, chemists, and engineers to provide innovative solutions to analytical problems. Biosensors are applicable to clinical diagnostics, food analysis, cell culture monitoring, environmental control, and various military situations. There is ever increasing demand for rapid and convenient analyses offered by this fusion of biology and electronics which mimics our principal concern: the effect of materials and environments on living systems. Biosensors are defined as analytical devices incorporating a biological material (e.g. tissue, microorganisms, organelles, cell receptors, enzymes, antibodies, nucleic acids, etc.), a biologically derived material, or biomimic intimately associated with or integrated within a physicochemical transducer or transducing microsystem, which may be optical, electrochemical, thermometric, piezoelectric, or magnetic. Biosensors usually yield a digital electronic signal which is proportional to the concentration of a specific analyte or group of analytes. While the signal may in principle be continuous, devices can be configured to yield single measurements to meet specific market requirements.

This series, *Advances in Biosensors*, presents a unique compendium of research-level publications that do not have a place in conventional journals, but have an increasingly important role to play in completing the primary research literature and offering a more incisive alternative to the full blown, exhaustive review article. In this second volume of a sub-series dedicated to regional activities, eminent authorities from China in the field of biosensors provide an up-to-date overview of their laboratory's contribution, summarizing the primary research as it has appeared, possibly scattered, in the journal and conference literature and reflecting on their findings. This produces an innovative synthesis of such smaller research efforts into an overall perspective on the topic, which is difficult for the reader to glean from the multifarious original publications often available only in Chinese. There is latitude for the inclusion of detail that may have been excised from the original publication and for speculation on future possibilities. The net result is intense, yet highly readable accounts of the state of the art at this leading edge of analytical technology in this key region of the world.

China, the Middle Kingdom, is a country with a wonderful history and amazing perspectives. The challenges for introducing novel bioanalytical tools to China are exciting and demanding: Imagine 1,300,000,000 people deserving not only proper diagnostic help and protection from disease, but also healthy food and clean water! China is not the only country in need of these technologies; the whole region of the Asian-Pacific Rim is waiting for solutions to these problems. This is both a gold mine for science and technology and a human challenge too. The creation of this book became an adventure far beyond simply compiling chapters from different scientific groups. It is the first book in the English language summarizing biosensor and bioelectronics efforts in China. We have included groups from Changchun, Beijing, Hangzhou, Xian, Jinan, Shanghai, Changsha, Hong Kong, and Wuhan.

Anthony P.F. Turner
Reinhard Renneberg
Editors

AMPEROMETRIC ENZYME ELECTRODES BASED ON CHEMICALLY MODIFIED ELECTRODES

Shaojun Dong and Qing Deng

OUTLINE

Advances in Biosensors
Volume 4, pages 1–39.

ISBN: 0-7623-0073-6

ABSTRACT

The state of and prospects for developing amperometric enzyme electrodes based on chemically modified electrodes (CMEs) are described. The important role played by CMEs in constructing these enzyme electrodes and the mechanisms of electrocatalysis are emphasized. Initial results in the field of organic-phase enzyme electrodes and preliminary studies on artificial enzymes are also discussed.

1. INTRODUCTION

Biosensors continue to arouse much attention in biochemistry and analytical chemistry. Although the routine use of biosensors has yet to be fully accepted, they have the potential to revolutionize analytical methodology and offer the promise of real-time analyses, which is particularly important for rapidly measuring body analytes.

Current research effort on biosensors is proceeding by using many different types of sensor principles. Electrochemical sensors, especially amperometric biosensors, hold a leading position among the systems presently available. Many significant developments have been realized, notably introducing surface modification in enzyme electrodes and the miniaturizing transducers. To tailor the electrode surface to meet the needs of the analytical systems in question, chemically modified electrodes (CMEs) can play an important role as substrates in enhancing sensitivity, selectivity, and stability of amperometric biosensors [1]. This paper is a review of the work done by our laboratory on amperometric enzyme electrodes based on CMEs.

2. FIRST-GENERATION ENZYME ELECTRODES

2.1 Oxygen Probe

The success of the Clark oxygen electrode as a reproducible probe for measuring oxygen led to the original concept of the first generation of amperometric enzyme electrodes, which is based on direct electrochemical detection of substrates or products of an enzymatic reaction. In the case of oxidases, oxygen and hydrogen peroxide are the substrate and product, respectively, based on this enzyme-catalyzed (EC) reaction:

$$\text{glucose} + O_2 \xrightarrow{\text{glucose oxidase}} \text{gluconic acid} + H_2O_2 \tag{1}$$

Amperometric detection of oxygen consumption at negative potentials (–0.6 V versus Ag/AgCl) is the simplest procedure. However, the reduction of hydrogen peroxide (H_2O_2) at similar potentials causes interference and the dialysis membrane used in a Clark oxygen electrode increases the response time [2]. The electrocatalytic activity of metalloporphyrin-modified glassy carbon electrodes toward oxygen reduction were studied systematically in our laboratory [3]. Accordingly, we designed and fabricated a series of enzyme glucose electrodes that rely on catalytic monitoring of oxygen consumption. These approaches greatly decrease the overpotential and increase the sensitivity of detection based on EC electrocatalysis. We chose the electrocatalyst cobalt tetraphenyl porphyrin (CoTPP) because of its higher catalytic activity in oxygen reduction and stability. Reactions (2) and (3) correspond to the reduction of O_2 by CoTPP.

$$\text{Co(III)TPP} + e \longrightarrow \text{Co(II)TPP} \tag{2}$$

$$\text{Co(II)TPP} + \tfrac{1}{2}O_2 + H^+ \longrightarrow \text{Co(III)TPP} + \tfrac{1}{2}H_2O_2 \tag{3}$$

CoTPP was simply adsorbed on a cleaned glassy carbon electrode [4]. However, the stability of the sensor was affected by the lifetime of the CoTPP adsorbed on the base electrode. Dong et al. [5] reported that the CoTPP-modified glassy carbon electrode exhibited good electrocatalytic activity and excellent stability after treatment at 750 °C for 1 h. The lifetime of the enzyme electrode based on this heat-treated CoTPP CME was hardly affected by thousands of continuous measurements or by storage over a period of three months. The operating potential was –300 mV versus Ag/AgCl [6]. Clearly, the overpotential was significantly decreased.

Many attempts have been made to miniaturize biosensors [7]. These are important not only for developing compact high performance autoanalyzing systems and also as sensing devices for artificial organs, especially for *in vivo* applications. More recently, microelectrodes and microelectrode arrays have gained attention in biosensor technology for use in miniaturization [8]. Microelectrodes show rapid mass transfer, small IR drop, and large signal-to-noise ratios. We found that the CoTPP-modified carbon-fiber microelectrode also showed catalytic activity in the electroreduction of oxygen at a negative potential of –400 mV versus Ag/AgCl [9]. The response time was less than 4 s. The microglucose sensor prepared was very stable and may be available for *in vivo* analysis. A carbon-fiber microelectrode is easily broken. Therefore, we prepared a CoTPP-modified microdisk-array, carbon-fiber glucose microenzyme sensor [10]. The characteristics of a microelectrode array are similar to those of a microelectrode. Furthermore, the sensitivity can be enhanced to almost three orders of magnitude, because the array is made of 1000 filaments of carbon fibers.

To reduce the overpotential of O_2 reduction and prevent the interference of electroactive anions, a water soluble catalyst, tetrakis (4-*N*-methylpyridyl) porphyrin Cobalt(III), was incorporated into Nafion, a cation-exchange polymer, to modify the microdisk-array electrode [11]. Figure 1 illustrates the cyclic voltammograms of the microenzyme electrode in a 0.1 M phosphate buffer (pH 7.0) at different glucose concentrations. From curves c to a, an obvious decrease reflects the consumption of O_2 by the enzymatic reaction of glucose oxidase, which varies linearly with increasing concentration of glucose.

2.2 Hydrogen Peroxide Electrode

The catalytic reaction of oxidases yields hydrogen peroxide (H_2O_2) as a by-product. This reaction can also be amperometrically monitored with a solid disc-electrode, such as platinum, gold, or glassy carbon, by oxidizing it at anodic potentials (> 0.6 V versus Ag/AgCl). These systems are more useful than oxygen probe-based devices because of their high sensitivity. Only one problem arises. All biological

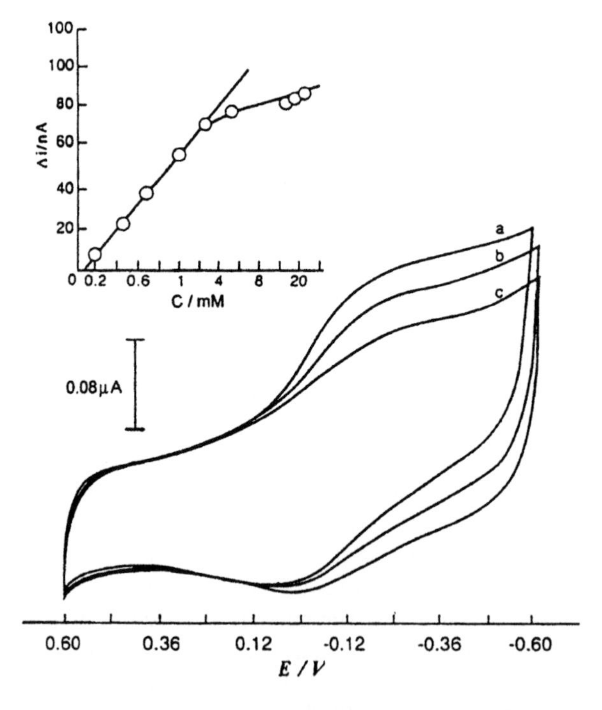

Figure 1. Cyclic voltammograms of a microenzyme electrode containing glucose oxidase prepared in 0.1 M phosphate buffer (pH 7.9) at a scan rate of 100 mv s^{-1}. (**a**) for blank solution; (**b**) 0.5 mM glucose; and (**c**) 1.0 mM glucose. Inset: Relationship of decrease in current due to reduction of oxygen to glucose concentration.

liquids contain reductants, such as ascorbate, uric acid, acetaminophenol, and bilirubin. Such reductants are oxidized at similar potentials and produce anodic current noise that interferes with analyte detection by biosensors. CMEs provide an effective means for the electrocatalytic oxidation of H_2O_2. Moreover, a polymer-film-modified-electrode greatly increases selectivity [12]. Cation-exchange polymer films, such as Nafion and Eastman Kodak AQ, can act as a membrane barrier, avoiding interference from electroactive anions in the application of sensors. Another kind of nonconducting polymer, poly(1,2-diaminobenzene) film, has permselectivity that allows the passage of H_2O_2, but prevents interference from reaching the electrode surface [13].

All electrodes covered with sputtered metal particles are much more efficient than the corresponding pure solid metal, even when careful electrochemical pretreatment has been applied [14]. Therefore, we electrodeposited a Pt particle layer on an AQ modified electrode [15]. The glucose oxidase (GOD) was cross-linked in the Pt/AQ CME. Galactose is also detected by using a similar strategy [16]. CMEs increased the sensitivity and selectivity. However, electrodeposited Pt cannot decrease the overvoltage. It was found that both the deposition of sputtered Pd and the mixture of sputtered Pd-Au on a carbon electrode results in a several hundred millivolts decrease of the large overvoltages for the electrochemical oxidation of H_2O_2 [17]. When appropriate apparatus is not available, electrodeposition is convenient. We found that electrodeposition of Pd on a carbon electrode decreases the overvoltage up to 300 mV. We cross-linked cholesterol oxidase with glutaraldehyde outside the electrodeposited Pd particle layer to prepare a biosensor [18]. A linear relationship was obtained in the range 0.05 to 4.5 mM using this Pd particle-modified cholesterol sensor.

In glucose sensor devices, GOD is usually immobilized by covalent cross-linking with glutaraldehyde and bovine serum albumin on the electrode surface. Recently, many new methods of enzyme immobilization have been developed, including the use of silicon dioxide [19] and the entrapment of GOD within electropolymerized films [20,21]. We developed a simple fabrication method based on the codeposition of Pd particles and GOD on a glassy carbon electrode (GCE) surface [22]. If deposited with Pd particles, the area of GCE surface is greatly increased by aggregates of fine Pd particles. When codeposited with Pd, GOD is strongly adsorbed on the "micropockets" formed by aggregates of Pd particles.

Typical cyclic voltammograms of H_2O_2 at the GCE and Pd/GCE substrate electrodes are shown in Figures 2(A) and 2(B), respectively. At the bare GCE, the oxidation of H_2O_2 is observed only above +0.7 V, and the current is small. In contrast, at the Pd/GCE current exhibits an anodic response to H_2O_2 starting at +0.2 V and reaching a maximum value at +0.7 V. The Pd deposition substantially decreases the overvoltage by about 500 mV, and increases the oxidation current more than 1000-fold. When GOD is present on the electrode surface, no influence on the electrocatalytic oxidation of H_2O_2 is observed, but the anodic peak shifts positively by about 50 mV (Figure 2C), which is due to the slight increase in

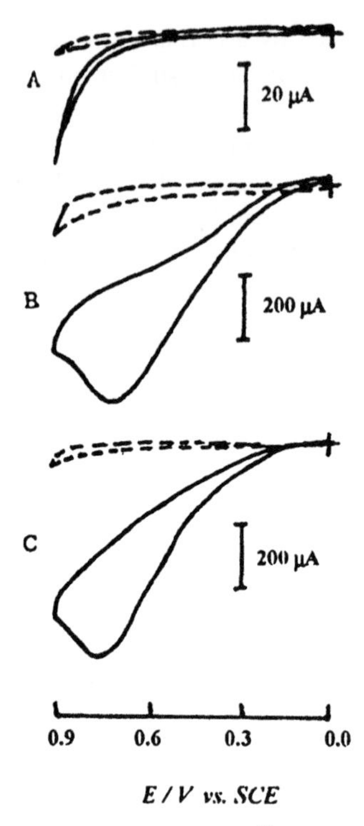

Figure 2. Cyclic voltammograms for 2.5×10^{-2} M hydrogen peroxide at (**A**) bare GCE, (**B**) Pd/GCE, and (**C**) GOD-Pd/GCE. The dashed lines indicate the response of the blank solution. Scan rate: 50 mV s^{-1}, electrolyte: 0.1M phosphate buffer (Ph 7.2).

resistance of the GOD film. Initial efforts to use the GOD-Pd/GCE in a flow-injection analysis system failed, mainly because GOD gradually leaves the surface of the modified electrode in a flow system. However, when a thin film of Nafion is coated on the outmost surface, the loss of GOD from the electrode surface is prevented effectively. The resulting sensor can be successfully applied to determine glucose in the range of 0.001–8 mM. Ascorbic acid or uric acid does not observably interfere with the response at +0.3 V because Nafion film is used.

Expensive noble metals increase the cost of biosensors. In addition, many low molecular mass substances are specifically absorbed on a Pt or Pd electrode, which significantly affects their electrochemical activity and dramatically reduces the biosensor response. The electrooxidation of H_2O_2 is suggested on the basis of the axial ligand exchange of protoporphyrin cobalt(II) and H_2O_2 as in this scheme [23]:

$$PCo(II)OH_2 + H_2O_2 \rightleftharpoons PCo(II)O_2H_2 + H_2O$$

$$PCo(II)O_2H_2 \xrightarrow{-2e} PCo(II)O_2 + 2H^+$$

$$PCo(II)O_2 + H_2O \rightleftharpoons PCo(II)OH_2 + O_2$$

The highest catalytic response was obtained at +0.6 V in a pH 7.0 phosphate buffer. A cholesterol sensor was constructed on the basis of this high electrocatalytic CME. A layer of poly(*o*-phenylenediamine) film was electropolymerized out of the enzyme film to avoid interference and fouling [24].

Above all, CMEs can reduce the overpotential and enhance the sensitivity of H_2O_2 oxidation effectively. However, the influence on the signal from other reductants is still a problem and the costs of noble metal and protoporphyrin are relatively high. On the other hand, during the last 15 years, inorganic compound-based CMEs have attracted much attention because of their potential use in electrocatalysis, electrochromism, and analysis. These inorganic substances include mainly iso- and heteropoly acids [25,26], zeolites, [27], and polynuclear transition metal hexacynometallates [28,29]. In particular, the latter group has become an important class of insoluble mixed-valent compounds. Among them, Prussian Blue (PB) is still the most important. The PB-modified electrode has extreme stability, well-defined redox transitions, and related color-switching properties. Hence, interest in using it in sensors, electrochromic displays, and various fundamental studies has been increasing. PB is easily deposited electrochemically on an electrode surface and produces a dense redox active layer. A completely reduced state (Prussian White) reduces both O_2 and H_2O_2, but the corresponding half-wave potentials are different. Therefore, one can try to detect H_2O_2 by electroreduction even in the presence of O_2. The completely oxidized state of Prussian Blue (Berlin Green) oxidizes H_2O_2. Thus, the H_2O_2 can also be detected by electroxidation at Prussian Blue-modified electrodes.

For the first time we found that GOD or amino oxidase can be incorporated into a PB film during its electrochemical growth process [30]. The glucose sensor was prepared in an unstirred potassium phosphate buffer solution containing 2 mM $K_3Fe(CN)_6$, 2 mM $Fe_2(SO_4)_3$, and GOD (10 mg mL^{-1}) by successive cyclic scanning. Figure 3 shows cyclic voltammograms of H_2O_2 oxidation at basal pyrolytic graphite (BPG) (A) and at PB/BPG (B) electrodes. At the bare electrode, H_2O_2 starts to oxidize only above +0.7 V and to reduce below –0.1 V. No redox peaks can be seen within the applied potential. In contrast, at the PB-modified electrode anodic and cathodic responses to H_2O_2 start to occur at +0.6 and +0.3 V, respectively. Moreover, well-defined redox peaks can be observed. There are marked decrease in the overvoltage and a large increase in the redox currents at the PB film electrode.

Figure 3 also demonstrates that electrocatalytic reduction of H_2O_2 is more effective. When the potential scan is limited to a narrower range, the corresponding cyclic voltammograms indicate that the two pairs of redox peaks of PB film catalyze both the oxidation and reduction of H_2O_2 independently. The redox peak at 0.85 V electrocatalyzes the oxidation of H_2O_2, whereas the redox peak at +0.2 V catalyzes the reduction of H_2O_2. At a more negative operating potential (–0.25 V), glucose

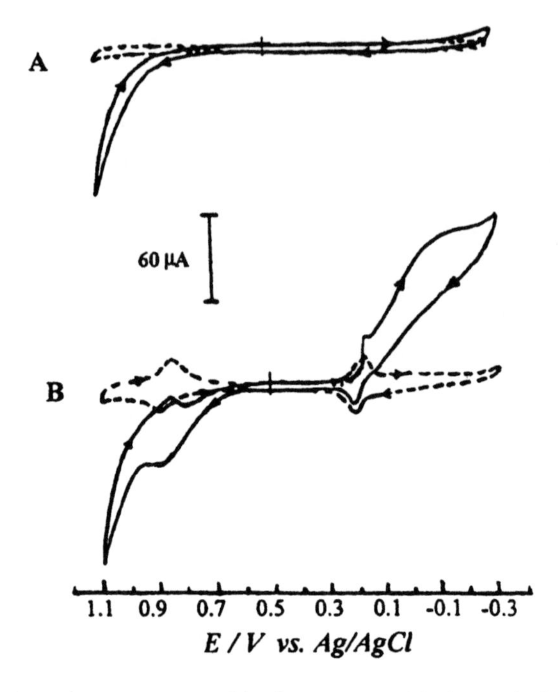

Figure 3. Cyclic voltammograms of hydrogen peroxide (3 mM) obtained with (**A**) BPG and (**B**) PB/BPG in phosphate buffered potassium salt (pH 6.4). Scan rate: 40 mV s^{-1}. The dashed lines indicate blank responses.

is amperometrically detected directly at PB/CME [31]. However, the PB/BPG electrode has only a negligible response to glucose at –0.05 V. Therefore, the enzymatically produced H_2O_2 is catalytically oxidized or reduced by the PB film. Meanwhile, in turn the PB film itself is electrochemically reduced or oxidized under the applied potentials. However, we found that the stability of such enzyme electrodes is limited by the detachment of enzyme from the PB film. About 40% of the initial response remains after four days. When a thin film of Nafion is coated on the outmost surface of GOD-PB/BPG electrode as Nafion/ GOD-PB/BPG, the stability clearly improves. Moreover, the interference is eliminated because of the low operating potential and Nafion coverage.

Many other first-generation biosensors based on enzymes other than oxidases, such as acetylcholinestease (AChE) and urease, have been reported. Because of the widespread concern about the health and environmental conditions surrounding humans, researchers are showing increasing interest in the AChE enzyme as a result of its important role in metabolizing the neurotransmitter and its blockage by some inhibitors (organo-phosphate and carbamate pesticides). Therefore, it is necessary to fabricate a new system to detect AChE activity. Amperometric detection of the acetyl- or butyryl-thiocholine (ATCh or BTCh) hydrolysis process catalyzed by AChE is the most commonly adopted strategy. The reaction scheme is as follows [32]:

$$CH_3COSCH_2CH_2N^+(CH_3)_3Cl^- + H_2O \xrightarrow{\text{A ChE}} CH_3COOH + HSCH_2CH_2N^+(CH_3)_3Cl^-$$

The products, thiocholine chloride or iodide, are electroactive. Their oxidation current can be recorded at a potential of +0.7 to +1.0 V at a glassy carbon electrode. The potential is too high and susceptible to interferences.

We found that CoTPP could be used as a good catalyst for oxidizing thiocholine [33]. Figure 4 shows the cyclic voltammograms of the AChE-CoTPP/GC electrode in a solution that contains 0 or 0.8 mM ATCh. When the potential scan is initiated from 0 V to +0.7 V, no current response occurs in a blank solution. When the enzyme

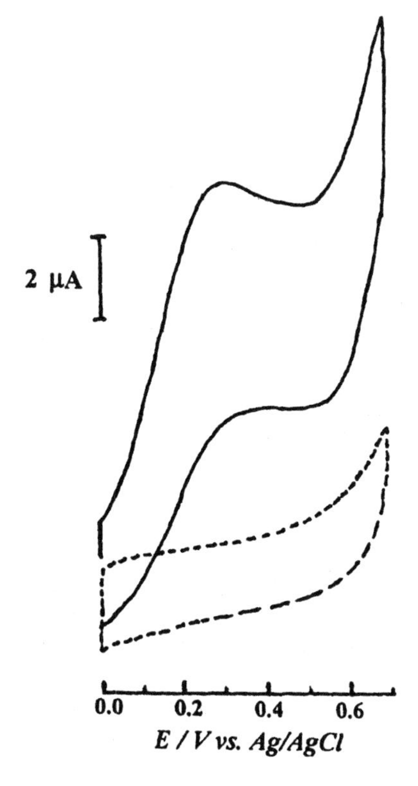

Figure 4. Cyclic voltammograms of the AChE-CoTPP/GC electrode in solution (pH 7.8) in the absence (dashed line) and presence (solid line) of 0.8 mM ATCh. Scan rate: 50 mV s^{-1}.

electrode is dipped into 0.8 mM ATCh solution for 10 minutes, a catalytic current is clearly obtained. The peak potential (E_p) is at +0.25 V. The peak current varies with different scan rates and ATCh concentrations, whereas the E_p value remains constant. In the concentration range from 5.0×10^{-6} to 8.0×10^{-4} M, the steady-state current increases linearly with ATCh concentration.

This system can be applied in detecting organophosphate and carbamate pesticides [34]. Measurements were carried out with acetylthiocholine as a substrate. Enzymatically produced thiocholine is oxidized at +250 mV on the CoTPP-modified electrode. The decreasing rate of the substrate steady-state current after the addition of pesticide was used for evaluation. Detection limits of 4.4, 52.1, and 11.1 $\mu g\ l^{-1}$ for dichlorvos, methylparathion, and carbofuran, respectively, were achieved within 5 minutes of inhibition time.

3. SECOND-GENERATION ENZYME ELECTRODE

Although the first-generation enzyme electrode that uses oxygen as a receptor is widely studied and applied to practical usage, problems arise from oxygen because of its limited solubility and fluctuation with surroundings.

The use of an alternative redox couple, such as an electron acceptor for shuttling electrons to the electrode surface, thereby conquering some of the problems of oxygen and hydrogen peroxide monitoring, is well established now.

Second-generation enzyme electrodes use relatively low molecular mass redox couples as alternatives to oxygen. A good mediator should [35] (1) be absorbed on the electrode surface and retained on it, (2) exhibit reversible kinetics, (3) react rapidly with the reduced enzyme, (4) have a low oxidizing potential and be pH independent, (5) be stable in both oxidized and reduced forms, and be unreactive toward oxygen, and (6) be nontoxic.

The potential at which these electrodes operate is only slightly positive of the formal potential of the mediator, and a costly noble metal electrode is not required for the reaction. Thus, spurious current due to competing species may be reduced. Nevertheless, in an oxygen-containing medium, there is competition between the oxidized form of the mediator and oxygen for the reduced form of the enzyme. Thus, the response current is independent of the oxygen concentration only insofar as the mediator competes effectively with O_2.

3.1 Oxidases-Mediator Systems

The most important mediator for oxidases is ferrocene (Fc) and its derivatives [36,37]. However, it has been found that ferrocene derivatives are not adsorbed very stably onto the electrode surface. In particular, the oxidized forms of ferrocenes are soluble in aqueous solution, indicating that the enzyme electrode is unstable for continuous application. To improve attachment to the electrode surface, a few ferrocene derivatives that have specific functional radicals have been synthesized, but the preparative methods are complicated [38]. We immobilized Fc with no substitutions by Nafion coating onto a glassy carbon electrode surface for use as an electron transfer mediator between the electrode and GOD [39,40]. Based on our previous work [41] in which an Fc-Nafion-modified electrode allows more reversible electrochemical reactions of the redox couple Fc^+/Fc, a sensor was constructed by immobilizing GOD on the Fc-Nafion electrode. Both the hydrophilic and hydrophobic domains in the Nafion structure contribute to keeping Fc/Fc^+ forms inside the film, which greatly improves the stability of the enzyme electrode.

Figure 5 shows the cyclic voltammograms of a freshly prepared Fc-Nafion-modified electrode immersed in a phosphate buffer (pH 6.90) during potential scanning between –0.6 V and –0.2 V versus SCE. In the first cycle of the anodic process, two peaks appear, then become one during subsequent scans. The anodic and cathodic currents rise continuously with potential scans, then reach a steady state after a few minutes. The cyclic voltammograms obtained in a phosphate buffer at the Fc-Nafion-modified electrode show that the electrochemical reaction of Fc in the film is diffusion-controlled. No catalytic oxidation current was observed at the Fc-Nafion-modified electrode when glucose was added to the phosphate buffer.

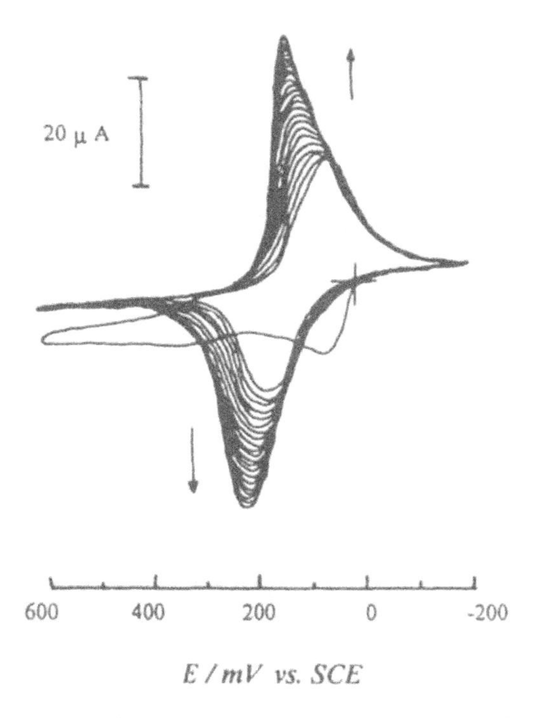

Figure 5. Cyclic voltammograms of a newly prepared Fc-Nafion-modified GC electrode in 0.1M phosphate buffer (pH 6.9). Coating film: 2 μl Nafion(1%)+2 μl 0.05 M Fc. Scan rate: 50 mV s^{-1}.

At the GOD-Fc-Nafion enzyme electrode, however, the anodic current increases rapidly, whereas the cathodic current reduces simultaneously with the addition of the glucose until the latter disappears, indicating that the catalytic reaction occurs at the enzyme electrode. This sensor showed good repeatability and a fast current response (< 20 s) to the variations of glucose concentration in a linear range up to 16 mM. The interference from some electroactive species can be efficiently limited at a lower applied potential (+0.25 V) and by using Nafion film. Another cationic exchange polymer, Eastman Kodak AQ, can also be used to achieve an effect similar to Nafion in biosensors [42,43].

Creating cheap, disposable, and highly stable biosensors is one goal of researchers in biosensors. In our laboratory a disposable biosensor was fabricated, as described schematically in Figure 6. The assembled disposable enzyme electrode is a tablet-like thin sheet. A metallic sheet was used as a substrate and covered with a mesh-print graphite layer and a mediator membrane. An enzyme layer was

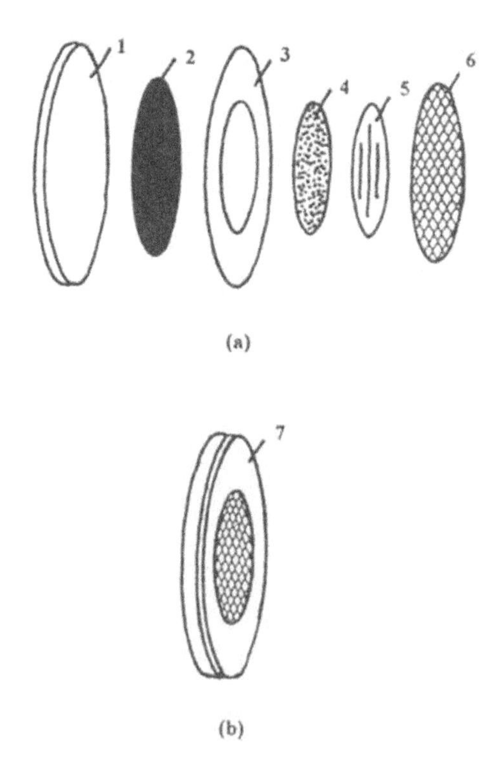

Figure 6. Construction of the disposable biosensor (**a**) and an assembled biosensor (**b**): 1, metallic substrate; 2, graphite layer; 3, isolating layer; 4, mediator; 5, immobilized enzyme membrane; 6, anti-interference layer; 7, the assembled disposable enzyme electrode.

assembled on this chemically modified base electrode, and finally an anti-interference and anti-fouling film covered the electrode. If the enzyme layer is GOD, the disposable sensor can be used to measure 0.5 to 25 mM glucose [44].

However, some oxidases do not react with the commonly used mediators. One of them is D-Amino acid oxidase, which cannot react directly with ferrocene. In this case, a double enzyme system was adopted [45]. Eastman Kodak AQ was used to immobilize two enzymes (D-amino acid oxidase and horseradish peroxidase) and 1,1′-bis(hydroxy ethyl)ferrocene (BHFc) and exhibited anti-interference characteristics. The basis for this system may be represented by these reactions:

$$\text{D-amino acid} + O_2 \xrightarrow{\text{D-amino acid oxidase}} H_2O_2 + NH_3 + \alpha\text{-Ketonic acid}$$

$$H_2O_2 + \text{BHFc} \xrightarrow{\text{horseradish peroxidase}} H_2O + \text{BHFc}^+$$

$$\text{BHFc}^+ + e \xrightarrow{+0.18\ \text{V(vs. SCE)}} \text{BHFc}$$

3.2 Dehydrogenase-Mediator Systems

More than 300 pyridine nucleotide-dependent redox dehydrogenases are known, but the number of publications on amperometric biosensors based on the use of dehydrogenase is much less than those based on other enzymes. One reason is the requirement of the soluble, low mass cofactor, $NAD(P)^+$, which makes constructing a practical device more complicated. Another reason is that the electrochemistry of reduced forms of the cofactor(NAD(P)H) is irreversible.

Electrocatalytic Oxidation of NADH at CMEs

The electrochemical reactions of pyridine coenzymes at metallic or carbonaceous electrodes are highly irreversible and take place at large overpotentials with the occurrence of side reactions. They are also complicated by adsorption (fouling) of cofactor-related products. Amperometric detection of NAD^+ will always suffer from the simultaneous reduction of oxygen because of the low reduction potential of NAD^+ and complications by dimer formation. Therefore, the electrochemical oxidation of NADH is the commonly adopted strategy for developing amperometric biosensors. Unfortunately, the direct oxidation of NADH at many bare electrodes is complex and requires a large overvoltage, such as 1.1 V at a carbon electrode [46] and 1.3 V at a platinum electrode [47]. This results in more easily oxidizable species that interfere with amperometric NADH detection in biological samples. Moreover, the adsorbed molecules of NAD^+ cause electrode fouling at NADH concentrations above 0.1 M. The electrocatalytic oxidation of NADH is an effective way to avoid these problems.

Ferrocene derivatives are the most effective mediator in oxidase systems. However, they are not as effective in the electrocatalytic oxidation of NADH. We determined the rate constant of electrocatalytic oxidation of NADH by acetoferrocene at a microdisk electrode [48]. Catalytic oxidation is obtained at +0.6 V with a rate constant of $4.68 \times 10^{-3}\ M^{-1}\ s^{-1}$.

Many other compounds, such as quinone [49], alkyphenezinium [50], phenoxazine [51,52] and phenothiazine [53,54] derivatives, have been tested as mediators. However, only a restricted number of organic quinoid redox compounds mediate the electron transfer with electrodes. Phenoxazine and phenothiazine derivatives that incorporate a charged *p*-phenylenediamine functionality are particularly appealing, because [55] (1) NADH is rapidly oxidized at these CMEs, (2) NADH oxidation does not cause electrode fouling, unlike the bare electrodes and CMEs with other mediators, and (3) the mediating properties are selective for NADH, unlike electrodes that incorporate other mediators.

We systematically compared the electrocatalytic ability of twelve mediators adsorbed on impregnated graphite electrodes (IGE) to oxidize NADH and confirmed that the electrocatalytic activity depends mainly on the formal potential and the molecular structure of mediators [56]. These mediators can be irreversibly adsorbed onto the IGE surfaces to yield stable CMEs. For some CMEs, the direct oxidation of NADH molecules at the IGE and the electrocatalytic oxidation of NADH coexist, and they compete with each other. More effective electrocatalysis is obtained at electrodes modified by mediators that have a formal potential of –0.1 V to –0.25 V at pH 7.0. On the other hand, the electrocatalytic ability of the CMEs is related to the mediator structure. The reaction mechanism between NADH and adsorbed organic dyes is similar to Michaelis–Menten kinetics [55]. The following electrocatalytic reaction scheme is assumed:

$$\mathrm{NADH} + \mathrm{M}^+ \underset{k_{-1}}{\overset{k_{+1}}{\rightleftharpoons}} \underset{\text{Complex I}}{[\mathrm{NADH}\cdot\mathrm{M}^+]} \rightleftharpoons \underset{\text{Complex II}}{[\mathrm{NADH}^+\cdot\mathrm{M}^{\bullet-}]} \xrightarrow{k_{+2}} \mathrm{NAD}^+ + \mathrm{MH}$$

where M^+ represents the oxidized mediator. First complex (I) forms, and then the charge exchange takes place between NADH and M^+ to generate complex (II). In complex (II) the mediator possesses a part negative charge. When the acceptor group exits in the mediator molecule, it enhances the stability of the complex (II). In contrast, the donor group decreases the stability of the complex (II). This may explain why Methylene Green (MG) and Meldola Blue are more effective catalysts, respectively, compared to Methylene Blue and Nile Blue A.

To study the catalytic mechanism and obtain kinetic information on the electrocatalytic process, Toluidine Blue O (TB) was adsorbed on an anodic GCE to give a stable CME [57]. The mechanism of electrocatalytic oxidation of NAD(P)H by TB was studied using a rotating-disk electrode. The catalytic reaction scheme is similar to the Michaelis–Menten expression. The coenzyme and the adsorbed mediator combine to form the charge transfer complex, and then the complex decomposes. The latter is a rate-limiting step in the overall catalytic reaction process for which the rate constant is 0.45 s^{-1}.

The oxidation current of NADPH at the bare and TB-modified electrodes as a function of pH has been examined. The oxidation current of NADPH at the bare electrode is nearly independent of solution pH in the range 2.0–10.0. However, the catalytic current at the TB-modified electrode markedly depends on the solution pH and reaches a maximum value at pH 6.0. It decreases drastically with increasing pH above 6.0. Similar results have been reported by Ni et al. [55] for the oxidation of NADH at a Nile Blue A-modified electrode. The main reason for this is that the orientation of the mediator on the electrode surface changes with pH, which affects the catalytic activity of the catalytic sites on the electrode surface. For adsorbed Nile Blue A, this has been confirmed by Raman spectra.

Amperometric Sensors Based on Electrocatalytic Oxidation of NAD(P)H

N-Methylphenazonium sulfate (PMS) is not an effective mediator in the direct electrocatalytic oxidation of NADH. However, when it is combined with hexacyanoferrate(III), the catalytic results become ideal. We prepared an alcohol sensor based on this double mediator system that has an operating potential of +300 mV [58]. Under the action of alcohol dehydrogenase, first alcohol is oxidized to aldehyde to form NADH, then the NADH is oxidized to NAD^+ by $K_3Fe(CN)_6$ in the presence of PMS. The chemically regenerated NAD^+ can continue an enzyme catalytic reaction. The newly formed $Fe(CN)_6^{4-}$ can be oxidized at the electrode surface. The anodic current produced depends on the concentration of alcohol. L-Malate can be detected by a similar strategy, but this system is costly because of the use of soluble NAD^+ and PMS [59].

Adsorption is the simplest method to obtain a dye-modified electrode. However, these CMEs are unsuitable as detectors in a flow system, mainly because of poor stability. Electrochemical copolymerization of pyrrole and MB solves the problem [60]. The resulting CME is electrocatalytically active in oxidizing NAD(P)H, and decreases the overvoltage more than 400 mV. When used as a detector for flow injection at a constant potential of 0.25 V versus SCE, it gives detection limits of 2.5×10^{-8} and 4.0×10^{-8} M for NADPH and NADH, respectively, and a linear concentration range of more than four orders of magnitude. The CME retains a stable response for more than 20 h in the flow system.

MG is an effective mediator in oxidizing NADH. The electrode modified by incorporating MG into carbon paste remains stable in electrocatalytic activity within a solution of pH 4–9. An alcohol dehydrogenase and its cofactor can be immobilized and retain their activity within carbon paste. When a poly(ester sulphonic acid) cation exchanger (Eastman Kodak AQ) is coated on the outside of the modified carbon-paste electrode to form a membrane, the resulting sensor responds rapidly to six alcohols because the membrane prevents the aqueous

Table 1. Parameters of the Alcohol Dehydrogenase Electrode Response to Various Alcohols

Alcohol	*Linear Range, mM*	*Sensitivity*[a] *nA/mM*	*Relative Response*[b]	K_m^{app}, *mM*
Ethanol	0.04–6.0	68	100	11.6
Propan-1-ol	0.09–7.5	54	90	14.1
Propan-2-ol	0.10–8.0	47	78	15.1
Butan-2-o1	0.40–6.5	39	76	15.6
1-Amyl alcohol	0.40–6.5	28	46	16.8
Butan-2-o1	1.0–16.5	13	35	32.2
Methanol	0	0	—	—

Notes: [a]Slope of linear portion.
[b]At 20 mM.

soluble species from dissolving out into the test solution and also prevents relevant negatively charged interfering compounds from reaching the electrode surface. This agrees with the fact that the yeast ADH readily oxidizes primary alcohols (except methanol) and slowly oxidizes secondary alcohols. The experiments show that 10% ADH and 15% NAD^+ contents are suitable for high sensitivity and a relatively wide linear range. The sensor characteristics of the response are summarized in Table 1. The trend in response sensitivity (ethanol > propan-1-ol > propan-2-ol > butan-1-ol > 1-amyl alcohol > butan-2-ol > methanol) differs partly from the known biospecificity of the solution-phase enzyme. This change may be caused by slight differences in the conformation between the immobilized enzyme and the solution-phase enzyme, but the exact reason for this change is not understood yet. A sensor covered with two layers of Eastman Kodak AQ polymer remains stable for ca. 15 days when it was stored in a dry state at 4 °C [61].

4. THE THIRD-GENERATION ENZYME ELECTRODE

Direct electrochemistry of enzymes has aroused increasing interest from researchers because it is a preferable way to produce a reagentless biosensor [62]. Horseradish peroxidase (HRP) has been studied extensively because of its relatively low molecular mass and its heme radical. Efficient electron transfer between HRP and different electrode materials (carbon black, graphite, gold, Pt, SnO_2, *etc.*) was achieved by several research groups [63–67]. We found that HRP reduces and oxidizes at a poly(*o*-phenylenediamine) (PPD) film- modified Pt electrode [68]. The polymer film mediates the reduction of HRP-Fe(III) to HRP-Fe(II). When *o*-phenylenediamine(*o*-PD) and HRP are coelectropolymerized, a reagentless H_2O_2 sensor is produced [69].

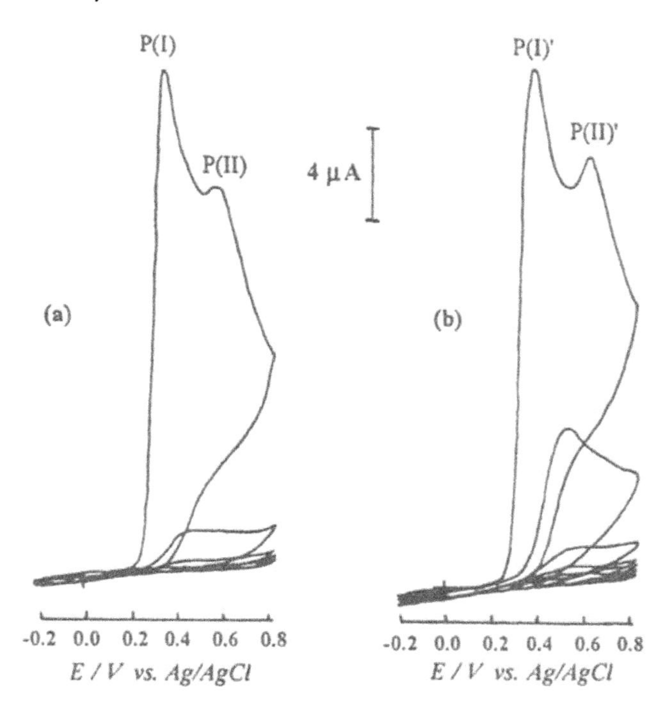

Figure 7. Cyclic voltammograms recorded continuously at a Pt electrode in a solution of (**a**) o-PD 5 mM, (**b**) o-PD (5 mM), and HRP (2 mg mL^{-1}) in pH 7.0 phosphate buffer (scan rate: 20 mV s^{-1}; time: 15 min)

Figure 7 shows cyclic voltammograms of the *o*-PD electropolymerization during cyclic potential scanning from –0.2 to +0.8 V. In the first scan, there are only two irreversible oxidation peaks, P(I) and P(II) at +0.39 and +0.6 V, respectively. With successive cyclic potential scans, the peak currents decrease significantly until almost no current appears. Compared with the result of codeposition with HRP, in a solution containing HRP, the second oxidation peak P(II)′ in the first scan is higher than that of the solution without HRP, and it decreases slowly in successive scans. During electropolymerization, the shape of cyclic voltammograms varies with different scan rates, but the effect of HRP on the peak currents is constant. The apparent activity of HRP entrapped in PPD film is affected mainly by the scan rate and electropolymerization time. The highest activity of HRP in PPD film on the electrode surface is obtained with a scan rate of 20 mV s^{-1} for 15 min. Apparently there is an interaction between HRP and *o*-PD during electropolymerization. The prepared enzyme electrode responds to H_2O_2 in less than 4 s and gives a current density of 73.3 nA μM^{-1} cm^{-2} at –100 mV. A linear calibration curve is obtained over the range of 1–130 μM.

Direct electron transfer is also realized on a highly oriented pyrolytic graphite (HOPG) surface preactivated by anodization as shown by imaging the native HRP structure with *ex situ* and *in situ* electrochemical scanning tunneling microscopy (STM). A cathodic response to H_2O_2 accompanied by a large increase in cathodic current starts to occur at +0.2 V at the HRP-adsorbed HOPG electrode. A linear range up to 2.5 mM is obtained for this enzyme electrode at −100 mV [70].

Glucose oxidase (GOD), a flavoprotein, is well known because of widespread use in biosensors. Recently, several people have attempted to achieve direct electron transfer between GOD and various electrodes. Yabuki et al. [71] claim that electrochemical polymerization of pyrrole in the presence of GOD produces an electroconductive enzyme membrane where GOD retains its bioactivity and could reversibly exchange electrons with the electrode. However, most other similar studies show that direct electron communication cannot occur [72,73] Koopal et al. [74] created conducting microtubles by synthesizing polypyrrole inside track-etch membranes to achieve direct electrochemistry of GOD and to fabricate the reagentless biosensor. The operating potential was 0.35 V versus Ag/AgCl. Unfortunately, the recent report by Kuwabate and Martin [75] demonstrates that the response of the sensor reported by Koopal et al. is caused by the direct electrochemical oxidation of glucose at the metal (Pt) film, rather than by the enzymatic oxidation of glucose. In our laboratory we developed a simple method to achieve direct electrochemistry of GOD in PPy film on a glassy carbon electrode (GCE) surface [76].

The sensor is prepared as follows: The preanodized GCE was immersed in an aqueous solution containing 0.15 mol/L pyrrole monomer. The electropolymerization was carried out by applying a potential of 0.75 V for 1–10 min. The resultant PPy film electrode is termed PPy/GCEa. Subsequently the PPy/GCEa was immersed in a GOD solution of 1000 U/ml for 3 h at 4 °C, called GOD-PPy/GCEa. Anodic pretreatment of GCE markedly improves the adsorption of GOD on the electrode surface. The GOD molecules are irreversibly adsorbed onto the anodized GCE to yield an extremely stable modified surface.

Direct electron transfer from the adsorbed enzyme to the electrode surface is demonstrated by the cyclic voltammogram shown in Figure 8 (solid line). A pair of redox peaks that have a formal potential of −0.41 V were observed. The cathodic and anodic peaks are symmetrical and appear at almost the same potential, demonstrating that the electron transfer is very fast. Another couple of broad peaks at about 0.05 V are usually considered quinone groups produced on the GCE surface by electrochemical oxidation. For comparison, the adsorption of GOD on the untreated GCE is not stable, and the direct electron communication cannot be measured. The result shows that anodic pretreatment is necessary to achieve direct electrochemistry of GOD on the GCE. However, once GOD is adsorbed on GCEa, it cannot catalyze glucose oxidation, indicating that the adsorbed enzyme has no bioactivity because the strong enzyme-surface interaction results in the GOD molecules extending to an unfolded structure. Direct evidence was provided by STM [77,78].

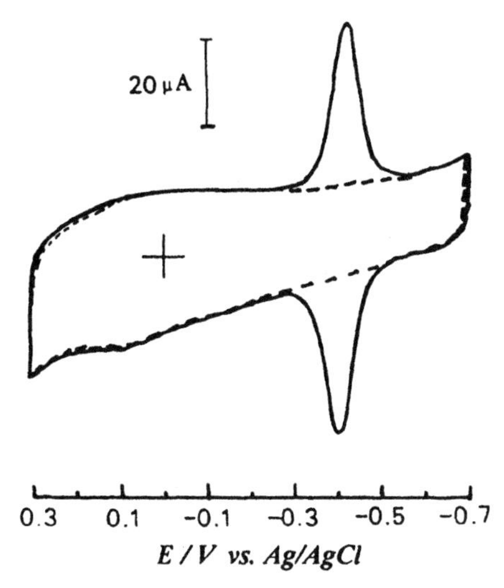

Figure 8. Cyclic voltammograms of an anodized GCE without (dashed line) and with adsorbed GOD (solid line) in 0.1M PBS (pH 6.9). Scan rate: 20 mV s^{-1}.

To prevent GOD denaturation, a thin film of PPy is coated on GCEa by electrochemical polymerization, and then GOD is adsorbed on the PPy layer. After anodic treatment, many etched pots are produced on the electrode surface. These pots vary from 1 μm to 5 μm in diameter and provide active sites for direct adsorption of the GOD molecule. After coating with the PPy film, the surface has a porous sponge-like structure that is beneficial to the adsorption of enzyme molecules by providing many open "micropockets." On the other hand, because PPy is positively charged in its oxidized form, the negatively charged GOD molecules in a neutral solution are adsorbed on the PPy film by electrostatic attraction. In this case, the adsorbed enzyme retains its bioactivity and directly electrocatalyzes glucose oxidation.

The chronoamperometric response to glucose increases with increasing glucose concentration at GOD-PPy/GCEa (Figure 9a), but not at PPy/GCEa (Figure 9b). These results indicate that direct electron transfer between GOD and PPy film electrode has actually occurred. As a glucose sensor, the GOD-PPy/GCEa responds linearly up to 20 mM glucose, showing a wider linear concentration range. Oxygen may play a mediator role and compete with the polypyrrole in accepting electrons from the redox centers of GOD if it is present in solution. However, all solutions in our experiments were purged continuously for a lengthy period with highly pure nitrogen or argon before electrochemical measurements, and an argon flow remained over the solutions during the measurement. In addition, the solution contained 10,000 U/ml catalase, which can rapidly decompose any H_2O_2 formed by enzymatic reaction. Comparative experiments with PPy/GCEa indicate that the

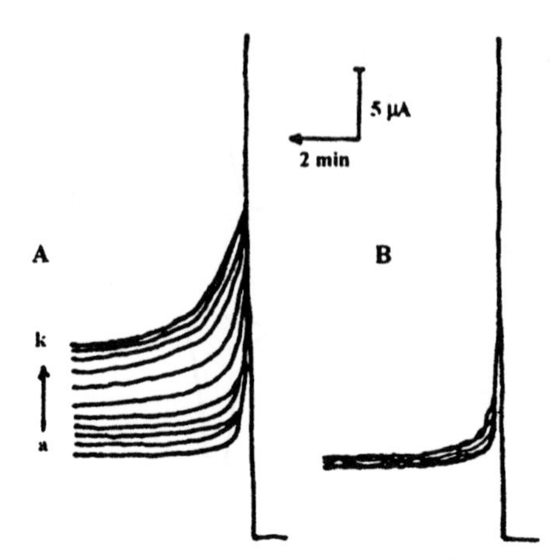

Figure 9. Chronoamperometric response of (a) GOD-PPy/GCEa and (**B**) PPy/GCEa to glucose of various concentrations: (**A**) a. 0; b. 3; c. 5; d. 7; e. 10; f. 15; g. 20; h. 25; i. 30; j. 35; and k. 40 mM (B) increments of 5 mM. operating potential, +0.3 V; supporting electrolyte: 0.1M PBS (pH 7.0) containing 10,000 U mL^{-1} catalase.

anodic response to H_2O_2 was observed only above +0.4 V. Therefore, we may conclude that oxygen mediation does not occur in our present experiments.

5. A NEW IMMOBILIZATION MATERIAL: CRYOHYDROGEL

One general area of biosensor research is immobilizing enzymes or other proteins. The major advantages of protein immobilization are close control of the reaction medium and conditions, prevention of bacterial and chemical degradation, cost-effective reusability of the protein, and enhanced bimolecular stability. However, proteins often fail to retain their native stability and reactivities upon immobilization, a flaw that results in low stability or altered functional responses of biosensors that incorporate them.

In general, there are three main ways to immobilize an enzyme onto an electrode surface:

1. Adsorption onto a graphite surface. These sensors show rapid decreasing signals caused by the washing out of the enzyme with continuous use.
2. Covalent binding of enzymes to the activated surface of the electrode by bifunctional reagents, such as glutaraldehyde.
3. Entrapment in a matrix. Enzymes were fixed within gels or polymers, such as poly(vinyl alcohol), carbon paste, gelatin, polyurethanes, or polypyrrole. With drying and polymerization, these materials form an enzyme layer that

has a specific diffusion resistance, depending on the polymer content within the film.

For optimum biostability and reaction efficiency, the preferred host matrix is one that isolates the biomolecule, preventing it from self-aggregation and microbial attack, while providing essentially the same local aqueous microenvironment as in biological media. Hence, a hydrogel, which is a cross-linked uniquely structured network of polymers swollen with water, is one of the ideal enzyme immobilization materials. Poly(vinyl alcohol) (PVA) has long been used as a support for immobilized enzymes because of its good biocompatibility, chemical stability, and inertness to microbial degradation. However, PVA is always polymerized by UV photopolymerization, and this process denatures enzymes.

In our laboratory, we created a new immobilization method, cryoimmobilization. We prepared a kind of polyhydroxy cellulose (PHC), which is a mixture of PVA and carboxymethyl hydroxyethyl cellulose (CMHEC) (with CMHEC/PVA from 1/5–1/3), the aqueous solution of which can be frozen to produce a hydrogel. This cryohydrogel has a semi-interpenetrating network, relatively high mechanical strength, and retains its water molecules even in organic solvents.

5.1 Cryohydrogel for Constructing Biosensors

Recently the amperometric determination of phenolic compounds and catecholamine neurotransmitters has received much attention in biosensor research. Tyrosinase (polyphenoloxidase, E.C. 1.14. 18.1) has broad specificity for *o*-diphenol compounds and catalyzes the oxidation of catechol to 1,2-benzoquinone. The direct amperometric detection of quinonoid compounds on carbon electrodes is usually adopted. This requires good immobilization methods because tyrosinase is not as robust as glucose oxidase and the stability of the electrode is not easy to control. A cryohydrogel-entrapped tyrosinase sensor was fabricated by freeze-thawing the mixture of tyrosinase and PHC solution three times on a glassy carbon electrode three times [79]. The gel solution forms a cross-linked, amorphous, netted polymer material through the reversed process of freeze-thawing. There are three kinds of water in this cryohydrogel [80]: unfreezing water, which strongly interacts with the polymer, does not freeze until –40 °C, bond water exits near the hydrophilic group of the polymer; the amount is related to the component of the polymer. Free water has exactly the same character as natural water. The amount of free water changes with the humidity of the environment, but the other kinds of water are stable. If the cryohydrogel is placed in a dry atmosphere, it shrinks about 40%, but when it comes into contact with water, the material recovers to its original state. When the cryodesiccated gel is dried with the enzyme, it stabilizes the activity of the enzyme. It is thought that the hydroxyl group in the polymer holds or substitutes for the "bound" water, which is necessary to retain the tertiary structure of the enzyme and the subsequent activity of the molecule [81], that is, the enzyme is retained in the network of the polymer by its perfect tertiary structure.

To obtain adequate mechanical robustness, the freeze-thawing process should be repeated more than four times. Adding protein makes the gel loosen. However, the repeated freeze-thawing process denatures some of the enzymatic activity. Therefore we chose to freeze-thaw the mixture only three times. The immobilized enzyme yields specific apparent activity that is more than 22% of the soluble enzyme based on the spectrophotometric results. In addition, the enzyme does not denature when the polymer shrivels. When supplied with some water, the enzyme layer rehydrates and the enzymatic activity is fully stimulated. This is quite a benefit in preserving and transporting the biosensor. The calibration curves of catechol, *p*-cresol, phenol, and dopamine were obtained under optimum conditions (in pH 6.9 phosphate buffer at –0.2 V) (shown in Fig. 10). The storage time of the sensor is satisfactory. The activity of the electrode (98%) was maintained after storage in a dry state at 4 °C for 3 months.

When glucose oxidase and hydroxyethylferrocene are mixed with the gel solution and stored in a refrigerator below –4 °C, a cryohydrogel glucose oxidase enzyme layer is formed [82]. Compared with physical adsorption and entrapping immobilization, the cryo-hydrogel immobilization distinctly improves the enzyme electrode's stability. Because the enzyme and mediator are physically and chemically entrapped in the three-dimensional interpenetrating network, they do not readily leach out of the hydrogel film. A linear range up to 10 mM is obtained, and the curve passes through zero. The apparent Michaelis–Menten constant is 27.5 mM,

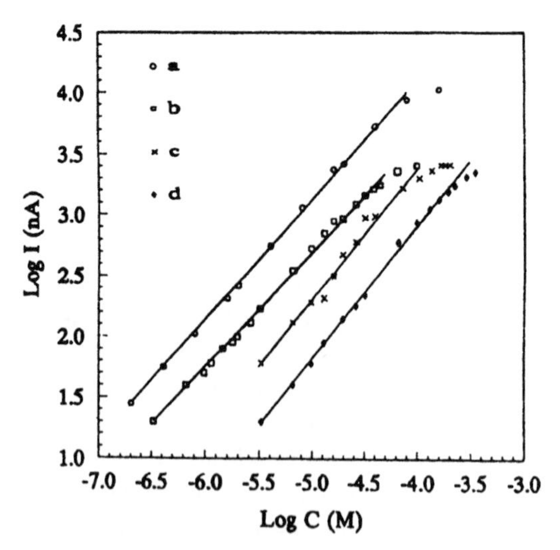

Figure 10. Calibration curve (log *I*–loc *C*) of catechol (**a**), p-cresol (**b**), phenol (**c**) and dopamine (**d**) obtained using a cryohydrogel-immobilized tyrosinase electrode.

which is somewhat higher than that of the soluble enzyme or that of an immobilized enzyme in a conducting polymer. The cryoimmobilized GOD electrode can be operated continuously in the flow system with 20 injections of 8 mM standard glucose solution, and the relative standard deviation is less than 1.5%. The sensor activity drops to 16.5% of its initial value after 300 analyses performed over 5 h, which implies that the enzyme and mediator do not readily leach out of the cryohydrogel and the response is stable and repeatable.

5.2 Cryohydrogel-Immobilized Protein Electrode for Direct Electron Transfer

This cryohydrogel immobilizes enzymes and also relatively low molecular mass proteins. We reported, for the first time, a cryohydrogel-immobilized protein electrode for direct electron transfer with horse heart myoglobin [83,84]. Rapid electron transfer at this electrode was achieved, showing a clear and stable redox wave of the commercially available myoglobin without further purification. The cryohydrogel membrane provides a suitable hydrophilic microenvironment for myoglobin to keep its activity.

Figure 11, the cyclic voltammograms of the horse heart myoglobin membrane electrode in a phosphate buffer solution at pH 5.3, shows a very stable and well-defined redox wave (solid line). No significant change in the voltammogram is observed during continuous measurement for several days. The protein membrane electrode remains stable for at least two months at 4 °C without decreasing its peak current or increasing its peak separation. This high stability shows the

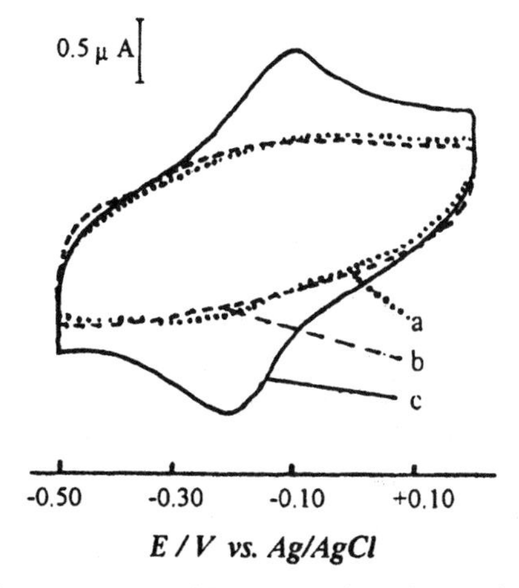

Figure 11. Cyclic voltammograms of the cryohydrogel-immobilized myoglobin electrode in a phosphate buffer (pH 5.3) solution at a glassy carbon electrode under a N_2 atmosphere at 20 ºC. Scan rate: 20 mV s[1]; .... bare glassy carbon electrode background, _ _ _ _ cryohydrogel membrane electrode background.

potential application of the protein electrode in the preparing of molecular/electronic devices. The formal redox potential was –0.158 V, which agrees well with that reported previously (–0.14–0.16 V vs. Ag/AgCl sat. KCl) [85,86]. The formal heterogeneous electron transfer rate constant is $k^{0\prime} = 5.7 \times 10^{-4}$ cm s^{-1} at pH 6.5, showing that rapid electron transfer is achieved. The hydrophilic environment is important for the rapid electron transfer of proteins. The three states of water in the hydrogel membrane provide an appropriate aqueous microenvironment for myoglobin to maintain its biological and electrochemical activities. During refrigeration, the interpenetrating network is formed gradually, and the active conformation of the myoglobin in aqueous solution is maintained in the hydrogel membrane because of its interaction with the polyhydroxyl compound. Therefore, the protein immobilized on the electrode surface has and retains relatively high and stable activity. The interaction between myoglobin and hydrogel also eliminates significant interference from the denatured apomyoglobin.

Other proteins also showed rapid electron transfer on the membrane electrode prepared this way [87]. From the cyclic voltammetric curves of cytochrome c (cyt. c) entrapped in the thin hydrogel membrane, a pair of well-defined redox peaks was obtained. The formal potential of +70 mV is similar to the result of +62 mV measured in solution by the spectrometric method, and to those observed on other chemically modified electrodes [88]. However, when cyt. c is immobilized in a thick hydrogel membrane, the voltammetric behavior differs slightly. The first pair of peaks appears at the same position as the peaks in the thin film. Interestingly, besides those two peaks, there were two other peaks that have formal potentials around +225 mV. The second pair may be caused by a surface process because cyt. c exists in at least three forms in the hydrogel membranes coated on the graphite electrode surface: immobilized molecules, "free" molecules, and directly adsorbed molecules. Hydrogel membranes have a polyhydroxyl structure, and there are many carboxyl groups and hydroxyl groups on the membrane backbones. They easily form hydrogen bonds with the carboxyl groups, amino groups, and phenol groups of cyt. c molecules and immobilize them. This elastic interaction maintains the natural conformations and activities of the proteins. The high water-containing and polyhydroxyl membranes also form a kind of hydrophilic environment around the surface which has been proved necessary for the direct electron transfer of cyt. c [89].

In the membranes are also some other cyt. c molecules that do not form hydrogen bonds with them. They exist in a "free" state, just as in the solution. These two kinds of cyt. c lead to the redox peaks at +70 mV. Some cyt. c molecules adsorb directly on the surface of graphite electrodes. This kind of adsorption usually is accompanied by the deformation and denaturation of the proteins. Because the redox peaks at around +225 mV are due to a surface reaction, there must be some cyt. c molecules that adsorb directly on the electrode surface. Moreover, the amount and extent of cyt. c molecules adsorbed on the surface of a graphite electrode are related to the slow kinetics of cyt. c denaturation on the electrode surface before they are frozen and fixed by the membranes.

6. PURE ORGANIC-PHASE ENZYME ELECTRODES

In the past 20 years, biocatalysis in media containing organic solvents has attracted considerable interest from classical enzymology [90], biosynthesis [91], and biosensors [92] because the reactivity and specificity of enzymes can be controlled by changing the physicochemical properties of the reaction medium. Since the pioneering work of Turner and co-workers [93], which showed that the enzyme electrode is stable for organic-phase detection of analytes, several so-called organic-phase enzyme electrodes (OPEE) have been developed [94–96]. Such novel sensor devices have received considerable attention because they can greatly expand the scope of biosensors in many previously inaccessible analytes (i.e., substrates with poor water solubility), and in many challenging sample matrices and hostile environments (e.g., butter, olive oil, and gasoline).

A wide variety of organic solvents are available, and an appropriate choice can improve sensor performance. Organic solvents also render the reaction species (i.e., the substrate and mediator), more soluble and increase their diffusion and partition coefficients across and within the immobilized enzyme layer. Organic-phase operation also offers improvements in the thermostability of enzymes, which consequently extends sensor lifetimes, eliminates microbial contamination, and reduces side reactions. Moreover, the immobilization schemes are greatly simplified because enzymes are insoluble in organic solvents. Low solubility of aqueous electroactive species decreases the interference of hydrophilic ionic species, which is a serious problem associated with amperometric enzyme electrodes in aqueous systems.

In the absolute absence of water, an enzyme ceases biocatalytic activity, although the exact role of water is not yet fully understood. Water is essential for the catalytic function of enzymes because it participates in all noncovalent interactions that maintain the native, catalytically active conformation of enzymes [97]. Moreover, water plays an important role in maintaining enzyme stability and kinetics. Although there is no doubt that water is necessary, the question is how much of it is necessary. Actually, only about a monolayer of water around the enzyme molecule is needed. The minimal amount of water required for biocatalytic activity is called the "essential water" layer. As long as this essential water is localized about the enzyme molecules, replacement of the rest of the water with an organic solvent should be possible without adversely affecting the enzyme.

In conclusion, the biocatalysis of enzymes in an organic solvent is a nonaqueous reaction macroscopically, but microscopically it is an aqueous reaction. However, in organic solvents, especially water-miscible solvents, this "essential" layer is easily disturbed or even lost, and the enzyme is deactivated. At present the only way of providing this essential hydration layer for enzymes immobilized on OPEEs is to add water deliberately to the organic solvent before use. Therefore, no OPEE reported previously operates strictly in pure nonaqueous media. The cryohydrogel created in our laboratory retains its water molecules to some extent in an organic solvent, which provides a water-containing microenvironment for the enzyme and allows it to function in pure organic solvents. Consequently, a new kind of OPEE—a pure organic-phase amperometric enzyme electrode—was developed.

6.1 Pure Organic-Phase Horseradish Peroxidase Electrode

We prepared a reagentless enzyme mediator system by coimmobilizing a water soluble mediator, potassium hexacyanoferrate(II), and HRP on a graphite electrode using a cryohydrogel [98,99]. The performance of the enzyme-mediator electrode was tested in different organic solvents in which the electrolyte tetrabutylammonium perchlorate (TBAP) readily dissolves. Table 2 shows that the HRP-mediator electrode operating in water-immiscible solvents has greater responses than in water-miscible solvents. This result is consistent with the stability of the hydrogel in different solvents. The water content of the hydrogel decreases gradually and the hydrogel is more difficult to dehydrate in water-immiscible solvents than in water-miscible solvents. Though no enzyme activity was observed in pure *N,N'*-dimethylformamide (DMF), responses of the enzyme electrode to H_2O_2 were found in DMF–chloroform and DMF–acetonitrile mixtures. Moreover, the enzyme-mediator electrode also has a relatively high response in water-containing organic phases and in aqueous solution. This demonstrates the flexibility of the cryohydrogel immobilization method that favorably facilitates the application of OPEEs in practical operation, especially monitoring *in situ* or on line.

A sensitive response of the enzyme electrode is obtained in the vicinity of 0.0 V. Figure 12 shows the current–time curves of H_2O_2 responses obtained at the enzyme electrode in chloroform. A stable base current is obtained after an equilibration time of 5–15 min, which is much more rapid than that previously described (1–1.5 h) [100] and has a shorter response time of 0.2–5 min. The useful measuring range is

Table 2. Response of the HRP-Mediator Electrode to H_2O_2 (0.25 mM) in Different Organic Solvents

Solvent	*Response, nA*	*Solvent*	*Response, nA*
Chlorobenzene	925	Methanol	34
Chloroform	740	Cyclohexane	30
Ethylene dichloride	705	Ethanol	6
Acetonitrile	100	Tetrahydrofuran	5
Acetone	100	DMF	0
Dioxane	50		

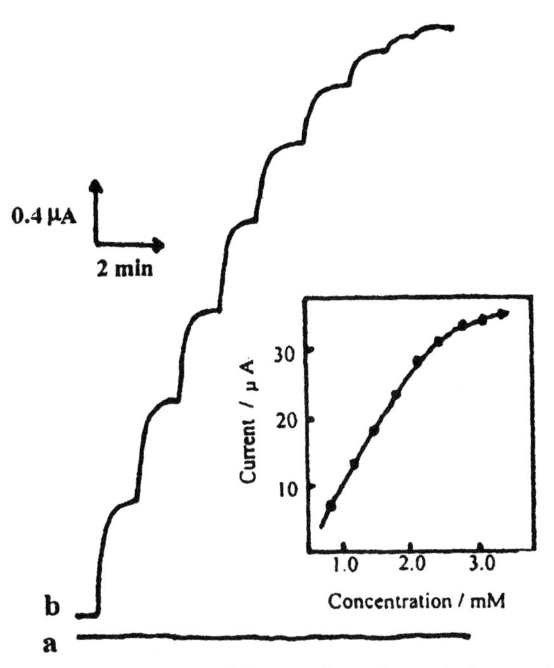

Figure 12. Amperometric response of the graphite electrode immobilized with PHC only (**a**) and HRP-mediator-PHC (**b**) to successive additions of 3.33×10^{-4}M H_2O_2. Also shown (inset) is the calibration plot. Conditions: solution of chloroform containing 0.1M TBAP; applied potential: +10 mV; stirring rate: 250 rpm.

up to 2.5 mM and has a detection limit of 5×10^{-7} M. Comparing the response of the enzyme electrode in chloroform and in acetonitrile, we can see that the enzyme electrode has an even shorter response time (within 10 s) and a wider measuring range (up to 7.5 mM) in acetonitrile than in chloroform, whereas the sensitivity decreases in acetonitrile possible because of the solubility of H_2O_2 in different solvents. The H_2O_2 standard solution is much more miscible with acetonitrile than with chloroform. Therefore, in acetonitrile, equilibration is more readily reached after injection of H_2O_2, and the response time is shorter. The improved measuring range and decreased sensitivity could be interpreted in the same way. In the chloroform background solution the preconcentration effect of H_2O_2 in the water phase inside the hydrogel contributes to the low detection limit and higher sensitivity in chloroform. The cryohydrogel enzyme electrode was used daily for 2 months, and the electrode retained a sensitivity of 60% of its initial value.

The direct electron transfer of HRP on a graphite electrode in aqueous solution is well known. Hence, we studied the response of a cryohydrogel-immobilized HRP graphite electrode in pure chloroform and pure chlorobenzene [101]. The apparent direct electron transfer between a spectrographic graphite electrode and immobilized HRP is obtained. When the applied potential is more than +100 mV, the current is small but increases upon reducing the applied potential. The current increases more rapidly at ca. 0.0 V and finally reaches a plateau at –100 mV. This behavior differs from that in aqueous solution, where the reduction of H_2O_2 starts at +600 mV and the current increases upon reducing the applied potential until it levels off at –100 mV [64]. Experiments under nitrogen show a two- to fourfold current increase. This indicates that functionalities on the electrode surface do not act as mediators and that direct electron transfer occurs from the active sites of the HRP to the graphite electrode.

A variety of organic solvents were investigated and no response to H_2O_2 was observed in water-miscible solvents, such as acetonitrile or acetone, whereas sensitive responses were obtained in water-immiscible solvents, such as chloroform and chlorobenzene. HRP was also immobilized on platinum and glassy carbon electrodes with a cryohydrogel. The modified electrodes show much larger responses to H_2O_2. This implies that unmediated catalytic reactions occur at both glassy carbon and platinum electrodes. It has been proposed that oxygen-containing functionality on the graphite electrode mediates electron transfer. However, on the surface of glassy carbon or platinum electrodes, especially the latter, without special treatment, such functionalities (e.g., phenolic or quinine groups) could be much lower than on graphite. Thus, it seems reasonable to propose direct electron transfer between the electrodes and HRP.

Using this enzyme electrode, a stable base current is obtained within 10 min, and the response time is 0.5–2 min the useful measuring range is up to 5.0 mM and has a detection limit of 1.3×10^{-6} M in chloroform. Corresponding values are 7.0 mM and 2.5×10^{-7} M in chlorobenzene. The enzyme electrode was used intermittently for a month without obvious deterioration in the sensing characteristics.

6.2 Tyrosinase-Based Enzyme Electrode in Pure Organic Phase

Tyrosinase is catalytically active in a number of organic solvents. A cryohydrogel-immobilized tyrosinase electrode was constructed in our laboratory [102]. The characteristics of this modified electrode were discussed and different phenolic substrates determined in different pure organic solvents. The effect of solvent on the OPEE can be divided into three aspects: (1) the effect of solvent hydrophobicity on the state of the enzyme layer; (2) the effect of the solution on the catalytic activity of the enzyme; and (3) the effect of solvent viscosity and the solubility of the conducting electrolyte on mass transport properties. In an organic solvent, the state and amount of water contained in the material depend on the characteristic of the solvent, especially the hydrophobicity. The more hydrophobic the solvent, the more water the cryohydrogel retains, the longer time the enzyme electrode can be successively used, and the easier the enzyme layer is to rehydrate. The recovery of enzymatic activity varies with solvent. After the enzyme electrode is used in chlorobenzene, chloroform, octanol, 1-butanol and ethanol, the activity is recovered fully by dipping in water. However, its recovery is only 30% in acetonitrile and 2-propanol, 50% in benzene, and 85% in toluene. Once the electrode has been used in *N,N'*-dimethylformamide, the activity is never recovered.

Although different organic systems have been adopted in research on OPEEs, the effect of organic solvents on the catalytic efficiency and substrate specificity of enzyme catalysis is unclear. This hinders researchers from taking full advantage of biosensing in nonaqueous media. Hence, we studied the effect of substrate and solvent properties on the response of a cryohydrogel-immobilized tyrosinase electrode in pure water-immiscible solvents [103]. The experimental strategy we adopted was to determine the steady-state current response of the sensor to different substrates in various solvents. The kinetic constants I_{max}, K_m^{app} and I_{max}/K_m^{app} were determined. The calibration curve of the enzyme electrode for the four substrates in pure chloroform—catechol, phenol, *p*-cresol, and *t*-butylcatechol—show that the sensitivity of the response decreases and the linear range enlarges gradually from catechol to *t*-butylcatechol. The catalytic efficiency (implied by the value of I_{max}/K_m^{app}) declines dramatically with increasing substrate hydrophobicity because of the profound increase in the K_m^{app} value. In chlorobenzene, the K_m^{app} values of catechol, phenol, *p*-cresol and *t*-butylcatechol are 86.05, 124.90, 287.02, and 1353.97 mM, respectively, which is up to three orders of magnitude greater than that of the aqueous buffer. These results can be explained by the partition of the substrate between the bulk organic solvent and the cryohydrogel enzyme layer. In our system the enzyme was immobilized inside a water-containing microenvironment. The more hydrophobic the substrate, the less it partitions into the hydrogel layer, and the larger concentration of substrate meets the need of the enzyme. Therefore, the apparent K_m increases, and the catalytic efficiency decreases.

If the hydrophobicity of the solvents (log *P* value) affects the biocatalytic activity of the enzyme solely, the I_{max}/K_m^{app} value should increase with increasing log *P* value of the solvents because a solvent that has a large log *P* value has much less propensity to strip the essential water from the enzyme molecule's microenvironment. The log *P* values of the three solvents we used here, chloroform, chlorobenzene, and 1,2-dichlorobenzene, are 2.0, 2.84, and 3.42, respectively. From our results shown in Figure 13(a), the catalytic efficiency in chlorobenzene is lower than that in chloroform, although the log *P* value of chlorobenzene is larger. Klibanov did not think that there is a direct correlation between the catalytic activity of an enzyme and the dielectric constant of a solvent. However he also claimed that enzymes generally should be more flexible in solvents that have high dielectric constants and that the higher flexibility would facilitate enzyme denaturation. The dielectric constants of chloroform, chlorobenzene, and 1,2-dichlorobenzene are 4.9, 5.6, and 9.8, respectively. Therefore, the catalytic efficiency in 1,2-dichlorobenzene should be lower than that in chlorobenzene. But the result in Figure 13 (b) shows just the opposite. On the basis of the results presented in Figure 13, we may conclude that both log *P* and dielectric constants affect on the catalytic efficiency of the enzyme, but that the log *P* value produces a greater effect.

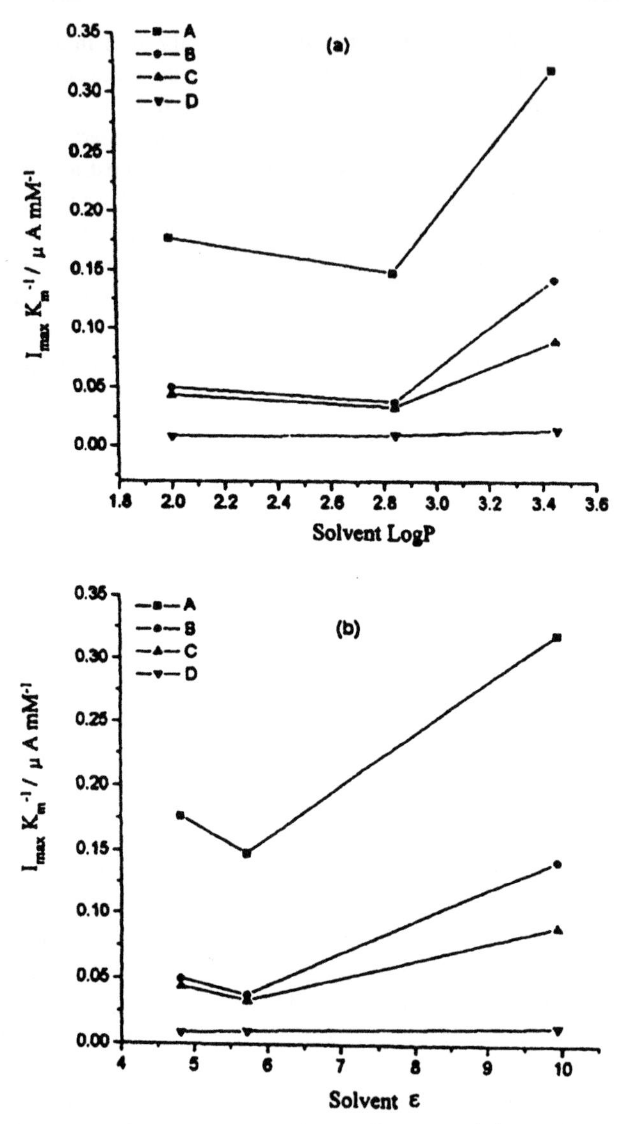

Figure 13. The dependence of the catalytic efficiency of the sensor with different substrates on solvent logP (**a**) and dielectric constant (**b**). A: catechol; B: phenol; C: *p*-cresol; and D: *t*-butylcatechol.

Recently, a new generation of organic-phase biosensors based on enzyme inhibition has been developed for monitoring pesticides and related compounds in organic media. This enlarges the practical applications of biosensors. Moreover, the study of the mechanism of enzyme inhibition in the organic phase will enhance the understanding of organic-phase enzymology. In our laboratory, thiourea, 2-mercaptoethanol, and benzoic acid are determined by exploiting their noxious effect on tyrosinase immobilized in cryohydrogel in pure water-immiscible solvents [104]. The kinetics of the inhibitory effects have also been studied. The amperometric detection of 2-mercaptoethanol, thiourea, and benzoic acid based on the modulated biocatalytic activity of tyrosinase in the presence of its substrate and inhibitors are depicted in Figure 14. Fast and sensitive responses are observed for the three inhibitors. The steady-state response of the electrode to phenol is obtained within one and a half minutes. The response decreases dramatically with successive additions of inhibitors. For 2-mercaptoethanol, the response time for inhibition is less than 20 seconds. The inhibitory effect of benzoic acid is very strong, and the response time is within one and a half minutes. Thiourea exhibits a relatively slow inhibition. The response time is between 2 and 3 minutes. The electrode responses to these three inhibitors need no incubation periods and the detection limits are 0.05, 0.1, and 0.5 μmol/L for benzoic acid, 2-mercaptoethanol, and thiourea, respectively, in chloroform. The characteristics of the enzyme for various inhibitors differ with the kind of organic solvents. In 1,2-dichlorobenzene they are similar to those in the aqueous phase because of the water-containing microenvironment of the cryohy-

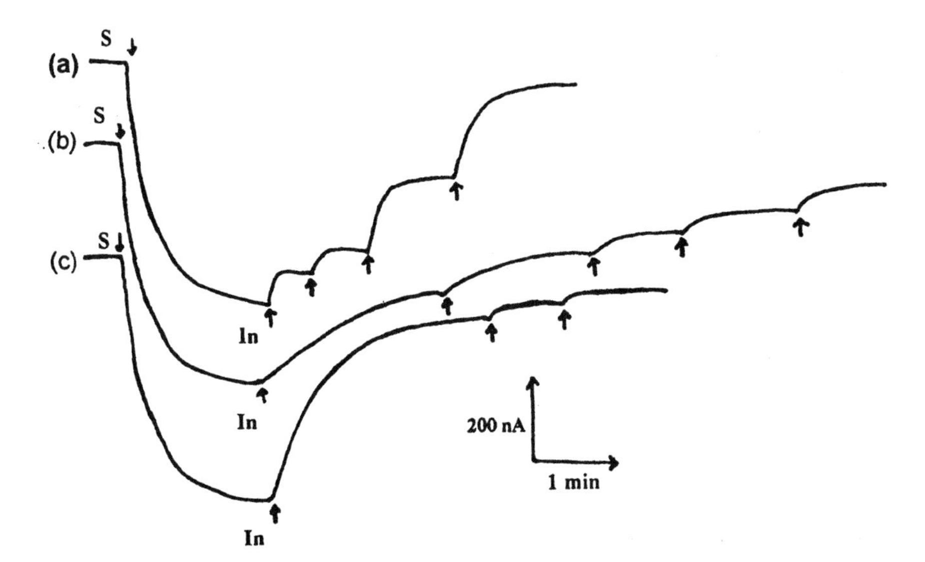

Figure 14. Steady-state current responses of the cryohydrogel tyrosinase electrode in the presence of 0.4 mmol/L phenol upon successive additions of inhibitors. Operating potential: -0.2 V; stirring rate: 300 rpm; solution: pure chloroform containing 0.1M TBAP. S: the addition of substrate (0.4 mM phenol); I_n: the addition of inhibitor; (**a**) 2-mercaptoethanol concentration: 10, 20, 40, and 80 μM (**b**) thiourea concentration: 10, 20, 30, 50, and 90 μM; (**c**) benzoic acid concentration: 8, 16, and 32 μM.

drogel in hydrophobic solvents. Different inhibitors have different inhibition mechanisms. Benzoic acid competitively inhibits tyrosinase in chlorobenzene.

7. THE PRELIMINARY STUDY OF AN ARTIFICIAL ENZYME IN BIOSENSOR

Enzymes have been widely used in biochemistry, chemical engineering, and clinical chemistry, and especially in biosensors. However, enzymes are expensive and unstable in solution, and enzyme electrodes are not suitable for use at high temperatures or in harsh environments. Therefore, enzyme mimics are an interesting trend in analytical biochemistry [105]. Artificial enzymes are synthetic polymer chains that have functional groups that mimic the biocatalytic activity of natural enzymes [106]. It was reported that metalloporphyrins show enzyme-like activity in fluorescent reactions [107]. Studies applying these artificial enzymes in biosensors are rare.

The following reaction of H_2O_2 by artificial (or mimetic) enzymes of catalase [cobalt or manganese- tetra(4-sulfophenyl) porphyrin; Co or Mn $TPPS_4$] was investigated [108,109]:

$$H_2O_2 \xrightarrow[\left(\begin{smallmatrix}Co\\Mn\end{smallmatrix} TPPS_4\right)]{\text{catalase}} H_2O + 1/2\,O_2$$

An anion-exchange resin modified with the artificial enzyme was immobilized on an oxygen electrode and supported by nylon mesh. The linear response of an Mn artificial enzyme ranges from 1 to 7 mM. The response time is within 15 min. The modified resin should be changed after 20 determinations.

The catalytic characteristics of $MnTPPS_4$ as an artificial peroxidase in the reaction of H_2O_2 with $K_4Fe(CN)_6$ have also been studied [110]. The concentration of $K_3Fe(CN)_6$ produced is directly proportional to the concentration of H_2O_2 in the sample, according to this reaction:

$$2Fe(CN)_6^{4-} + H_2O_2 + 2H^+ \underset{(Mn_{TPPS_4})}{\overset{HRP}{\rightleftharpoons}} 2Fe(CN)_6^{3-} + 2\,H_2O$$

$$Fe(CN)_6^{3-} + e \xrightarrow{+80\text{ mV vs. SCE}} Fe(CN)_6^{4-}$$

$K_3Fe(CN)_6$ is measured amperometrically by reduction at a glassy carbon electrode at 80 mV vs. SCE. The linear relationship is in the range of 0.3 to 1 mM. However, ascorbic acid and DL-cysteine interfere with the reaction, and the response time is 15 min. To reduce the response time, we changed the mediator to ferrocene derivatives. The response time was 240 s with dihydroxyethyl ferrocene, whereas with monohydroxythyl ferrocene it took only 90 s to obtain a steady response [111].

The field of synthesizing and studying artificial enzymes is in its infancy [112]. Already it is already clear that by appropriate molecular design it will be possible to achieve very large rate accelerations, comparable to those typical of enzymatic processes. However, there are still a lot of questions that must be studied further.

8. A NEW KIND OF BIOSENSOR: MONENSIN INCORPORATED PHOSPHOLIPID/ALKANETHIOL BILAYERS

Monensin is a naturally existing antibiotic compound. In solution monensin exists in a ring configuration due to hydrogen bonding. It is well known that monensin transports alkali cations across biological membranes and exhibits a strong preference for the sodium ion, compared with the other alkali metals. Earlier results [113] obtained from model membranes revealed characteristics of the interaction of monensin with biological membranes, but the stability of these model systems, especially the free-standing lipid film is poor. They last only a few hours. A membrane formed by self-assembled lipid bilayers on a rigid support improves the stability up to as much as 36 hours [114]. Bilayers consisting of self-assembled monolayers (SAMs) [115] of an alkanethiol and lipid monolayer on solid support show extraordinary stability because of the additional energy of the exposure of the hydrophobic surface. In our laboratory, monensin was incorporated into phospholipid/alkanethiol bilayers on a gold electrode surface by a new method, the painted-frozen method, to deposit a lipid monolayer of alkanethiol [116] on SAMs. This method was based on previous reports [117] that the organic solvent could be frozen out of the lipid films at a lower temperature. The advantages of this assembly system that has a suitable function for investigating ion-selective transfer across a mimetic biomembrane are based on the characteristics of SAMs made from alkanethiols and monensin. On one hand, the SAMs of alkanethiols offer efficiency of packing and coverage of the metal substrate and relatively long-term stability. On the other hand, monensin clearly improves the ion selectivity. The selectivity coefficients K_{Na^+,K^+}, K_{Na^+,Rb^+}, and K_{Na^+,Ag^+} are 6×10^{-2}, 7.2×10^{-3}, and 30, respectively. However, the selectivity coefficient K_{Na^+,Li^+} could not be obtained by potentiometric methods, because of the specific interaction between Li^+ and phospholipid and the lower degree of complexation between Li^+ and monensin. The potential response of this bilayer system to a monovalent ion is fairly good. For example, the slope of the response to Na^+ is close to 60 mV per decade, its linearity range is from 10^{-1} M to 10^{-5} M, and the detection limit is 2×10^{-6} M. The bilayer is stable for at least two months without changing its properties. This monensin-incorporated lipid/alkanethiol bilayer is a good mimetic biomembrane system that provides great promise for investigating the ion transfer mechanism across a biomembrane and developing a practical biosensor.

Valinomycin incorporated in phospholipid/alkanethiol bilayers was also prepared similarly to produce a successful K^+ sensor that has high selectivity and stability [118].

9. CONCLUSION

In principle, and electrode that has an appropriate chemical or physical surface modification that alters the electrochemical response of the electrode can be included among CMEs, if the concept of CMEs is treated broadly. Therefore, all electrochemical biosensors can be regarded as bio-component-containing CMEs. In our laboratory, different types of three generations of biosensors, aqueous or organic phase, and enzyme- or artificial enzyme-based amperometric biosensors were developed. This work clearly indicates that the studies on CMEs contributed to expanding the field of biosensors. We are sure that the results of studies on CMEs will continue to lead the advance in theoretical and practical studies of biosensors, especially up to the level of routine applications.

ACKNOWLEDGMENTS

The financial support of the National Natural Science Foundation of China is gratefully acknowledged.

REFERENCES

[1] Dong, S.J., Che, G.L., Xie, Y.W. (1995). *Chemically Modified Electrodes*, Science Press, Beijing.

[2] Zhang, Y.H., Men, G. (1989). Advances in biosensors. *Fenxi Huaxue(Chin. J. Anal. Chem.)*, **17**, 469.

[3] Jiang, R.Z., Dong, S.J. (1990a). Rotating ring disk electrode theory dealing with nostationary electrocatalysis: Study of the electrocatalytic reduction of dioxygen at cobalt protoporphyrin modified electrode. *J. Phys. Chem.*, **94**, 7471.

[4] Zhang, Y.H., Feng, L.Y. (1990). Glucose enzyme electrode based on cobalt prophrin modified electrode. *Yingyong Huaxue(Chin. J. Appl. Chem.)*, **7**(4), 93.

[5] Dong, S.J., Jiang, R.Z. (1987). Researches on chemically modified electrode XVIII. Heat treatment of cobalt tetraphenyl porphyrin modified electrode and catalytic reduction of dioxygen on its surface. *Huaxue Xuebao(Acta Chimca Sinica)*, **45**, 865.

[6] Feng, L.Y., Zhang, Y.H. (1991). Study on durable glucose sensor based on a cobalt porphyrin modified glassy carbon electrode. *Fenxi Huaxue(Chin. J. Anal. Chem.)*, **19**, 22.

[7] Frew, J.E., Hill, H.A.O. (1987). Electrochemical biosensors. *Anal Chem.*, **59**, 933A.

[8] Morita, K., Shimizu, Y. (1989). Microhole array for oxygen electrode. *Anal. Chem.*, **61**, 159.

[9] Che, G.L., Feng, L.Y., Zhang, Y.H., Dong, S.J. (1991). A study of glucose microsensor based on cobalt prophyrin chemically modified carbon fiber microelectrode. *Fenxi Huaxue(Chin. J. Anal. Chem.)*, **19**, 650.

[10] Feng, L.Y., Che, G.L., Dong, S.J., Zhang, Y.H. (1991). Glucose micro-enzyme sensor based on cobalt porphyrin modified microdisk array carbon fiber electrode. *Yingyong Huaxue(Chin. J. Appl. Chem.)*, **8**(3), 86.

[11] Dong, S.J., Kuwana, T. (1991). Cobalt-porphyrin-Nafion film on carbon microarray electrode to monitor oxygen for enzyme analysis of glucose. *Electroanalysis*, **3**, 485.

[12] Dong, S.J. (1994). Surface modification of electrodes for amperometric biosensors in analytical chemistry. *Anal. Sci.*, **10**, 175.

[13] Ji, X.F., Zhang, Y.H. (1993a). A glucose sensor based on poly-1,2-diaminobenzene modified platinized carbon electrode. *Yingyong Huaxue (Chin. J. Appl. Chem.)*, **10**(2), 97.

[14] Lingane, J.J., Lingane, P.J. (1963). Chronopoteniometry of hydrogen peroxide with a platinum wire electrode. *J. Electroanal. Chem.*, **5**, 411.

[15] Ji, X.F., Zhang, Y.H. (1993b). A glucose sensor based on a platinized Eastman AQ modified glassy carbon electrode. *Shengwu Huaxue Shengwu Wuli Jinzhan (Chin. Prog. Biochem. Biophys.)*, **20**, 395.

[16] Ji, X.F., Zhang, Y.H. (1993c). A galactose sensor based on the platinized glassy carbon electrode modified with National film. *Fenxi Huaxue(Chin. J. Anal. Chem.)*, **21**, 519.

[17] Gorton, L., Csorei, E., Dominguez, E., Emneus, J., Jonsson-Pettersson, G., Marko-Varga, G., Persson, B. (1991). Selective detection in flow analysis based on the combination of immobilized enzyme and chemically modified electrodes. *Anal. Chim. Acta*, **250**, 203.

[18] Dong, S.J., Deng, Q., Cheng, G.J. (1993). Cholesterol sensor based on electrodeposition of catalytic palladium particles. *Anal. Chim. Acta*, **279**, 235.

[19] Yokoyama, K., Tamiya, E., Karube, I. (1991). Amperometric glucose sensor using silicon oxide deposited gold electrodes. *Electroanalysis*, **3**, 469.

[20] Foulds, N.C., Lowe, C.R. (1986). Enzyme entrapment in electrically conducting polymers immobilises glucose oxidase in polypyrrole and its application in amperometric glucose sensor. *J. Chem. Soc., Farady Trans., 1*, **82**, 1259.

[21] Bartlett, P.N., Whitaker, R.G. (1987). Electrochemical immobilisation of enzymes. *J. Electroanal. Chem.*, **224**, 27.

[22] Chi, Q.J., Dong, S.J. (1993). Flow-injection analysis of glucose at an amperometric glucose sensor based on electrochemical codeposition of palladium and glucose oxidase on a glassy carbon electrode. *Anal. Chim. Acta*, **278**, 17.

[23] Jiang, R.Z., Dong, S.J. (1990b). Electrocatalytic oxidation of H_2O_2 at a cobalt protoporphyrin modified electrode and its application in analysis. *J. Electroanal. Chem.*, **292**, 11.

[24] Deng, Q., Dong, S.J. (1993). Cholesterol oxidase sensor based on cobalt protoporphyrin modified electrode. *Gaodeng Xuexiao Huaxue Xuebao(Chem. J. Chin. Univ.)*, **14**, 1214.

[25] Wang, B.X., Song, F.Y., Dong, S.J. (1993). The preparation of electrodes modified with isopolymolybdic + polyaniline film and its catalytic effects on chlorate ions. *J. Electroanal. Chem.*, **353**, 43.

[26] Liu, M, Dong, S.J. (1995). Electrochemical behavior of molybdosilicic heteropoly complex with dysprosium and its doped polpyrrole film modified electrode. *Electrochim. Acta*, **40**(2), 197.

[27] Murray, C.G., Nowak, R.J., Polison, D.R. (1984). Electrogenerated coatings containing zeolites. *J. Electroanal. Chem.*, **164**, 205.

[28] Dong, S.J., Ji, Z. (1989). Electrochemistry of indium hexacyanoferrate film modified electrodes. *Electrochim. Acta*, **34**, 963.

[29] Cheng, G.J., Liu, C.W., Li, J.H., Jin, J.G., Dong, S.J. (1996). Probe beam deflection study of the ion exchange between Prussian blue film or indium hexacyanoferrate film and electrolyte solutions. *Chem. J. Chin. Univ.*, **17**(2), 196.

[30] Chi, Q.J., Dong, S.J. (1995a). Amperometric biosensors based on the immobilization of oxidases in a Prussian blue film by electrochemical codeposition. *Anal. Chem. Acta*, **310**, 429.

[31] Zhou, J.X., Wang, E.K.J. (1992). Sensitive amperometric detection of glucose by reversed phase liquid chromatography at a Prussian blue chemically modified electrode of novel construction. *J. Electroanal. Chem.*, **331**, 1029.

[32] Stoytcheva, M. (1995). Acetylcholinesterase-based amperometric sensor. *Electroanalysis*, **7**, 560.

[33] Deng, Q., Dong, S.J. (1996a). Acetylcholinesterase amperometric detection system based on a cobalt (II) tetraphenyl porphyrin-modified electrode. *Analyst*, **121**, 1123.

[34] Deng, Q., Dong, S.J. (1996b). Amperometric determination of pesticides using an acetylcholinesterase biosensor based on tetraphenylporphyrin cobalt(II) modified electrode, submitted to the *Analyst*.

[35] Mascini, M., Palleschi, G. (1989). Design and applications of enzyme electrode probes. *Selective Electrode Rev.*, **11**, 191.

[36] Cass, E.G., Davis, G., Francis, G.D., Hill, H.A.O., Aston, W.J., Higins, I.J., Plotkin, E.V., Scott, L.D., Turner, A.P.F. (1984). Ferrocene-mediated enzyme electrode for amperometric determination of glucose. *Anal. Chem.*, **56**, 667.

[37] Foulds, N.C., Lowe, C.R. (1988). Immobilization of glucose oxidase in ferrocene-modified pyrrole polymers. *Anal. Chem.*, **60**, 2473.

[38] Jonsson, G., Gorton, L., Pettersson, L. (1989). Mediated electron transfer from glucose oxidase at a ferrocene modified graphite electrode. *Electroanalysis*, **1**, 49.

[39] Dong, S.J., Wang, B.X. (1991). Glucose sensor based on ferrocene mediator. *Kexue Tongbao(Chin. Sci. Bull.)*, **11**, 877.

[40] Dong, S.J., Wang, B.X., Liu, B.F. (1991). Amperometric glucose sensor with ferrocene as an electron transfer mediator. *Biosensors & Bioelectronics*, **7**, 215.

[41] Dong, S.J., Lu, Z.L. (1990). Ferrocene-Nafion modified electrode and its catalysis for Cerium(IV). *Cryst. Liq. Cryst.*, **190**, 197.

[42] Dong, S.J., Tian, M., Liu, B.F. (1993). Glucose microsensor based on ferrocene-AQ chemically modified carbon fiber. *Fenxi Huaxue(Chin. J. Anal. Chem.)*, **31**(3), 255.

[43] Tian, M., Dong, S.J. (1993). Electrochemical behavior of Eastman AQ polymer/DMFc modified microdisk array electrode and its application to glucose analysis. *Yingyong Huaxue(Chin. J. Appl. Chem.)*, **10**(6), 6.

[44] Yu, P.G., Dong, S.J. (1995). A disposable biosensor. *Chin. Pat.* CN 2195757Y.

[45] Ji, X.F., Zhang, Y.H. (1993d). Construction of medicated amperometric bienzyme D-amino acid sensor. *Fenxi Huaxue(Chin. J. Anal. Chem.)*, **21**(6), 625.

[46] Moirox, J., Elving, P.J. (1978). Effect of adsorption, electrode material and operations variables on the oxidation of dihydronicotinamine adenine dinucleotide at carbon electrodes. *Anal. Chem.*, **50**, 1056.

[47] Jaegfelet, H. (1980). Adbsorption and electrochemical oxidation behavior of NADH at a clean platinum electrode. *Electroanal. Chem.*, **110**, 295.

[48] Tian, M., Dong, S.J. (1995). Study on electrocatalytic oxidation of NADH by ferrocene derivation and determination of catalytic rate constant at microdisk electrode. *Electroanalysis*, **7**, 1063.

[49] Ueda, C., Tse, D.C.-S., Kunans, T. (1982). Stability of catechol modified carbon electrodes for electrocatalysis of dihydronictinaminde adenine dinucleotide and ascorbic acid. *Anal. Chem.*, **54**, 850.

[50] Torstensson, A., Gorton, L. (1981). Catalytic oxidation of NADH by surface-modified graphite electrodes. *J. Electroanal. Chem.*, **130**, 119.

[51] Persson, B., Gorton, L. (1990). A comparative study of some 3,7-diaminophenoxazine derivative and related compounds for electrocatalytic oxidation of NADH. *J. Electroanal. Chem.*, **292**, 115.

[52] Kulys, J., Gleixner, G., Schubmann, W., Schmidt, H.L. (1993). Biocatalysis and electrocatalysis at carbon paste electrodes doped by diaphorase-methylene green and diaphorase-meldola blue. *Electroanalysis*, **5**, 201.

[53] Gorton, L., Torstensson, A., Jaefeldt, H., Johansson, J. (1984). Electrocatalytic oxidation of reduced nicotinaminde coenzymes by graphite electrodes modified with an adsorbed phenoxazinium salt, meldola blue. *J. Electroanal. Chem.*, **161**, 103.

[54] Persson, B. (1990). A chemistry modified graphite electrode for electrocatalytic oxidation of reduced nicotinamide adenine dinucleotide based on a phenothiazine derivative, 3-β-naphthoyl-toluidine blue O. *J. Electroanal. Chem.*, **287**, 61.
[55] Ni, F., Feng, H., Gorton, L., Cotton, T.M. (1990). Electrochemical and SERS studies of chemically modified electrodes: Nile Blue A, a mediator for NADH oxidation. *Langmuir*, **6**, 66.
[56] Chi, Q.J., Dong, S.J. (1996a). A comparison of electrocatalytic ability of various mediators absorbed onto paraffin impregnated graphite for oxidation of reduced nicotinamide coenzymes. *J. Mol. Cat. A: Chem.*, **105**, 193.
[57] Chi, Q.J., Dong, S.J. (1995b). Electrocatalytic oxidation of reduced nicotinamide coenzymes at organic dye-modified electrodes. *Electroanalysis*, **7**, 147.
[58] Ji, X.F., Zhang, Y.H. (1993e). Study of ethanol dehydrogenase electrode. *Fenxi Huaxue(Chin. J. Anal. Chem.)*, **21**, 267.
[59] Ji, X.F., Zhang, Y.H. (1993f). L-malate-sensing electrode based on malate dehydrogenase with the mediation of hexacyanoferrate. *Shengwu Huaxue Shengwu Wuli Jinzhan (Chin. Prog. Biochem. Biophys.)*, **20**, 301.
[60] Chi, Q.J., Dong, S.J. (1994a). Electrocatalytic oxidation and flow-injection determination of reduced nicotinamide coenzyme at glassy carbon electrode modified by a polymer thin film. *Analyst*, **119**, 1063.
[61] Chi, Q.J., Dong, S.J. (1994b). Electrocatalytic oxidation of reduced nicotinamide coenzymes at methylene green modified electrode and fabrication of amperometric alcohol biosensors. *Anal. Chim. Acta*, **286**, 125.
[62] Chi, Q.J., Dong, S.J. (1994c). Direct electrochemistry of enzymes and the third generation biosensors. *Fenxi Huaxue(Chin. J. Anal. Chem.)*, **22**, 1065.
[63] Yaropolov, A.I., Malovik, V., Varfolomeer, S.D., Berrzin, I.V. (1979). Bioelectrocatalysis. Direct electron transfer from peroxidase active site to electrode. *Dokl. Akad. Nank SSSR*, **249**, 1399.
[64] Jonsson, G., Gorton, L. (1989). An electrochemical sensor for hydrogen peroxide based on peroxidase absorbed on a spectrographic graphite electrode. *Electroanalysis*, **1**, 465.
[65] Zhao, J., Henkens, R.W., Stonehuerner, T., O'Daly, J.P., Grumbliss, A.L. (1992). Direct electron transfer on horseradish peroxidase-modified colloidal gold electrodes. *Electroanal. Chem.*, **327**, 109.
[66] Wollenberger, U., Bogdanovskaya, V., Bobrin, S., Scheller, F., Tarasevich, M. (1990). Enzyme electrodes using bioelectrocatalytic reduction of hydrogen peroxide. *Anal. Lett.*, **23**, 1795.
[67] Tatsuma, T., Gondaira, M., Watanabe, T. (1992). Peroxidase-incorporated polypyrrole membrane electrodes. *Anal. Chem.*, **64**, 1183.
[68] Deng, Q., Dong, S.J. (1994a). Redox reaction of peroxidase at poly (*o*-phenylenediamine) modified electrode. *Electroanalysis*, **6**, 878.
[69] Deng, Q., Dong, S.J. (1994b). Mediatorless hydrogen peroxide electrode based on horseradish peroxidase entrapped in poly (*o*-phenylenediamine). *J. Electroanal. Chem.*, **377**, 191.
[70] Zhang, J.D., Chi, Q.J., Dong, S.J., Wang, E.K. (1996). *In situ* electrochemical scanning tunnelling microscopy investigation of structure for horse radish peroxidase and its electrocatalytic property. *Bioelectrochem. Bioenerg.*, **39**, 267.
[71] Yabuki, So-ichi, Shinohara, H., Aizawa, M., (1989). Electro-conductive enzyme membrane. *J. Chem. Soc., Chem. Commun.*, 945.
[72] Rishpon, J., Gottesfeld, S. (1991). Investigation of polypyrrole/glucose oxidase electrodes by ellipsometric, microgravimetric and electrochemical measurement. *Biosensor & Bioelectronics*, **6**, 143.
[73] Sun, Z., Tachikawa, H. (1992). Enzyme-based bilayer conducting polymer electrodes consisting of polymetallophthalocyanines and polypyrrole-glucose oxidase thin films. *Anal. Chem.*, **64**, 1112.

[74] Koopal, C.G.J., Feiters, M.C., Nolte, R.J.M., de Ruiter, B., Schasfoort, R.B. (1992). Glucose sensor utilizing polyporrole incorporated in track-etch membranes as the mediator. *Biosensors & Bioelectronics*, **7**, 461.

[75] Kuwabata, S., Martin, C.R. (1994). Mechanism of the amperometric response of a proposed glucose sensor based on a polypyrrole-tubule-impregnated membrane. *Anal. Chem.*, **66**, 2757.

[76] Chi, Q.J., Dong, S.J. (1996b). Direct electron communication between glucose oxidase and carbon electrode covered with polypyrrole. *Chem. Res. Chin. Univ.*, **12**, 37.

[77] Chi, Q.J., Zhang, J.D., Dong, S.J., Wang, E.K. (1994a). Direct electrochemistry and surface characterization of glucose oxidase adsorbed on anodized carbon electrodes. *Electrochem. Acta*, **39**, 2431.

[78] Chi, Q.J., Zhang, J.D., Dong, S.J., Wang, E.K. (1994b). Direct observation of native and unfolded glucose oxidases structures by scanning tunnelling microscopy. *J. Chem. Soc., Faraday. Trans.*, **90**, 2057.

[79] Deng, Q., Guo, Y.Z., Dong, S.J. (1996). Cryo-hydrogel for the construction of tyrosinase-based biosensor. *Anal. Chim. Acta*, **319**, 71.

[80] Feng, Y.D. (1989). Study on state of water in cryo-hydrogel. *Sichuan Daxue Xuebao(J. Sichuan University Nat. Sci. Ed.)*, **26**, 470.

[81] Gibson, T.D., Woodward, J.R., Edelman, P.G., Wang, J. (Eds.). (1992). In *Biosensors and Chemical Sensors*. American Chemical Society, Washington, DC., Chap. 5.

[82] Wang, Y., Guo, Y.Z., Zhu, G.Y., Dong, S.J. (1996). Cryoimmobilized enzyme for biosensors construction. *Sensors and Actuators B*, **31**, 193.

[83] Niu, J.J., Guo, Y.Z., Dong, S.J. (1995a). A Cryo-hydrogel immobilized protein electrode for direct electron transfer of myoglobin. *Chin. Chem. Lett.*, **6**, 421.

[84] Niu, J.J., Guo, Y.Z., Dong, S.J. (1995b). The direct electrochemistry of cryohydrogel immobilized myoglobin at a glassy carbon electrode. *J. Electroanal. Chem.*, **399**, 41.

[85] Taniguchi, I., Watanabe, K., Tominaga, M. (1992). Direct electron transfer of horse heart myoglobin at an indium oxide electrode. *J. Electroanal. Chem.*, **333**, 331.

[86] Dong, S.J., Chu, Q.H. (1993). Study on electrode process of myoglobin at a polymerized toluidine blue film electrode. *Chin. J. Chem.*, **11**, 12.

[87] Chen, T., Guo, Y.Z., Dong, S.J. (1995). Voltammetry of cytochrome c entrapped in hydrogel membrane on graphite electrode. *Bioelectrochem. Bioenerg.*, **37**, 125.

[88] Heineman, W.R., Norris, B.J., Godz, J.F. (1975). Measurement of enzyme E^0 values by optically transparent thin layer electrochemical cells. *Anal. Chem.*, **47**, 79.

[89] Bowden, E.F., Hawkridge, F.M. (1984). Interfacial electrochemistry of cytochrome c at tin oxide, indium oxide, gold and platinum electrodes. *J. Electroanal. Chem.*, **161**, 355.

[90] Klibanov, A.M. (1990). Asymmetric transformations catalyzed by enzymes in organic solvents. *Acc. Chem. Res.*, **23**, 114.

[91] Homandberg, G.A., Mattis, J.A., Laskowski, M., Jr., (1978). Synthesis of peptide bonds by proteinases. Addition of organic cosolvents shifts peptide bond equilibriums toward synthesis. *Biochemistry*, **17**, 5220.

[92] Saini, S., Hall, G., Down, M., Turner, A.P.F. (1991). Organic phase enzyme electrodes. *Anal. Chim. Acta*, **249**, 1.

[93] Hall, G., Best, D., Turner, A.P.F. (1988). The determination of *p*-cresol in chloroform with an enzyme electrode used in the organic phase. *Anal. Chim. Acta.*, **213**, 113.

[94] Hall, G., Turner, A.P.F. (1991). An organic phase enzyme electrode for cholestrol. *Anal. Lett.*, **24**, 1375.

[95] Wang, J., Lin, Y., Chen, W. (1993). Organic phase biosensors based on the entrapment of enzymes within ply(ester-sulfonic acid) coatings. *Electroanalysis*, **5**, 23.

[96] Wang, J., Lin, Y. (1993). On-line organic phase enzyme detector. *Anal. Chim. Acta*, **271**, 53.

[97] Klibanov, A.M. (1986). Enzymes that work in organic solvents. *Chemtech*, **6**, 354.

[98] Dong, S.J., Guo, Y.Z. (1994a). A novel enzyme electrode for the water-free organic phase. *J. Electroanal. Chem.*, **375**, 405.
[99] Dong, S.J., Guo, Y.Z. (1994b). Organic phase enzyme electrode operated in water-free solvents. *Anal. Chem.*, **66**, 3895.
[100] Schubert, F., Saini, S., Turner, A.P.F., Scheller, F. (1991). Mediated amperometric enzyme electrode incorporating peroxidase for the determination of hydrogen peroxide in organic solvents. *Anal. Chim. Acta*, **245**, 133.
[101] Dong, S.J., Guo, Y.Z. (1995). An organic-phase enzyme electrode based on an apparent direct electron transfer between a graphite electrode and immobilized horseradish peroxidase. *J. Chem., Soc., Chem.. Commun.*, 483.
[102] Deng, Q., Dong, S.J. (1995). Construction of a tyrosinase-based biosensor in pure organic phase. *Anal. Chem.*, **67**, 1357.
[103] Deng, Q., Dong, S.J. (1996c). The effect of substrate and solvent properties on the response of an organic phase tyrosinase electrode. *J. Electroanal. Chem.*, in press.
[104] Deng, Q., Dong, S.J. (1996d). Amperometric biosensor for tyrosinase inhibitors in pure organic phase. *Analyst*, **121**, 1979.
[105] Breslow, R. (1982). Artificial enzymes. *Science*, **218**(5), 532.
[106] Ho, M.Y.K., Rechnitz, G.A. (1987). Highly stable biosensor using an artificial enzyme. *Anal. Chem.*, **59**, 536.
[107] Ci, Y.X., Wang, F. (1990). Studies on catalytic fluorescence formation with peroxidase-like metallotetrakis-(*N*-methlpyridiniumyl) porphyrins. *Talanta*, **37**(12), 1133.
[108] Xi, X.L., Zhang, Y.H. (1993a). Preliminary study of mimetic catalase $CoTPPS_4$ for electrochemical sensor. *Yinyong Huaxue(Chin. J. Appl. Chem.)*, **10**(2), 102.
[109] Xi, X.L., Zhang, Y.H., Wu, J.L., Zhang, Y.S. (1993). Synthesis of mimetic catalase and its catalytic activity—manganese porphyrin acts as mimetic enzyme. *Fenxi Huaxue(Chin. J. Anal. Chem.)*, **21**, 500.
[110] Xi, X.L., Zhang, Y.H. (1993b). Application of mimetic enzyme of peroxidase for biosensor. *Fenxi Huaxue(Chin. J. Anal. Chem.)*, **21**, 838.
[111] Zhang, Y.H., Yang, S.P., Wang, E.K. (1995). Application of mimetic peroxidase for biosensor. *Fenxi Huaxue(Chin. J. Anal. Chem.)*, **23**, 9.
[112] Xi, X.L., Zhang, Y.H. (1993c). A review of the current research status of artificial oxidoreductase. *Huaxue Chuanganqi(Chin. Chem. Sensor)*, **13**(2), 1.
[113] Tien, H.T. (1974). *Bilayer Lipid Membranes: Theory and Practice*, Marcel Dekker, New York.
[114] Wardak, A., Tien, H.T. (1990). Cyclic voltammetry studies of bilayer lipid membranes deposited on platinum by self-assembly. *Bioelectrochem. Bioenerg.*, **24**, 1.
[115] Ulman, A. (1991). *An Introduction to Ultrathin Organic Films: From Langmuir–Blodgett to Self-Assembly*. Academic, New York.
[116] Li, J.H., Ding, L., Wang, E.K., Dong, S.J. (1996). The ion selectivity of monensin incorporated phospholipid/alkanethiol bilayers. *J. Electroanal. Chem.*, **414**, 17.
[117] White, S.H. (1974). Temperature-dependent structural changes on planar bilayer membranes: Solvent "freeze-out." *Biochim. Biophys. Acta*, **356**, 8.
[118] Ding, L., Li, J.H., Wang, E.K., Dong, S.J., (1997). K^+ sensors based on supported alkanethiol/phospholipid bilayers. *Thin Solid Film*, **293**, 153.

ELECTROCHEMICAL BIOSENSORS BASED ON CHEMICALLY MODIFIED ELECTRODES

Litong Jin, Pingang He, Jiannong Ye, and Yuzhi Fang

OUTLINE

Advances in Biosensors
Volume 4, pages 41–58.

ISBN: 0-7623-0073-6

ABSTRACT

Electrochemical biosensors are usually certain response systems consisting of electrotransducers that have immobilized biomaterials. Ordinary electrochemical biosensors often suffer from short lifetimes, poor stability, low sensitivity, and slow response time. In addition, dissolved oxygen, temperature change, and coexistent electroactive materials often interfere with the required response. To improve the characteristics of biosensors, we have worked for years in our laboratory to develop several electrochemical biosensors based on chemically modified electrodes (CMEs). Our work includes the following: (1) immobilizing enzymes to electropolymerized conductive macromolecules to prolong the lifetime of the biosensors; (2) improving the detection sensitivity by using Nafion/methyl viologen, which induces catalytic reduction of dissolved O_2, as the substrate electrode to mobilize enzymes; (3) developing an enzyme-modified electrode using tetrathiafulvalene (TTF) and ferrocene carboxylic acid (FCA) as the electron transfer media. In doing so, the rate of electron transfer between the enzyme and substrate electrode is enhanced, and the detection error caused by variable dissolved O_2 and coexistent electroactive species is reduced; (4) using the avidin-biotin technique to prepare a self-assembled monolayer enzyme electrode, thereby allowing the study of enzyme sensors at the molecular level.

1. INTRODUCTION

Electrochemical biosensors are usually response systems consisting of electrotransducers that have immobilized biomaterials. In 1967, S.J. Updike and G.P. Hicks [1] prepared a glucose biosensor by immobilizing glucose oxidase on a platinum (Pt) electrode surface. Since then, numerous biosensors have been developed, and today biosensors are an active research area [2,3]. Unfortunately, the ordinary electrochemical biosensor often suffers from a short lifetime, poor stability, low sensitivity, and slow response time. In addition, dissolved oxygen, temperature change, and coexistent electroactive materials often interfere with the response. We have attempted to improve the characteristics of biosensors, based on chemically modified electrodes (CMEs). Our work includes the following: (1) immobilizing enzymes to electropolymerized conductive macromolecules to prolong the lifetime of the biosensors; (2) improving the detection sensitivity by using Nafion/methyl viologen (MV), which induces catalytic reduction of dissolved O_2, as the substrate electrode to mobilize enzymes; (3) developing an enzyme-modified electrode using tetrathiafulvalene (TTF) and ferrocene carboxylic acid (FCA) as the electron transfer media. In doing so, the rate of electron transfer between the enzyme and substrate electrode is enhanced, and the detection error caused by variable dissolved O_2 and coexistent electroactive species is reduced; (4) using the avidin-biotin technique to prepare a self-assembled monolayer enzyme electrode, thereby allowing the study of enzyme sensors at the molecular level.

2. BIOSENSORS USING NAFION-MV CHEMICALLY MODIFIED ELECTRODES AS THE SUBSTRATE ELECTRODE

Using noble metals as the working electrodes [4], conventional electrolytic oxygen sensors suffer from high potentials required for oxygen reduction and from low sensitivity and poor selectivity. Biosensors using Nafion-MV CME as the substrate electrode for enzyme immobilization have the advantages of high sensitivity, fast response, and good selectivity. With this approach, we prepared glucose, ascorbic acid, acetonic acid, uric acid, and cholesterol biosensors [5–9].

2.1 Preparation of Enzyme-Immobilized Nafion-MV CME

A clean glassy carbon electrode was dried under an IR lamp. Nafion was coated on the electrode by pipetting 1.4 μL of 0.1% Nafion onto the Nafion-coated electrode surface and then air-dried. The Nafion-MV CME was prepared by immersing it in 0.1 mol/L MV solution for 20 min and then rinsing it with doubly distilled water. Bovine serum albumin (3 mg), and 5% glutaraldehyde (20 μL) were added to a mixture of enzymes (40 μL) and mixed quickly. Seven μL of this mixture was spread on the Nafion-MV CME surface and dried at 4 °C. The enzyme-immobilized Nafion-MV CME was immersed in pH 7.2 phosphate buffer and stored in a refrigerator.

2.2 Response Mechanism of Biosensors Using Nafion-MV CME

As the substrate electrode, Nafion-MV CME can catalyze oxygen reduction [10]:

$$-[-(SO_3^-)_2MV^{2+}] + e = -[-(SO_3^-)_2MV^{+}]$$

$$-[-(SO_3^-)_2MV^{+}] + H^+ + 1/2O_2 = -[-(SO_3^-)_2MV^{2+}] + 1/2\,H_2O_2$$

The experimental results shown in Figure 1 demonstrate that, with Nafion-MV CME, oxygen reduction occurs at –0.55 V (vs. Ag/AgCl). At an unmodified electrode, this reduction occurs at –0.75 V. In addition, the oxygen reduction peak current increases by a factor of 7 at this mobilized electrode. Therefore, because of the catalytic effect of Nafion-MV CME on oxygen reduction, oxygen detection sensitivity is enhanced. If an oxidase is immobilized on the Nafion-MV CME, the resultant biosensor detects the corresponding substrate.

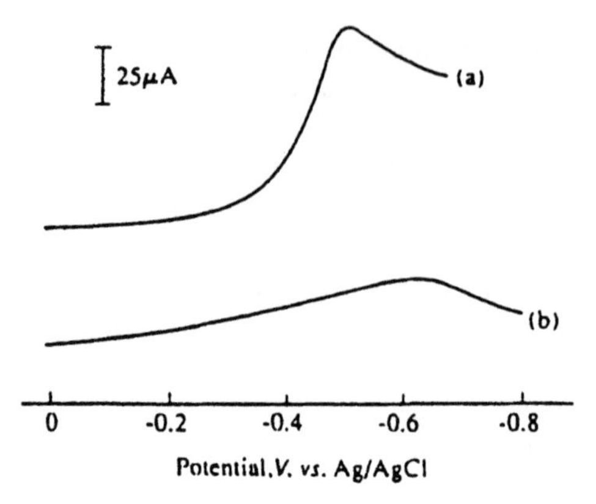

Figure 1. Voltammogram of oxygen reduction at both Nafion-MV CME (curve **a**) and bare glassy carbon electrodes (curve **b**) in pH 3, 0.1 M NaCl/HCl solution (DO concentration: 7.5 ppm) Scan rate: 80 mV/s.

For example, the Nafion-MV CME/ketonic acid biosensor can be used to detect glutamic pyruvic transaminase (GPT). In the presence of GPT, alanine reacts with α-ketoglutarate to produce acetonic acid, as shown in reaction(1):

$$\alpha\text{-ketoglutarate} + \text{L-alanine} \xrightarrow{\text{GPT}} \text{L-glutamic acid} + \text{acetonic acid} \quad (1)$$

Then, the acetonic acid produced in this reaction reacts with dissolved oxygen in the presence of acetonic acid oxidase:

$$\text{Acetonic acid} + H_3PO_4 + O_2 \rightarrow H_3OPO_3H_2 + \text{Ethyl acetate} + CO_2 + H_2O_2 \quad (2)$$

Thus, by measuring the concentration change of dissolved oxygen, the acetonic acid and subsequently the GPT activity are obtained.

2.3 Characteristics of Biosensors Using Nafion-MV CME

Response Range of Biosensors That Use Nafion-MV CME

Table 1 lists the linear response range of biosensors that use Nafion-MV CME as the substrate electrode.

Response Time of Biosensors That Use Nafion-MV CME

The response time of the above biosensor ranges from 5 to 50 s, depending on the individual enzyme employed.

Table 1. The Detection Range of Several Nafion-MV CME Biosensors

Enzymes	*Samples*	*Linear Range*
Glucose oxidase	Glucose	5.0×10^{-5}–2.5×10^{-3} mol/L [6]
Ascorbic acid oxidase	Ascorbic acid	7.5×10^{-7}–7.5×10^{-4} mol/L [5]
Pyruvate oxidase	GPT	0–110 μg/L [8]
Urate oxidase	Uric acid	5.0×10^{-6}–1.0×10^{-4} mol/L [7]
Cholesterol oxidase	Cholesterol	2.5×10^{-7}–1.0×10^{-4} mol/L [9]

Lifetime of Biosensors That Use Nafion-MV CME

Depending on the type of enzyme immobilized on the Nafion-MV CME, the lifetime of the resultant biosensor varies from more than two and half months for glucose biosensors to eight days for keto acid biosensors. The typical lifetime of other biosensors is about two weeks. Figure 2 illustrates the response of the uric acid biosensor versus time at 20 °C.

Ability of Biosensors to Resist Interference

Oxygen concentration should remain constant when using Nafion-MV CME biosensors and other coexisting electroactive species. For example, K^+, Na^+, Cl^-, and HPO_4^{2-}, often found in serum or urine, do not significantly interfere, even at concentrations 1000 times higher than the analyte. Other coexisting electroactive species do not interfere at concentrations 50–100 times higher than the analyte.

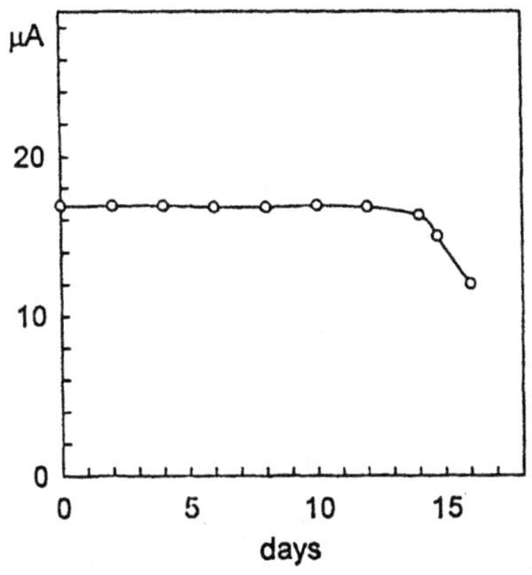

Figure 2. Lifetime of the Nafion-MV CME uricase oxidase biosensor. Solution: pH 7.4 citric acid-dibasic sodium phosphate buffer with 5.0×10^{-5} mol/L of uric acid; scan rate 100 mV/s; temperature 20 °C.

Table 2. Cholesterol Measurements in Human Serum Samples

No.	*Cholesterol in Serum Samples (mg/100 mL)*	*Cholesterol Added (mg/100 mL)*	*Cholesterol in Diluted Serum Samples after Standard Addition (mg/100 mL)*	*Recovery (%)*	*Cholesterol in Serum Samples Before Dilution (mg/100mL)*	*Cholesterol in Serum Samples by Spectrophotometry (mg/100 mL)*
1	1.36	1.0	2.31	101	130	131
2	1.69	1.0	2.69	100	169	165
3	1.42	1.0	2.41	99	142	141
4	1.98	1.0	2.96	98	198	195
5	2.42	1.0	3.40	99	241	240
6	2.67	1.0	3.62	95	267	262
7	3.34	1.0	4.32	98	334	330
8	3.15	1.0	4.12	97	315	312

2.4 Application

Nafion-MV CME biosensors satisfactorily detect blood sugar, ascorbic acid, GPT, uric acid, and cholesterol in human blood serum. Table 2 shows the determination of cholesterol in human blood serum, proving the usefulness of the Nafion-MV CME cholesterol biosensor.

3. BIOSENSOR BASED ON POLY(*o*-PHENYLENEDIAMINE)

Enzyme immobilization is often achieved by trapping it in conductive or redox polymer films during electropolymerization. This immobilization method is simple and allows control of the thickness of polymer film, improving the stability of the resultant electrode and prolonging its lifetime. Moreover, because of the change in selectivity often associated with the polymer film and the consequent selective penetration, electrode fouling and possible interference is reduced greatly [11,12]. Such examples include the glucose biosensor and the ascorbic acid biosensors, both of which utilize poly(*o*-phenylenediamine) as the substrate electrode [13,14].

These biosensors are prepared by immersing a glassy carbon electrode in a solution containing *o*-phenylenediamine and a particular oxidase and then applying a constant potential of +0.65 V (vs. Ag/AgCl) for 20 min. The oxidase is immobilized during polymerization.

The cyclic voltammogram of poly(*o*-phenylenediamine) (PPD) CME itself shows a reduction peak at –0.65 V and an oxidation peak at –0.47 V (vs. Ag/AgCl) in an oxygen-free HAc/NaAc buffer, corresponding to the reaction:

$$PPD_{ox} + e = PPD_{red}$$

As shown in Figure 3, this PPD CME catalyzes the reduction of dissolved oxygen:

$$PPD_{red} + H^+ + 1/2O_2 = PPD_{ox} + 1/2H_2O_2$$

Figure 3 suggests that the PPD CME has a catalytic effect on the reduction of dissolved oxygen, and the peak potential is at a plain carbon electrode. The peak current is increased by a factor of 2.5. Thus, the detection sensitivity of dissolved oxygen is improved by using a PPD CME.

If an oxidase is entrapped in PPD film, a corresponding biosensor is formed. For example, when ascorbic acid oxidase is entrapped in PPD, the resultant biosensor selectively catalyzes the oxidation of L-ascorbic acid. In the same reaction dissolved oxygen is consumed quantitatively, as follows:

$$\text{L-ascorbic acid} + O_2 \xrightarrow{\text{ascorbic acid oxidase}} \text{L-dehydroascorbic acid} + H_2O_2$$

Using this biosensor, two catalytic processes are operating, the catalytic reduction of oxygen and the catalytic oxidation of L-ascorbic acid. Therefore, both the selectivity and sensitivity of ascorbic acid detection can be improved. Based on our experiments, the linear response range of ascorbic acid detection is from 2.5×10^{-5} to 1.0×10^{-2} mol/L, with a response time of 10 s. K^+, Na^+, and Ac^- do not significantly interfere, even at concentrations 5000 times higher than the analyte.

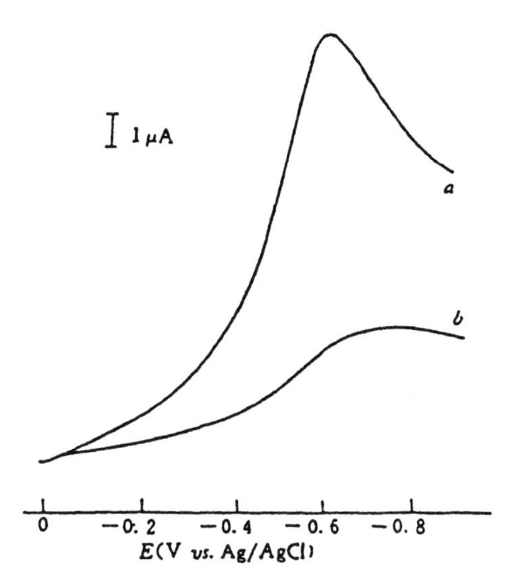

Figure 3. Comparison of oxygen reduction peaks of both poly(*o*-phenylenediamine) modified electrode and glassy carbon electrode peaks: (**a**) poly(*o*-phenylenediamine) modified electrode; (**b**) glassy carbon electrode. Solution: pH 5.6 HAc-NaAc buffer saturated with air.

For Ca^{2+}, glucose, uric acid, urea, L-proline, L-lysine, glycine, and fructose, concentrations 100 times higher than the analyte do not significantly interfere.

We also tested the *p*- and *m*- isomers of *o*-phenylenediamine. The experimental results indicate that *o*-phenylenediamine is the best choice based on the quality of the resultant polymer film. This conclusion is consistent with the results of quantum chemical calculations and surface IR spectra [15].

4. BIOSENSOR BASED ON A CHEMICALLY MODIFIED ELECTRODE WITH ELECTRON TRANSFER MEDIA

In the case of amperometric enzyme-based biosensors, attention has focused on the use of electron transfer media to eliminate the effects caused by unstable oxygen concentration and the decreased activity of enzymes caused by the enzymatic reaction product H_2O_2 [16,17]. Tetrathiofulvalene (TTF) and ferrocene derivative (Fc) are two commonly used electron transfer media. In this section we describe the preparation and characteristics of a glucose oxidase biosensor based on carbon paste CMEs with Nafion-TTF and carboxylforrecene.

4.1 Glucose Oxidase Biosensor with Tetrathiafulvalene (TTF) CME [18]

As a cation exchanger, the structure of Nafion can be expressed as

$$(CF_2CF_2) \times (CFCF_2)_y$$
$$\qquad | $$
$$O\text{-}(C_3F_6)\text{-}O\text{-}CF_2CF_2\text{-}SO_3\text{-}Na^+$$

Therefore, Nafion film coated on a substrate electrode adsorbs TTF rather than exchanging ions with TTF^+ and $TTF2^+$. In this way, the electron transfer medium is attached firmly to the electrode surface.

A glucose biosensor based on Nafion-TTF was prepared as follows: 2 μL of 0.1% Nafion/methanol solution was applied to a clean and dry glassy carbon surface, the surface was dried at room temperature, and then 5 μL of 0.2% TTF/ethanol solution was applied before drying again at room temperature. Finally, 8 μL of freshly prepared glucose oxidase-bovine albumin-pentyl dialdehyde was applied evenly to the surface of the Nafion-TTF coating, then it was stored at 4 °C for enzyme immobilization.

Figure 4 shows the cyclic voltammogram obtained with a Nafion-TTF CME. The two waves correspond to the electrochemical process $TTF \rightarrow TTF^+ \rightarrow TTF^{2+}$. Both ΔE_{p1} and ΔE_{p2} equal approximately 60 mV, indicating two reversible one-electron processes.

Both TTF^+ and TTF^{2+} function as glucose oxidase (GOD) oxidizing agents, and the biosensor response mechanism is expressed as follows (take TTF^{2+} as an example):

$$\text{Glucose} + \text{GOD(FAD)} \rightarrow \text{GA} + \text{GOD(FADH}_2)$$

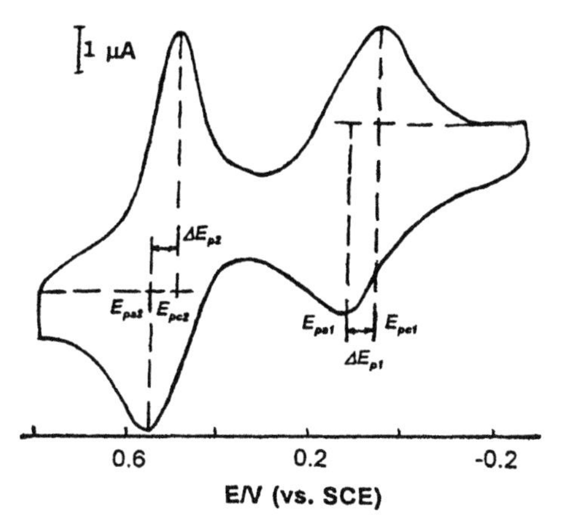

Figure 4. Cyclic voltammogram of FFT buffer: 0.02 mol/L HAc/NaAc (pH 5.0).

$$\mathrm{GOD(FADH_2) + TTF^{2+} \rightarrow TTF + GOD(FAD)}$$

$$\mathrm{TTF \rightarrow TTF^{2+} + 2e}$$

where GA is gluconic acid, FAD is flavine-adenine dinucleotide, the redox center of GOD, and $FADH_2$ is the reduced form of FAD. In these processes, TTF functions only as an electron transfer medium, and the consumption of glucose is quantitatively determined by the increase of the TTF oxidation peak at +0.55 V (vs. SCE). This glucose biosensor has a linear response in the glucose concentration range from 4.0×10^{-5} to 2.0×10^{-3} mol/L and a response time of about 20 s. In 4.0×10^{-4} mol/L glucose, none of the following significantly interfere: K^+, Na^+, or Ca^{2+} at 1000-fold concentrations; sucrose, urea, citric acid, L-glutamine, or L-hydroxyproline at 100-fold concentrations; glutathione at 50-fold concentrations; lysine at 10-fold concentrations; or ascorbic acid at twofold concentrations.

4.2 Glucose Biosensor Based on Carboxyl Ferrocene Carbon Paste CME [19]

Carboxyl ferrocene (Fc) carbon paste was made by mixing 0.5 g of Fc with 3.5 g of graphite powder and paraffin oil. The hydrophobicity of the carbon paste means that it is stable in aqueous solution. Glucose oxidase is immobilized simply by applying a mixture of glucose oxidase/bovine albumin/pentyl diadhydeon Fc carbon paste to the electrode surface. To prevent the enzyme film from becoming detached and to reduce the contamination of the biosensor electrode surface, a nylon gauze (300-mesh) coated with Nafion was used to cover the biosensor electrode surface.

Figure 5 is the cyclic voltammogram of this glucose biosensor based on a Fc carbon paste CME. The wave at +0.28 V (vs. SCE) is the Fc oxidation peak. Following the addition of glucose, this oxidation peak increased to become propor-

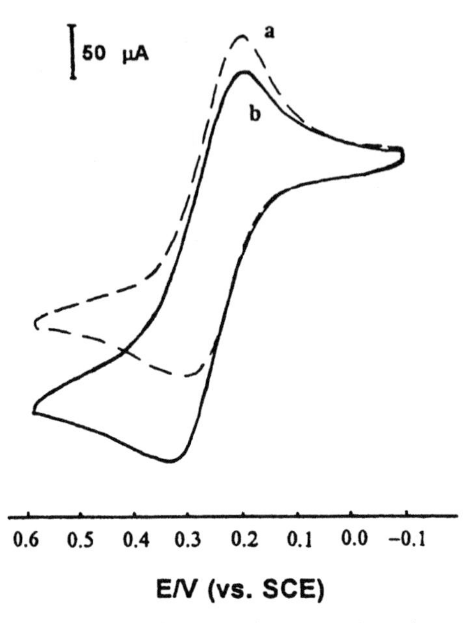

Figure 5. Cyclic voltammogram of glucose biosensor based on carboxyl ferrocence CME: (**a**) 0.1 mol/L KH_2PO_4-Na_2HPO_4 Buffer solution (pH 7.0); (**b**) a + 9.0×10^{-3} mol/L glucose.

tional to the glucose concentration. The linear response range is from 5.0×10^{-4} to 1.4×10^{-2} mol/L of the glucose concentration with a response time of 40 s. Molecules in blood sera, such as alanine, methionine, c-tyrosine, lysine, phenylalanine, ascorbic acid, uric acid, L-hydroxyproline, glycine, sucrose, lactose, urea, cysteine, citric acid, and glutathione, do not significantly interfere with glucose determination.

5. APPLICATION OF AN AVIDIN-BIOTIN SYSTEM FOR ENZYME SENSORS

Recently, much attention has been devoted to modifying electrode surfaces at the molecular level with enzymes [20–22]. The concept of modification at the molecular level has made it possible to fabricate quickly responding biosensors by removing conventional types of thick membranes from the electrode surface. These membranes often disturb the smooth transport of analytes and reaction products and result in slow sensor response. However, the magnitude of the output signal of such sensors is not always high enough because a limited amount of enzymes are available on the thin layer of the modified surface. Therefore, regulating the enzyme load on the surface of an electrode is crucial to developing biosensors with high and rapid response rates.

The use of an avidin-biotin system reported recently immobilizes enzymes on the electrode surface by taking advantage of the strong affinity between avidin- and biotin-labeled enzymes [23,24]. These studies demonstrate the usefulness of the avidin-biotin system for modifying electrode surfaces at the molecular level to prepare enzyme sensors. However, the avidin-biotin-based sensors reported so far often suffer from low response. This arises from an insufficient enzyme load caused by the molecular-level (or monolayer-level) modification of the electrode surface. To improve the response, a large amount of enzyme should be immobilized on the electrode.

We used a self-assembled monolayer of aminoethanethiol at an Au or indium tin oxide (ITO) surface. This monolayer would react further with glutaraldehyde to form an active surface for avidin. An alternative method is to use a sulfur-containing biotin derivative to form a self-assembled monolayer on a gold surface, and then to prepare a monolayer or multilayer enzyme biosensor using the avidin-biotin technique. The present techniques may help to regulate performance characteristics of enzyme sensors, such as lower and higher detection limits and the magnitude of the output current, by simply changing the number of deposits of enzyme layers on the electrode surface [25–27].

The procedure for modifying the electrode surface is illustrated schematically in Figure 6. The ITO electrode was washed successively with sulfuric acid, chloroform, water, and methanol, and then the pretreated ITO electrode was immersed in 1×10^{-4} mol/L 2-aminoethanethiol/ethanol for 16 h and then in 30% glutaraldehyde (in water) for 2 h at room temperature (Electrode 1a). An alternative approach is to use an electrochemically activated gold electrode. The gold electrode was immersed in 1×10^{-4} mol/L bis(*N*-biotinylhydrazidoethyl) disulfide (BDS) solution for 14 h (Electrode 1b). Electrodes 1a and 1b were immersed in avidin (50 μg/mL) for 30 min, and in biotin-labeled glucose oxidase (B.GOD) (50 μg/mL) for 30 min to form electrode 2. Electrode 1a or 1b was immersed in a mixed solution of 50 μg/mL avidin and 25 μg/mL B.GOD (called ABC) for 2 h to form Electrode 3. Electrode 2 was immersed in 50 μg/mL B.GOD for 30 min to form Electrode 4. Electrodes 5 and 6 were prepared by repeating the procedures for making Electrodes 3 and 4 twice and three times, respectively.

Figure 7 shows the response obtained at electrodes 2–6 for 1.0×10^{-2} M glucose. In the case of Electrode 2, the B.GOD monolayer was immobilized via avidin. From Figure 7a we can see that some GOD has been immobilized at the electrode surface. Therefore, electrode 2 has some response to glucose. However, the response shown in Figure 7a is rather small because it is only a monolayer of GOD, a technique called the avidin-biotinylated enzyme complex or ABC technique. Figure 7b shows a much larger response than that in Figure 7a. If we let the ABC membrane at the electrode surface interact further with B.GOD, more B.GOD can be immobilized, because active biotin exists at the ABC surface. This explains why the response of

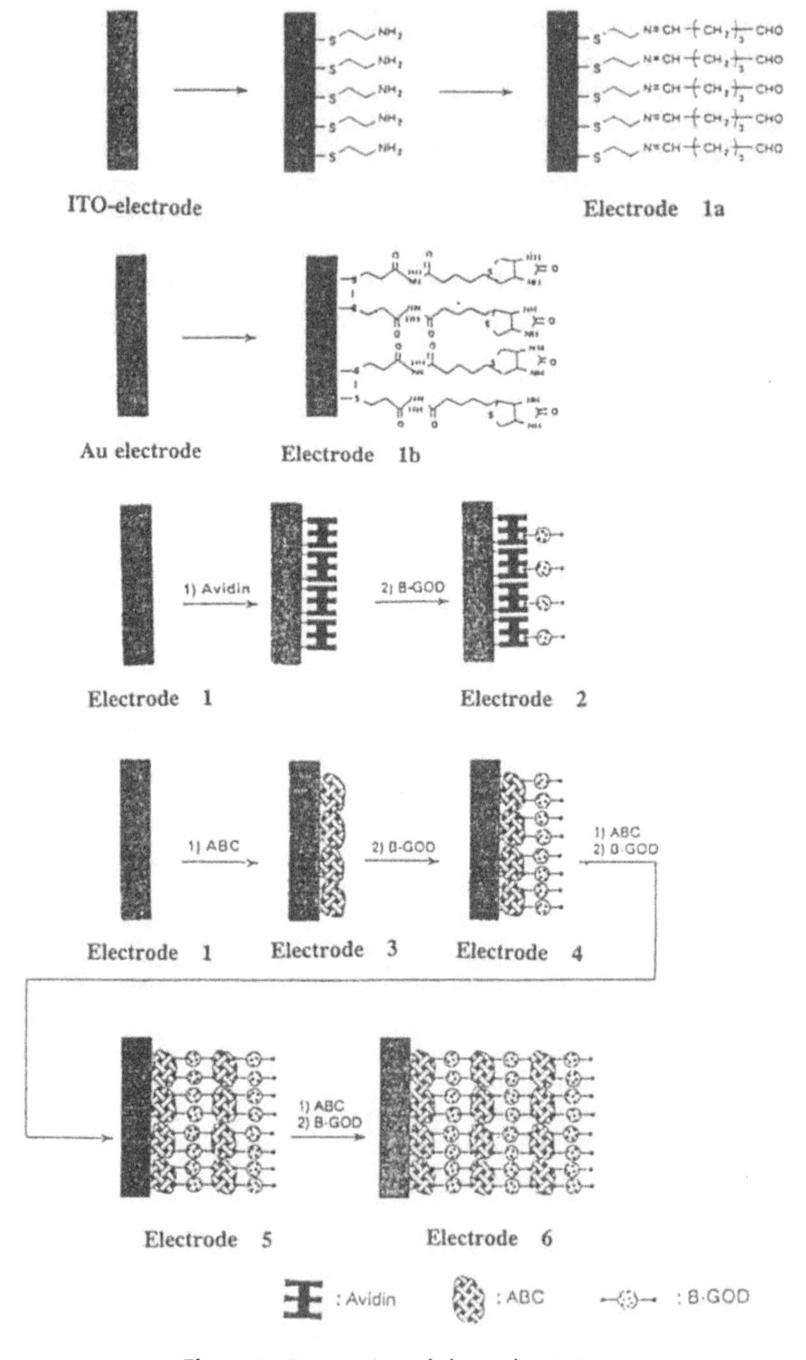

Figure 6. Preparation of electrodes **1–6**.

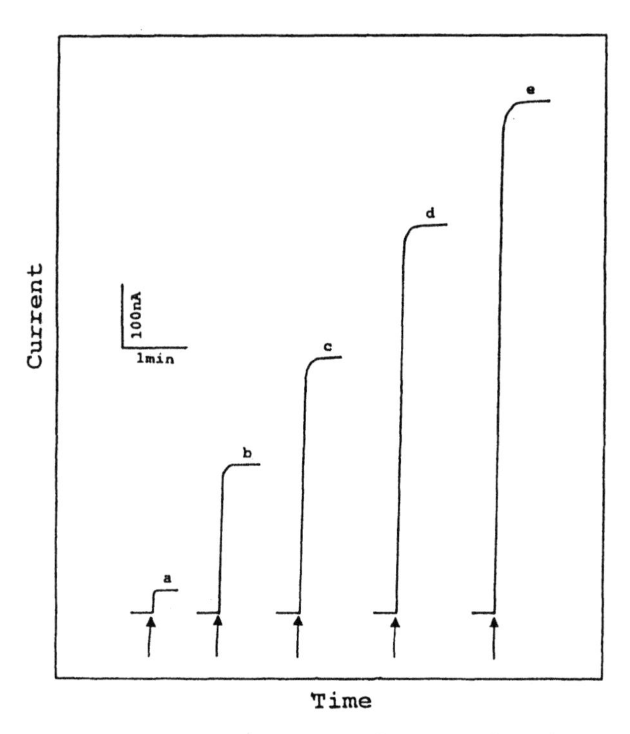

Figure 7. Typical response of electrode 2(**a**), 3(**b**), 4(**c**), 5(**d**), and 6(**e**) to 10 mmol/L glucose. Glucose was added to the sample solution at the time indicated by the arrow.

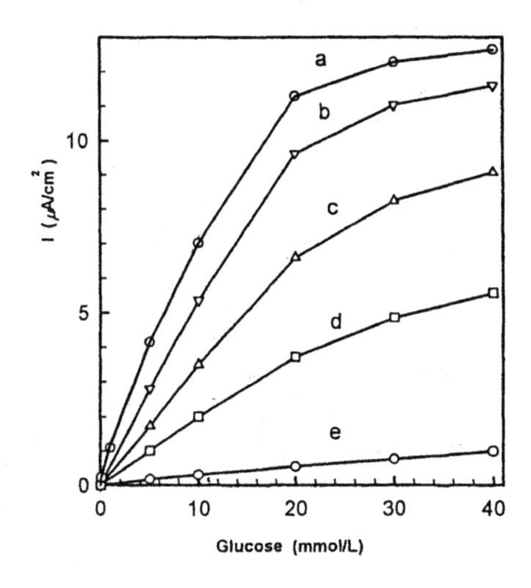

Figure 8. Calibration curves of electrodes 2–6. Electrode: 6(**a**), 5(**b**), 4(**c**), 3(**d**), and 2(**e**).

7c is higher than that of 7b. In the case of Electrodes 5 and 6, multi ABC-B.GOD layers were covered at the electrode surface. The amount of fixed GOD is even greater. Therefore, the responses of 7d and 7b are even higher. These phenomena give us an opportunity to study the relationship between the electrode response and the amount of GOD fixed at the molecular level.

Figure 8 shows the concentration dependence obtained at electrodes 2–6. As expected, Electrode 6 is the most sensitive. It responds linearly in the glucose concentration range of 6×10^{-6} to 5×10^{-3} mol/L. Although Electrodes 2–5 are less sensitive than Electrode 6 if the glucose concentration is relatively high, say 1–40 mmol/L, the blood glucose concentration levels for most diabetes patients usually fall in this range. This means that the technique may help to control the performance characteristics of enzyme sensors simply by changing the thickness of the enzyme layer deposited on the electrode surface.

6. OTHER BIOSENSORS

6.1 Ascorbic Acid Biosensor with Dual Enzymes [28,29]

Conventional biosensors that have noble metals as the substrate electrodes can be used to measure ascorbic acid at relatively high positive potentials. However, interference caused by uric acid or other species often raises problems. The dual-enzyme biosensor was designed to overcome these problems. For example, both ascorbic acid oxidase and peroxidase can be simultaneously immobilized on a glassy carbon electrode surface. The resultant dual-enzyme biosensor can be used to determine ascorbic acid at about 0 V (vs. Ag/AgCl) if a small amount of $K_4Fe(CN)_6$ is added to the sample solution [28].

Ascorbic acid oxidase catalyzes the oxidation of ascorbic acid, producing H_2O_2. In the presence of peroxidase, H_2O_2 oxidizes $Fe(CN)_6^{4-}$ to $Fe(CN)_6^{3-}$ and the $Fe(CN)_6^{3-}$ produced is easily reduced at a low potential:

$$\text{L-ascorbic acid} + O_2 \xrightarrow{\text{ascorbic acid oxidase}} \text{L-dehydroascorbic acid} + H_2O_2$$

$$H_2O_2 + 2\,Fe(CN)_6^{4-} + 2\,H^+ \xrightarrow{\text{peroxidase}} 2\,Fe(CN)_6^{3-} + 2\,H_2O$$

$$Fe(CN)_6^{3-} + e \xrightarrow{0.0\text{V (vs. Ag/AgCl)}} Fe(CN)_6^{4-}$$

At 0.0 V (vs. Ag/AgCl), interference from uric acid and other electroactive species can be eliminated, improving the selectivity of the biosensor. Experimental data indicate that this dual-enzyme biosensor responds linearly in the ascorbic acid concentration range of 5.0×10^{-4} to 1.5×10^{-6} mol/L. In a 1.0×10^{-4} mol/L ascorbic acid solution, neither uric acid nor urea at 100-fold concentrations interfere detectably.

Based on the same principle, GOD and peroxidase can be simultaneously immobilized on a glassy carbon electrode surface to form another dual-enzyme glucose biosensor. This biosensor shows excellent selectivity against uric acid, ascorbic acid, tryptophan, lysine, and cysteine and thus is more selective and sensitive than the corresponding single-enzyme biosensor. Its lifetime is also longer [29].

6.2 Glucose Oxidase Biosensor with Casein as the Carrier [30]

Enzyme immobilization is a key issue in preparing an enzyme-containing biosensor. The normal cross-linking approach employs bovine serum albumin (BSA) as the carrier [31]. Then, enzyme is immobilized in the presence of glutaraldehyde. Because the BSA membrane can block H_2O_2, penetration can be improved greatly. Therefore, the sensitivity of the resultant biosensor can also be improved. Here we describe an electrochemical approach to adsorb casein onto a Pt electrode surface. Then, in the presence of glutaraldehyde, GOD is immobilized by cross-linking to form the glucose biosensor. In practice,a pretreated Pt electrode is immersed in a 20 mg/mL casein phosphate buffer (pH 6) solution. A constant potential of +0.6 V (vs. Ag/AgCl) is applied to the Pt electrode for 10 min, and it is transferred to a 25% aqueous glutaraldehyde solution and left in the solution for 45 min. Then it is transferred to pH 7.4 phosphate buffer containing 5 mg/mL GOD for 1 h.

Casein molecules exist as anions when the pH > 4.6. If a specific positive potential is applied to the Pt electrode, casein is adsorbed on the Pt electrode surface to form a uniform and firm casein film, which in turn functions as the BOD carrier. Figure 9 compares the response currents of two glucose biosensors. The glucose biosensor

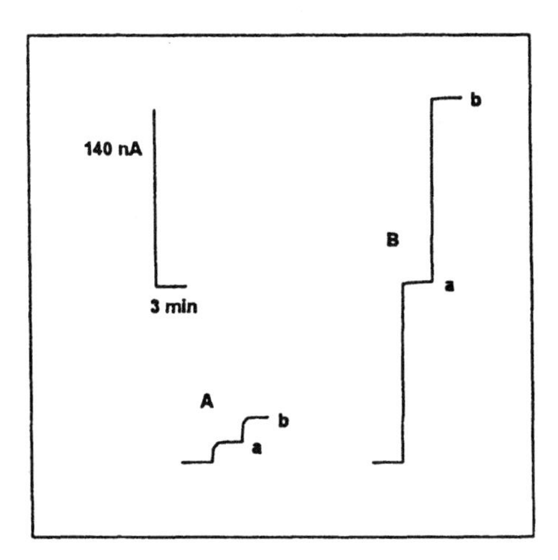

Figure 9. Current response of glucose biosensor: **A**: Using BSA as the carrier; **B**: Using casein as the carrier; (**a**) 1×10^{-4} mol/L; (**b**) 2×10^{-4}mol/L glucose.

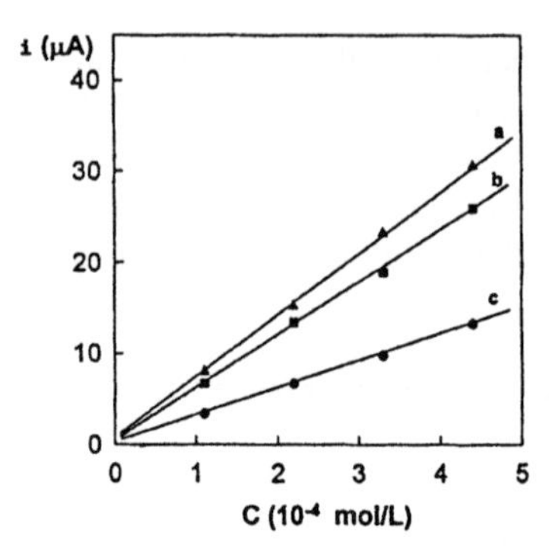

Figure 10. H_2O_2 penetration at various electrodes: (**a**) Pt electrode; (**b**) Pt-casein electrode; (**c**) Pt-BSA electrode.

using casein as the BOD carrier has a much higher response than that of the glucose biosensor using BSA. One important reason for this is the difference with which H_2O_2 is free to penetrate between the two glucose biosensors, as described previously. Further evidence can be found in Figure 10. Compared with a bare Pt electrode, the casein film causes only a 17% decrease in H_2O_2 reduction current, whereas the BSA membrane causes a 59% decrease.

At 4 °C, the casein glucose biosensor can be stored for at least one month without significant loss of sensitivity. The linear glucose response range is from 1.0×10^{-5} to 1.7×10^{-3} mol/L.

7. CONCLUSION

Several types of biosensors based on chemically modified electrodes have been described. One type uses Nafion-MV CME as the substrate electrode, in which the Nafion-MV catalyzes the reduction of oxygen, improving both sensitivity and selectivity. Biosensors of this type detect blood sugar, ascorbic acid, and cholesterol in human sera with satisfactory results. Biosensors can also be prepared by immobilizing suitable enzymes during the process of electropolymerizing *o*-phenylenediamine. The resultant biosensor has good sensitivity and selectivity and a relatively long lifetime. Biosensors using TTF or FCA as the electron transfer media between enzyme and substrate improve the detection sensitivity, reducing the interference caused by dissolved oxygen. The avidin-biotin technique and self-assembled membrane technique can be used to prepare a monolayer or multilayer avidin-biotin labeled glucose oxidase biosensor. As a result, the relationship between the amount of immobilized enzyme and the detection sensitivity can be studied at the molecular level, and the response range of the biosensor can be adjusted and controlled by varying the number of layers that cover the electrode surface. Finally, a dual-enzyme biosensor was prepared by immobilizing peroxidase and oxidase on the electrode surface. The resultant biosensor responds to H_2O_2 at low potential, effectively eliminating the interference caused by uric acid and resulting in better sensitivity and selectivity. Furthermore, using casein instead of bovine serum albumin (BSA) in the process of enzyme immobilization via cross-linkage facilitates H_2O_2 penetration and enhances sensitivity.

REFERENCES

[1] Updike, S.J., Hicks, G.P. (1967). The enzyme electrode. *Nature*, **214**, 986.

[2] Janata, J. (1990). Chemical sensors. *Anal. Chem.*, **62**, 33R.

[3] Cass, A.E.G. (ed.). (1990). *Biosensors, A Practical Approach.* Oxford University Press, New York.

[4] Oudelka, M.K. (1989). *Sensors and Actuators.* **18**, 157.

[5] Jin, L.T., Liu, H.Y., Fang, Y.Z. (1992). Ascorbic acid oxidase biosensor based on the glassy carbon electrode modified with Nafion and methyl viologen. *Chin. J. Anal. Chem.*, **20**, 515.

[6] Jin, L.T., Liu, H.Y., Fang, Y.Z. (1991). A study of glucose oxidase biosensor based on glassy carbon electrode modified with Nafion and methyl viologen. *J. East China Normal Univ.*, **9**, 50.

[7] Jin, L.T., Ye, J.S., Tong, W., Fang, Y.Z. (1993). A study of uricase biosensor based on glassy carbon electrode modified with Nafion and methyl viologen. *Mikrochim. Acta.*, **112**, 71.

[8] Jin, L.T., Ye, J.S., Fang, Y.Z. (1993). Pyruvate oxidase biosensor for the determination of GPT based on glassy carbon electrode modified with Nafion and methyl viologen. *Chem. J. Chin. Univ.*, **14**, 1210.

[9] Jin, L.T., Zhao, G.Z., Fang, Y.Z. (1994). Cholesterol oxidase biosensor based on glassy carbon electrode modified with Nafion and methyl viologen. *Chin. J. Chem.*, **12**, 343.

[10] Jin, L.T., Jin, P., Ye, J.N., Fang, Y.Z. (1992). Determination of dissolved oxygen by catalytic reduction on Nafion-methyl viologen chemically modified electrode. *Talanta*, **39**, 145.

[11] Sasso, S.V., Pierce, R.J., Walla, R., Yacynych, A.M. (1990). Electropolymerized 1,2-diaminobenzene as a means to prevent interferences and fouling and to stabilize immobilized enzyme in electrochemical biosensors. *Anal. Chem.* **62**, 1111.

[12] Geise, R.J., Adams, J.M., Barone, N.J., Yacynych, A.M. (1991). *Biosensors & Electronics*, **6**, 151.

[13] Jin, L.T., Zhao, G.Z., Fang, Y.Z. (1994). A study of ascorbic acid sensor based on ascorbic oxidase immobilized in an electropolymerized Poly)o-phenylenediamine) film. *Chem. J. Chin. Univ.*, **15**, 189.

[14] Liu, H.Y. (1992). *M.S. Thesis, East China Normal University.*

[15] Zhao, G.Z. (1993). *M.S. Thesis, East China Normal University.*

[16] Williams, D.L., Doig, A.R., Korosi, J.A. (1970). Electrochemical enzymatic analysis of blood glucose and lactate. *Anal. Chem.*, **42**, 118.

[17] Palleschi, G., Turner, A.E. (1990). Amperometric tetrathiafulvalene-mediated lactate electrode using lactate oxidase absorbed on carbon foil. *Anal. Chim. Acta*, **234**, 459.

[18] Jin, L.T., Mao, Y.P., Sun, X.Y., Fang, Y.Z. (1994). A study of glucose biosensor using tetrathafulvalene as a mediator. *Anal. Lab.*, **13**, 13.

[19] Mao, Y.P. (1994). *M.S. Thesis, East China Normal University.*

[20] Bourdillon, C., Bourgeoie, D., Tomas, D. (1980). Covalent linkage of glucose oxidase on modified glassy carbon electrodes. Kinetic phenomena. *J. Am. Chem. Soc.*, **102**, 4231.

[21] Johnson, K.W. (1991). Reproducible electrodeposition of biomolecules for the fabrication of miniature electroenzymic biosensors. *Sensors and Actuators B.*, **5**, 85.

[22] Bianco, P., Haladjian, J., Bourdillon, C. (1990). Immobilization of glucose oxidase on carbon electrodes. *J. Electroanal. Chem.*, **293**, 151.

[23] Lee, S., Anzai, J., Osa, T. (1993). Enzyme-modified Langmuir-Blodgett membranes in glucose electrodes based on avidin-biotin interaction. *Sensors and Actuators B.*, **12**, 153.

[24] Pantato, P., Kubr, W.G. (1993). Dehydrogenase-modified carbon-fiber microelectrodes for the measurement of neurotransmitter dynamics. 2. Covalent modification utilizing avidin-biotin technology. *Anal. Chem.*, **65**, 623.

[25] He, P.G., Anzai, J., Osa, T. (1994). A facile method to regulate enzyme load on biosensor electrode based on avidin/biotin complexation. *Pharmazie*, **49**, 8.

[26] He, P.G., Anzai, J., Osa, T. (1994). Preparation of enzyme multilayers on electrode surface by use of avidin and biotin-labeled enzyme for biosensor applications. *Mater. Sci. Eng.*, **2**, 103.

[27] He, P.G., Fang, Y.Z., Anzia, J., Osa, T. (1997). Self-assembled biotinylated disulfide derivative monolayer on gold electrode for immobilizing enzymes. *Talanta*, **44**, 885.

[28] Jin, L.T., Liu, H.Y., Fang, Y.Z. (1993). Bienzyme electrode for determination of ascorbic acid in blood serum sample. *Chin. J. Anal. Chem.*, **21**, 584.

[29] Jin, L.T., Zhao, G.Z., Fang, Y.Z. (1994). A study of bienzyme electrode for determination of glucose in blood serum samples. *J. East China Normal Univ.*, **2**, 60.

[30] Lu, X., Li, Z.,He, P.G.,Fang, Y.Z. (1996). Study of glucose biosensor using casein as carrier. *J. East China Normal Univ.*, **4**, 63.

[31] Pandey, P.C. (1988). A new conducting polymer-coated glucose sensor. *J. Chem. Soc., Faraday Trans.*, **84**, 2259.

ASPECTS OF THE DEVELOPMENT OF TISSUE AND IMMUNOSENSORS

Guo-Li Shen, Xia Chu, and Ru-Qin Yu

OUTLINE

Advances in Biosensors
Volume 4, pages 59–78.

ISBN: 0-7623-0073-6

ABSTRACT

This chapter discusses biosensors that have tissue slices and immuno elements incorporated in their transduction scheme, and reviews their design, characterization, and application. The chapter covers biosensors that incorporate immobilized mammalian and plant tissue slices and immunosensors based on potentiometric and piezoelectric detection.

1. INTRODUCTION

Biosensor technology represents one of the most exciting new trends in the development of chemical sensors. Incorporating a biological component as a special functional element, which has selectivity and other valuable characteristics, in the overall transduction sequence adds new dimensions and analytical capability to chemical sensors. Tissue slices and immuno elements are examples of such biological components. This review covers some aspects of developing these types of biosensors based on research work undertaken in the authors' laboratory.

2. TISSUE-BASED BIOSENSORS

Following the pioneering work of Rechnitz et al. [1] that reported a biosensor immobilizing a thin slice of porcine kidney at the surface of an ammonia gas sensing probe, the use of different animal and plant tissue materials has been the subject of many researchers' investigations [2–8]. This kind of biosensor can compensate for the unavailability of isolated high-purity enzyme preparations for constructing a desired sensor. The use of mammalian and plant tissues provides relatively high enzyme loading retained in the natural, "immobilized" state that is presupplied with the requisite cofactor or related agents. The lifetime of a biosensor prepared this way is usually longer than that of the conventional isolated, enzyme-based sensor. Tissue-based biosensors are relatively simple and inexpensive devices for routine use. In this section, the preparation and characterization of some tissue-based biosensors studied in the authors' laboratory are discussed.

2.1 Biosensors Using Mammalian Kidney Slices

Rechnitz et al. [1] used porcine kidney slices and the mitochondria fraction-flow porcine kidney cells to prepare sensors for glutamine. Ma et al. [9,10] used porcine and sheep kidneys to prepare sensors for glucose-6-phosphate and D-amino acid, respectively. The porcine kidney contains Cytochrome c (cyt. c) oxidase and succinic dehydrogenase. The latter catalyzes the following reaction:

$$\begin{matrix} CH_2COOH \\ | \\ CH_2COOH \end{matrix} \xrightarrow[\text{dehydrogenase}]{\text{succinic}} \begin{matrix} CH_2COOH \\ \| \\ CH_2COOH \end{matrix} + H_2$$

An equilibrium exists between the oxidized and reduced forms of central iron metal in Cytochrome c:

$$\text{Cyt. c (Fe III)} \underset{-e}{\overset{+e}{\rightleftharpoons}} \text{Cyt. c (Fe II)}$$

The oxidized form reacts with hydrogen:

$$2\ \text{Cyt. c (Fe III)} + H_2 \rightarrow 2\ \text{Cyt. c (Fe II)} + 2\ H^+$$

The reduced form reacts with oxygen in the presence of cytochrome c oxidase:

$$2\ \text{Cyt. c (Fe II)} + 1/2\ O_2 \underset{\text{oxidase}}{\overset{\text{Cyt. c}}{\longrightarrow}} 2\ \text{Cyt. c (Fe III)} + O^{2-}$$

The combination of H^+ and O^{2-} forms H_2O and the overall reaction is formulated as

$$\begin{matrix} CH_2COOH \\ | \\ CH_2COOH \end{matrix} + 1/2\ O_2 \xrightarrow[\text{dehydrogenase}]{\text{succinic}} \begin{matrix} CH_2COOH \\ || \\ CH_2COOH \end{matrix} + H_2O$$

This overall reaction takes place only in the presence of both succinic dehydrogenase and Cytochrome c oxidase. Porcine kidney tissue is an ideal source of two enzymes necessary for the reaction sequence. By using the reagentless monitoring of oxygen consumption at an oxygen probe, a tissue-based Cytochrome c sensor can be realized [11]. The optimum reaction medium for Cytochrome c determination is a pH 7.3 phosphate buffer to which is added appropriate amounts of succinates as the substrate. The thickness of the tissue slice substantially affects the oxygen consumption rate (V):

$$V_{200\text{–}250\ \mu m} > V_{100\text{–}150\ \mu m} > V_{300\text{–}350\ \mu m}$$

The enzyme loading and diffusion rate are two factors that influence the oxygen consumption rate in opposite ways. An intermediate thickness must be selected as a compromise. It takes time for the sensor readings to reach steady state. A dynamic approach is recommended, and reproducible results are obtained in 1–2 min. The dynamic response of the sensor is linearly related to the logarithm of the concentration of Cytochrome c in the range of 10–227 μg/mL. The sensor is usable for up to 10 days.

2.2 Biosensors Using Mammalian Liver Tissues

Rechnitz et al. [12] used urease together with a bovine liver slice to prepare a sensor for arginine, and Mascini et al. [13] used bovine liver for an H_2O_2 sensor. The mitochondrial fractions of rabbit and porcine livers were used to prepare sensors for tyrosine, serine, and phenylalanine, respectively [14]. A comparative study has been undertaken in the authors' laboratory on using different mammalian liver slices to prepare sensors for hydrogen peroxide, L-ascorbic acid, and uric acid.

Hydrogen Peroxide Sensor [15]

Porcine, rabbit, and chicken liver tissues were taken to prepare a H_2O_2 sensor, using an oxygen electrode as the basic probe. All three types of liver tissue show a well-defined dynamic response to variations in H_2O_2 concentration, with a relatively wide linear response range and fairly good sensitivity (slope). Mammalian liver tissues are more active than, for example, slices of potato or sweet potato. The best response characteristics were obtained with the porcine liver slice. The variation of the slice thickness from 200 to 1000 μm does not substantially affect the sensor performance. Moderate thickness would guarantee sufficient enzyme loading, mechanical strength of the slice, and a longer sensor lifetime. To prevent the loss of enzymes, immobilization by glutaraldehyde cross-linking and the addition of NaN_3 is recommended for extending the sensor lifetime to about one month. The sensor shows a linear response range extending from 6.4×10^{-5} to 1.2×10^{-3} mol/L. A $1/V$ versus $1/C$ plot according to the Linewearer equation gives an estimate of the Michaelis constant (K_m) for the enzyme-catalyzed reaction as 0.543×10^{-3} mol/L. The sensor has been used to determine the H_2O_2 content in milk samples, because interference studies show that species coexisting in milk do not interfere with the H_2O_2 determination. The reaction kinetics of the H_2O_2 sensor were studied [16] by using Kalman filtering. The enzyme (E) participates in the reaction with the substrate (S) according to the well-known mechanism:

$$S + E \underset{k_{-1}}{\overset{k_1}{\rightleftharpoons}} ES \overset{k_2}{\longrightarrow} E + P,$$

where ES and P are the intermediate and final products, respectively. The reaction rate v can be expressed as

$$v = k_2[\mathrm{ES}] = \frac{k_2[\mathrm{E}][\mathrm{S}]}{K_m} = \frac{k_2\,[\mathrm{E_o}][\mathrm{S}]}{(K_\mathrm{m} + [\mathrm{S}])}$$

Discretizing from state $(k-1)$ to k and taking the limit of [S] approaching zero, one obtains

$$\int_{k_{-1}}^{k} -d[\mathrm{S}] = \int_{k_{-1}}^{k} \left(\frac{k_2[\mathrm{E_o}][\mathrm{S}]}{K_\mathrm{m}}\right) dt$$

$$[\mathrm{S}]_k = \varphi[\mathrm{S}]_{k_{-1}} + a_{k_{-1}}$$

$$\varphi = \exp\left(-\frac{k_2[\mathrm{E_o}]}{K_\mathrm{m}}\Delta t\right)$$

Here a is the parameter to be estimated and Δt is the sampling interval. It is assumed that the reaction rate is linearly related to [S] at the beginning of the reaction:

$$\frac{d[\mathrm{O_2}]}{dt} = k[\mathrm{S}] + b$$

The state model and measurement model can be formulated respectively as

$$\begin{pmatrix} [\mathrm{E}]_k \\ b_k \\ a_k \end{pmatrix} = \begin{pmatrix} \varphi & 0 & 1 \\ 0 & 1 & 0 \\ 0 & 0 & 1 \end{pmatrix} \begin{pmatrix} [\mathrm{S}]_{k_{-1}} \\ b_{k_{-1}} \\ a_{k_{-1}} \end{pmatrix} + Q_{k_{-1}}$$

$$Z_{k_{-1}} = (k \ \ 1 \ \ 0) \begin{pmatrix} [\mathrm{S}]_{k_{-1}} \\ b_{k_{-1}} \\ a_{k_{-1}} \end{pmatrix} + X_{k_{-1}}$$

where Z is the measured oxygen consumption rate, $d[\mathrm{O_2}]/dt$, Q is the model noise, and X is the measurement noise. The covariances of the state model and measurement errors are denoted as W and R, respectively. For Kalman iteration one takes $R = 0.01$ and $W_{11} = 0.001$ for iterations. Using Kalman filtering, the Michaelis constant was estimated as $K_\mathrm{m} = 0.553$ mol/L. Moreover, a modification of the Kalman filter algorithm provides the estimate of initial concentration of H_2O_2, which is the analyte concentration that is being sought. The Kalman filtering approach estimates the value of this concentration more accurately.

L-Ascorbic Acid Sensor [17]

Porcine liver tissue contains L-ascorbic acid oxidase, which can be utilized to prepare a sensor for the acid. In phosphate buffer solutions the variation of pH between 5.0 and 8.0 does not significantly influence the oxygen consumption rate. Because the ascorbic acid is more stable in a medium that has lower pH, a pH value of 6.1 is recommended for the sensor operation. The sensitivity and selectivity of the sensor for ascorbic acid is improved substantially by adding Cu^{2+} ions (up to 0.64 mmol/L) as an activator. A comparison of different antiseptic agents, such as chloroform, phenol, and NaN_3 shows that NaN_3 least affects sensor sensitivity. The tissue slice can be presented in 0.2% NaN_3 solution as long as 15 days. A newly prepared sensor can be used at least 100 times. The sensor response has been tested for possible interferents, including pyridoxine, glucose, L-lysine, L-serine, L-proline, L-tryptophan, valine, citrate, L-glutamic acid, DL-histidine, L-tyrosine, and DL-glutamine. Only the latter two substances interfere. In the presence of Cu^{2+}, the sensor shows a linear relationship between the logarithm of the ascorbic acid concentration and the dynamic response in the range 21.1–352.3 μg/mL. The Michaelis constant of the enzyme-catalyzed reaction was estimated at 0.8×10^{-3} mol/L. The sensor has been used to determine ascorbic acid in pharmaceutical preparations, and the results agree with those obtained by the pharmacological procedure.

Sensor For Uric Acid [18]

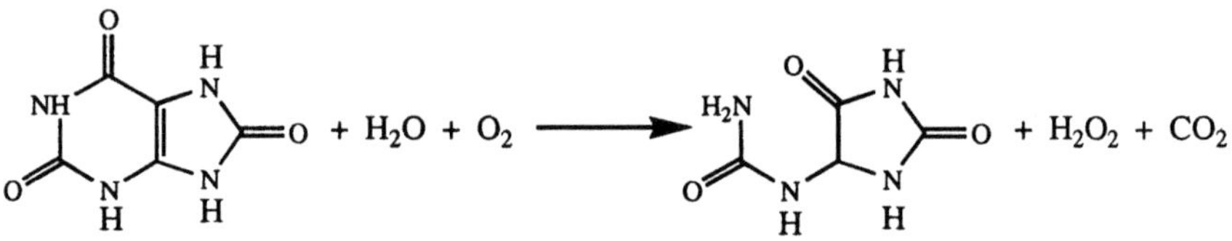

$$+ H_2O + O_2 \longrightarrow \quad + H_2O_2 + CO_2$$

The uricase in porcine liver tissue catalyzes the reaction
The reaction product, hydrogen peroxide, is in turn decomposed in the presence of peroxidase contained in the liver tissue:

$$H_2O_2 \rightarrow H_2O + 1/2\ O_2$$

Both of these reactions would lead to a decrease in oxygen concentration in the reaction medium. A pH 8.0 tris buffer is recommended as an optimum reaction medium. Immobilization by glutaraldehyde cross-linking extends the sensor lifetime. However, adding antiseptic agents, such as NaN_3, is not recommended, because these agents inhibit the enzyme-catalyzed reaction. The linear dynamic response covers the range of 0.2–61.6 μg/mL. The sensor has been used to determine uric acid in clinical urine and blood samples. The results agree fairly well with those obtained by the standard method using the phosphotungstate agent.

2.3 Banana-Pulp-Based Biosensors

Sidwell et al. reported the preparation of a biosensor for the neural transmitter dopamine using banana pulp [19]. Fonong [20] used banana tissue to prepare a sensor for oxalic acid. In the authors' laboratory, banana pulp tissue has been used to prepare several ordinary and flow-through biosensor devices.

Oxalate Sensor [21]

Banana pulp coupled with an oxygen probe was used to prepare the ordinary and the flow-through types of oxalate sensors. Banana pulp is a soft mass. After coating a layer of the pulp mass on the gas-permeable membrane of the oxygen probe, the layer is covered with a dialysis membrane. The optimum reaction medium for oxalate determination is a butane diacid solution at pH 2.5. Among the 20 possible interferents tested, including benzoic acid, ascorbic acid, citric acid, glucose, uric acid, and different amino acids, only catechol significantly interferes. The average analysis time is about 3–5 minutes. The sensor has been used to determine oxalate in tea, spinach, celery, radish, pear, apple, cucumber, and tomato samples.

Sensor for Catechol [22]

As mentioned earlier, interference from catechol is caused by the presence of

$$2\ \text{catechol} + O_2 \xrightarrow[\text{oxidase}]{\text{polyphenol}} 2\ \text{o-benzoquinone} + 2\,H_2O$$

polyphenol oxidase in the banana pulp, which causes the oxidation of catechol: The optimum reaction medium to determine catechol is a phosphate buffer at pH 7.7. *p*-Dihydroxybenzene and pyrogallol interfere somewhat.

Sensor for Epinephrine [23]

The epinephrine molecule contains phenolic hydroxy groups, which are oxidized by oxygen in the presence of polyphenol oxidase. This is the basis for using the banana pulp tissue sensor to determine epinephrine. A comparison of using different parts of banana pulp tissue shows that the central transparent part of the pulp is more active than the surface or the inner white parts. The optimum reaction medium is a phosphate buffer at pH 6.5. Although the sensor produces maximum current output at about 50 °C, a temperature of 30 °C is recommended for routine operation to avoid pulp deterioration. *m*-Aminophenol interferes somewhat.

Sensor for Methyl Dopa [24]

Methyl dopa or L-3-(3, 4-dihydroxyphenyl)-2-methylalanine (sesquihydrate) is an antihypertensive that has phenolic hydroxyl groups which oxidize in the presence of polyphenol oxidase:

$$2\ (HO)_2C_6H_3{-}CH_2{-}C(CH_3)(NH_2){-}COOH + O_2 \xrightarrow[\text{oxidase}]{\text{polyphenol}} 2\ O_2C_6H_3{-}CH_2{-}C(CH_3)(NH_2){-}COOH + 2\,H_2O$$

The banana-tissue-based biosensor is suitable for determining methyldopa. Because the polyphenol oxidase is not an enzyme specific to a unique reaction, it may catalyze the oxidation of many types of polyphenols. One way to improve the selectivity is to add activators to the reaction medium. Adding Cu^{2+} up to 1×10^{-4} mol/L substantially activates the sensor performance.

2.4 Biosensors Based on Jack Bean Tissue

Smit et al. [25] used jack bean meal coupled with an ammonia probe to construct urea sensors. We tried to immobilize a jack bean slice on an ammonia probe using a polyvinylchloride (PVC) membrane pH electrode that has a graphite inner contact [26]. A phosphate buffer solution is the reaction medium. The pH of the medium should be higher than 8.0 to meet the conditions required for ammonia sensor operation. At pH 9.0 the jack bean tissue biosensor shows satisfactory response characteristics, including linear range, slope, and reproducibility. The response properties deteriorate when the medium's pH is above 9.0, apparently because of decreased urease activity. Jack bean slices can be preserved for about 2 months in 0.02% chloroform solution at 4 °C. The operative lifetime of a newly prepared sensor is about two weeks. The linear response range extends from 4.6×10^{-5} to 1.0×10^{-2} mol/L. The experimentally estimated Michaelis constant is 3.39×10^{-3} mol/L.

2.5 Biosensors Based on Potato Slices

Schubert et al. [27] used potato slices combined with glucose oxidase to prepare sensors for phosphate and fluoride. Using potato slices containing polyphenol oxidase, we tried to prepare a sensor for catechol [28]. Usually the optimum pH range for plant tissues extends from 4.5 to 6.5. For the catechol sensor the maximum oxygen consumption rate (V) was obtained in a pH 5.6 phosphate buffer. The following rate sequence was observed for different slice thicknesses: $V_{100\,\mu m} > V_{200\,\mu m} > V_{300\,\mu m}$. Diffusion is the dominant factor affecting the reaction rate. Because the slice that is 100 μm thick gave relatively unstable current readings, a thickness of 200 μm is an optimum choice. Among the possible interferents tested, only *p*-dihydroxybenzene interfered. *m*-Dihydroxybenzene, α-naphthol, β-naphthol, trihydroxybenzene, *m*-aminophenol, *m*-nitrophenol, *p*-nitrophenol and common amino acids did not interfere with the catechol determination.

3. IMMUNOSENSORS

It is estimated that about half the world's diagnostics industry falls into the immuno-chemical sector. Chemical measurements based on immuno reactions are used to determine protein in test samples. The proteins, called antibodies or immunoglobulins, are produced in animal bodies to neutralize and help destroy invading foreign species, known as antigens. A given antibody structure has specific binding sites for corresponding antigenic determinants formed on the surface of an antigen. This is the basis of the specificity of immunoassay. An important category of biologically sensitized sensors involves sensor structures that use antibodies and antigens. Because the specificity of such systems is relatively high, promising biosensors could be designed and prepared by using appropriate measurement devices that have the necessary sensitivity.

3.1 Potentiometric Immunosensors with Polymeric Membranes

Janata [29] reported an immunosensor using concanavalin A, which is covalently attached to the surface of a PVC membrane deposited on a platinum electrode. Indirect measurements of immunoagents using ion-selective electrodes have been reported by D'Orazlo et al. [30]. Yeating et al. [31] reported on the so-called potentiometric ionophore-modulation immunoassay for measuring antibodies to digoxin with conjugate-based membrane electrodes. Attempts were made in the authors' laboratory at direct potentiometric detection of analyte by its binding to an affinity surface. Either antigens or antibodies have a net electrical polarity that is correlated to the isoelectric points of the species involved. If antibodies are bound covalently on a polymeric membrane complex with corresponding antigens, the electrical charge of such an immunocomplex membrane will differ from that of the membrane functionalized with the antibodies alone. Attaching the antibody-functionalized membrane or the immunocomplex membrane to the tip of a basic electrode would alter the charge transport process and the electrode membrane potential itself, which can be measured against a reference electrode in solution.

Immunosensors Using Acetylcellulose Membranes

Sensors for α-fetoprotein (AFP) [32], immunoglobulin M (IgM) [33], and C-reactive protein (CRP) [34] have been prepared using acetylcellulose membranes. By mixing acetylcellulose, hexanediamine, and glutaraldehyde dissolved in CH_2Cl_2 and C_2H_5OH and allowing the solvent to evaporate to form a transparent, flexible film on a glass plate. Disks of appropriate diameter (7 mm) are cut from the "master membrane" and soaked with 20% glutaraldehyde solution at 30 °C for several hours. The unreacted hexanediamine on the membrane surface reacts with glutaraldehyde. The number of aldehyde groups on the surface can be increased to enhance its binding ability. After thorough washing with water, the membranes were readied for immobilization of the antibodies by soaking them in appropriately diluted antiserum solutions. The membranes were washed with water, and the free active sites on the membranes were blocked by soaking in 0.25% bovine serum albumin (BSA) solution. Then, the membranes prepared this way were immersed in standard serum or analytical samples to enable the antigens to bind to the antibodies bound on the membrane. The reaction time needed varies for different antigen-antibody complexes. The optimum reaction times were 1 h, 2.5 h, and 2 h for AFP, IgM, and CRP, respectively. The membranes were washed sequentially with a 0.025 % Tween 20 solution and water.

The membrane bound with the antigen-antibody complex was attached to the sensitive tip of a chloride electrode, and an equivalent chloride electrode lacking an attached membrane was the reference. The potential of the electrode couple was measured in 10^{-3} mol/L NaCl solution. The potential readings (E) were related to the concentration (C) of antigen to be determined. For AFP, a linear response ($E = 1.8 + 0.012C$) was established in the concentration range 0–600 ng/mL. For IgM, the linear response ($E = -1.1 + 0.014C$) covers the range extending from 47 to 850 ng/mL. The CRP sensor's linear response range covers the range 1.8–60 μg/mL, and the regression equation is $E = 0.2 + 0.091C$.

Comparing the chloride electrode with other basic electrodes, such as the sulfide and Ag/AgCl electrodes, tested for CRP sensor preparation, shows that a solid-state chloride electrode pair coupled with the 10^{-3} mol/L NaCl electrolyte medium is the best choice for immunosensor design.

Immunosensors Using Hydroxyethyl Methacrylate (HEMA)/Methyl Methacrylate (MMA) Copolymer Membranes

The HEMA/MMA copolymer has been used as the membrane matrix to prepare immunosensors for thyroid-stimulating hormone (TSH) [35] and 3, 5, 3′, 5′-tetraidothyronine (T_4) [36]. A comparison of the HEMA/MMA copolymer with hydroxyethyl acrylate (HEA)/MMA at various copolymerization ratios shows that HEMA/MMA is preferred. Relatively high HEMA (or HEA) content enhances sensor sensitivity. This phenomenon is related to the increasing content of hydroxyl groups in the membrane. A 3:1 HEMA/MMA copolymer is an optimum choice. HEMA/MMA copolymer possesses a block structure. MMA can improve the membrane strength and enlarge the distance between two HEMA molecules and their reactive sites, reducing the steric hindrance and facilitating the reaction of the antibodies with HEMA.

The HEMA/MMA copolymer is easily prepared by reacting HEMA and MMA in 1, 4-dioxane in the presence of 2, 2′-azobis (isobutyronitrile) at 110 °C, sequentially adding *p*-methoxyphenol. The glue-transparent copolymer solution in 1, 4-dioxase is diluted with tetrahydrofuran to prepare the transparent membrane on a glass plate, according to the conventional procedure for preparing polymeric membranes for electrodes. The membrane is treated with cyanogen bromide solution to convert the hydroxy groups into active sites capable of coupling with antibodies.

The analytical procedure is similar to that of the immunosensors for AFP. The TSH and T_4 sensors show linear response ranges covering 0–17.5 μIU/mL and 0–20 ng/mL, respectively. The order of the potential change for the TSH sensor is about 30 mV.

3.2 Piezoelectric Immunosensors

The use of piezoelectric (Pz) devices in sensor technology is attributed to Sauerbrey [37], who formulated the relationship of the crystal's resonant frequency to the mass deposited on its electrodes:

$$\Delta F = -2.3 \times 10^{-6} F^2 \, \Delta M / A \tag{1}$$

where ΔF is the change in frequency, F is the resonant frequency of the Pz crystal, A is the area coated, and ΔM is the mass change due to surface deposition. Following the pioneering work of Shons et al., who immobilized BSA onto the crystal's surface and determined the activity of anti-BSA antibody in solution [38], Pz quartz crystals have been introduced into immunosensor design. Most early applications were limited to measurements in the gas phase. In 1985, Kanazawa and Gordon [39,40] and Bruckenstein and Shay [41] derived a theoretical equation describing the oscillation frequency of a quartz crystal in solution. The oscillation of a quartz crystal in air is taken as a reference, and the difference of the frequencies in solution and in air, ΔF_S, is given by:

$$\Delta F_S = -k(\rho_S \eta_S)^{1/2} \tag{2}$$

where ρ_S and η_S are the density and viscosity of the solution, respectively. This electrical foundation and subsequent technological advances that facilitate oscillation in the liquid phase have led to the rapid development of piezoimmunosensors for liquid immunoassay. In this section, some piezoimmunosensors developed recently in the authors' laboratory are reviewed briefly.

Microgravimetric Immunoassay

Analytes in solutions were determined early by comparing the resonant frequency of the crystal before and after exposure to the test sample. The frequencies were measured in the gas phase. The general way to use a Pz quartz crystal for microgravimetric immunoassay involves coating the crystal with a specific film to which either an antigen or an antibody is fixed by chemical treatment. An immunoreaction increases the mass on the quartz crystal's surface, which is recognized by the decrease in oscillating frequency. Immunoglobulins were detected by several workers using antibody- [42] or protein A-coated [43,44] crystals. Certain microbes [45], haptens [46], and protein antigens [47] were also detected by similar methods. Based on the aforementioned studies, piezoimmunosensors for detecting IgM and AFP and a Pz immunosensor array for simultaneously determining IgM and CRP have been developed in the authors' laboratory.

Piezoelectric immunosensor for detecting immunoglobulin M [48]. A Pz immunosensor to detect human IgM has been developed. Selecting coatings that bind with antibodies (antigens) is of primary importance in applying Pz quartz crystals to microgravimetric immunoassay. In the literature, the organic coatings employed have mainly been silane coupling agents, such as glycidoxypropyltrimethoxysilane (GOPS), (3-aminopropyl)triethoxysilane, polyamide 6, and polyethyleneimine. One drawback of this type of coating is the irreversibility, because it is difficult to remove from the crystal's surface. Moreover, its ability to bind antibodies (antigens) is greatly limited by the steric hindrance caused by the very short reaction arm. In the IgM piezoimmunosensor, the HEMA/MMA copolymer is utilized to immobilize an anti-IgM antibody. In contrast to silane coupling agents, the HEMA/MMA copolymer coating binds effectively with antibody after activation of CNBr by enlarging the distance between the quartz crystal surface and the binding group. In addition, the copolymer coating can be removed easily after rinsing with tetrahydrofuran (THF), and each crystal can be used repeatedly more than 20 times without a detectable loss of sensitivity. The IgM Pz immunosensor can be used for human IgM determination in the range of 5–93 μg/mL. The analytical results of eight clinical samples given by this approach agreed satisfactorily with those given by single radial immunodiffusion (SRID). Furthermore, the valence of goat antihuman IgM antibody and human IgM antigen and the affinity constant of immunoreaction in this experimental system were estimated. For nonrigid molecules, such as antigens, antibodies, and their complexes, Sauerbrey's equation can be modified as

$$\Delta F = -2.3 \times 10^{-6} b F^2 \frac{\Delta M}{A} \tag{3}$$

where b is the modification factor. Then, the frequency shifts caused by immobilizing the antibody on the crystal (ΔF_{Ab}) and by the interaction of the antigen with the antibody-immobilized crystal (ΔF_{Ag}) can be used to calculate the total amount of bound antibody and interacted antigen by using this equation. Because the molecular masses of human IgM antigen and IgM antibody are about 900,000 and 160,000, respectively, the ratio of the total amount of antibody immobilized on the crystal (N_{Ab}, mol) to the amount of antigen interacted with the crystal (N_{Ag}, mol) is obtained as follows:

$$\frac{N_{Ab}}{N_{Ag}} = \frac{90}{16} \times \frac{\Delta F_{Ab}}{\Delta F_{Ag}} \tag{4}$$

One can plot the ratio of the amount of antibody to antigen, calculated using this equation, versus the concentration of IgM antigen. When the concentration of antigen increases to a certain value, the formation of complexes reaches a maximum, and therefore, the ratio of N_{Ab} to N_{Ag} is minimal. This minimal ratio, estimated at 4:1, is the most suitable for forming antigen-antibody complex. In other words, one molecule of human IgM can bind with four molecules of IgM antibody. Consequently, the valence of IgM antigen is four in this experiment. The affinity constant (K_{ass}) of this immunoreaction can be calculated by the following equation:

$$\frac{1}{r} = \frac{1}{n} \times \frac{1}{C \times K_{ass}} + \frac{s}{n} \tag{5}$$

where $r = x/B$ and $C = A - x$. The term x represents the concentration of the bound antigen, B the total antibody concentration, A the total antigen concentration, n the valence of antibody, and s the valence of antigen. In this experimental system, $r = x/B = N_{Ag}/N_{Ab}$, $1/r = N_{Ab}/N_{Ag}$, and C was taken as approximately equal to A. One can plot $1/r$ versus $1/C$, and a linear relationship is obtained when the concentration of antigen lies in a certain range. Therefore, the valence of antibody is approximately 2 and the affinity constant is 1.91×10^6 L/mol, calculated, respectively, from the intercept and slope of the regression line in the linear range.

Piezoelectric immunosensor for α–fetoprotein [49]. A Pz immunosensor for detecting AFP has also been developed. Two methods were employed for immobilizing antibody, covalent binding and entrapping. These two methods have comparable sensitivities. However, the covalent binding method has a lower limit of determination and better linearity, so it was used for immobilizing the antibody. The effect of the amount of copolymer coating on the frequency change has been investigated. The less the copolymer coating, the smaller the decrease in resonant frequency. When the frequency shift caused by the copolymer coating is less than 1900 Hz, the amount of bound antibody increases with the amount of copolymer coating. When the frequency shift caused by the copolymer coating is more than 1900 Hz, the amount of bound antibody no longer depends on the amount of copolymer coating, and it becomes constant. This may result from the fact that, under the experimental conditions, when the amount of copolymer coating is sufficient to decrease the frequency more than a certain value (1900 Hz in this experiment), the electrode surface is completely covered with copolymer, and large molecules generally couple with the surface molecules of the copolymer, because of the steric hindrance. The frequency shifts are linearly dependent on the AFP concentrations in the range of 100–800 ng/mL. Other antigens in human serum do not interfere with the sensor's response. The valences of AFP and anti-AFP antibody were estimated at 2 and the affinity constant of immunoreaction is 6.31×10^8 L/mol.

Piezoelectric immunosensor array for simultaneously determining IgM and CRP [50]. A simultaneous immunoassay technique has been developed for determining the dual analytes, IgM and CRP, by constructing a Pz quartz crystal array system composed of five Pz quartz crystals. Each crystal was immobilized with an antibody mixture, which had a fixed ratio of anti-IgM and anti-CRP antibodies, but worked under different detection conditions, specifically in five buffer solutions of different pH. The crystals react independently either with IgM or with CRP and provide excellent linear responses to IgM and CRP. Because different pH values make the crystal's response characteristics for IgM and CRP differ from each other, the constructed array system gives different response patterns for IgM and CRP. This can be used to determine these dual analytes simultaneously. The frequency responses of the five sensor elements, each working at a different pH, can be combined to generate the characteristic array response for each pure analyte. Such a response pattern of the Pz immunosensor array is unique for each analyte involved and thus is a "fingerprint" identifying the individual components in a sample. The response patterns for IgM and CRP differ significantly from each other, which indicates that there is no significant collinearity in the array system. This helps with the problem of estimating concentration. The five-dimensional response of the array system to individual pure analytes whose concentrations are in the linear response range is governed merely by one free parameter, the concentration of the pure component. One could predict that all the responses to an individual pure analyte would lie on a line in the five-dimensional response space. Therefore, all the data points in the calibration set of the pure component lie on two lines in the five-dimensional response space, each associated with one analyte. This can be confirmed by the two-dimensional projections of the data set along the first two principal components (PCs).

By applying principal component analysis (PCA) to the calibration set of the pure component, it was observed that the first two PCs accounted for almost all of the variation. One could conclude that all of the data points lie on the plane generated by the first two PCs, the plane on which all the data are projected. Therefore, the two-dimensional projections of the data onto the first two PCs can reflect almost entirely the actual distribution of all the data points. Such a two-dimensional representation of the array's responses provide a straightforward approach to qualitative analysis, which can identify whether an unknown sample contains only IgM, only CRP, or both analytes. With a pure-component calibration set of various concentrations, eight samples of unknown concentration were determined with the aid of several linear regression methods, including the ordinary least squares (OLS), partial least squares (PLS), and least trimmed squares (LTS) estimators. All of these regression methods correctly estimated the concentration of the majority of unknown samples. The concentrations of the remaining samples, whose array responses were contaminated by some outliers, were also predicted with satisfactory accuracy by LTS.

Immunoassay in the Liquid Phase

In general, immunoassays in the liquid phase have been carried out directly in solutions that contain analytes using antibody- or antigen-immobilized quartz crystals. Roederer and Bastiaans reported the first assay in a solution phase for detecting IgG [51]. Subsequently, human granulocytes [52] and herbicide atrazine [53], among others, were detected by similar methods. In another study, a method for the immunoassay of CRP was developed by Kurosawa et al. [54]. In this method, designated as latex piezoelectric immunoassay (LPEIA), the frequency shift is observed by using an antibody-bearing latex without any film on the crystal. The mechanism proposed for the frequency change is that the crystal acts as a sensor for viscosity or density changes in the solution because latex particles aggregate. Later, Muratsugu et al. [55] used it to detect anti-streptolysin O antibody. Ghourchian et al. [56] detected the rheumatoid factor by using a similar method. Studies of these two kinds of liquid-phase immunoassay have been undertaken recently in the authors' laboratory.

Kinetic studies of immunoglobulin M immunoreaction using quartz crystal microbalance methodology [57]. Kinetic studies of the immunoreaction between the IgM and anti-IgM antibody based on the quartz crystal microbalance methodology have been reported. Via a CNBr-activated HEMA/MMA copolymer coating, the anti-IgM antibodies are immobilized on the crystal's surface. By continuously monitoring the adsorption and desorption processes of the IgM antigen on the (anti-IgM antibody)-modified crystal surface, the kinetics of the immunoreaction between IgM and its antibodies is investigated. For an immunoreaction, the affinity constant (K_{ass}) can be defined as

$$K_{ass} = \frac{\frac{sx}{A}}{\left[\frac{s(N_{Ag} - x)}{V}\right]\left[\frac{n(N_{Ab} - y)}{A}\right]} \tag{6}$$

where s and n are the valences of antigen and antibody, respectively, N_{Ag} and N_{Ab} are the total amounts (mol) of antigen and antibody, respectively, x and y are the amounts (mol) of reacted antigen and antibody, respectively, A is the area of the silver electrode, and V is the volume of the reaction cell. Because the frequency shift-mass equation implies that the frequency shift of a quartz crystal in a solution is proportional to the mass of materials adsorbed onto the crystal surface, $\Delta F = -k\,\Delta M$, then x, y, and N_{Ab} can be represented by $\Delta F/k\,M_{Ag}$, $s\,\Delta F/n\,k\,M_{Ag}$ and $s\,\Delta F^{o}/n\,k\,M_{Ag}$, respectively. Here ΔF is the frequency shift response to the antigen solution of a certain concentration after the immunoreaction has reached the equilibrium, ΔF^{o} is the frequency shift response to an antigen solution of sufficiently large concentration, in which all reactable sites of antibodies on the crystal's surface are assumed to be occupied by the antigens, and M_{Ag} is the molecular weight of antigen. Therefore, the previous equation can be transformed to

$$\frac{1}{K_{ass}C_{F}s} + 1 = \frac{\Delta F^{o}}{\Delta F} \tag{7}$$

where C_F is the concentration of the free antigens in the solution. Because the concentration of antigen reacted is small enough to be neglected compared with the initial concentration of antigen, C_F approximately equals C_P. From Eq. (7),

$$\frac{C_F}{\Delta F} = \frac{1}{K_{ass}\Delta F^o s} + \frac{C_F}{\Delta F^o} \tag{8}$$

that is, $C_F/\Delta F$ linearly correlates with C_F and has a slope of $1/\Delta F^o$ and an intercept of $1/(K_{ass}\,\Delta F^o s)$. Therefore, the affinity constant can be calculated according to the regression equation of $C_F/\Delta F$ on C_F. For the immunoreaction studied, the affinity constant is estimated at 7.10×10^7 L/mol. On the other hand, the kinetic equation of an immunoreaction is

$$r = \frac{d(sx)}{Adt} = k_f s\left(C_P - \frac{x}{V}\right) n\left(\frac{N_{Ab} - y}{A}\right) \tag{9}$$

where k_f is the rate constant of the forward reaction. In this reaction system, the amount of antigen in the solution is significantly greater than the amount of antibody on the crystal's surface, that is, the decrease in the amount of antigen in the solution is negligible. Therefore, one can assume that the immunoreaction studied is a quasi-first-order-reaction. Therefore, Eq. (9) can be rewritten as

$$r = \frac{d(sx)}{Adt} = k'n\left(\frac{N_{Ab} - y}{A}\right) \tag{10}$$

where $k' = k_f\, s\, C_P$. Equation (10) can be solved easily, and one obtains

$$\ln\left(1 - \frac{\Delta F}{\Delta F^o}\right) = -k't + C \tag{11}$$

where ΔF is the frequency shift at a specific time and ΔF^o is the frequency shift after the immunoreaction reaches equilibrium. A linear relationship between ln $(1 - \Delta F/\Delta F^o)$ and t is obtained. Therefore, according to the regression equation of ln $(1 - \Delta F/\Delta F^o)$ on t, the reaction rate constant can be estimated by using the regression slope. In this system, the reaction rate constants at 10 °C, 27 °C, and 40 °C are 0.8257×10^4, 1.2396×10^4, and 1.7599×10^4 $(\text{mol/L})^{-1}$ s^{-1}, respectively. The Arrhenius activation energy (E_a) of an immunoreaction can be estimated from the following equation:

$$\ln k = -\frac{E_a}{RT} + \ln A \tag{12}$$

This equation implies a linear correlation between ln k_f and $1/T$. Therefore, if the reaction rate constants at different temperatures are available, the Arrhenius activation energy is easily estimated by a linear regression. According to the reaction rate constants at 10 °C, 18 °C, 27 °C, 34 °C, and 40 °C calculated as described previously, the calculated Arrhenius activation energy of this immunoreaction is 18.052 kJ/mol. The activation energy is used mainly to overcome the hydration energy of the antigen or antibody molecules during the formation of the antigen-antibody complex.

Polymer agglutination-based piezoelectric immunoassay for determining complement III [58]. A Pz immunoassay technique, based on detecting agglutination of antibody- or antigen-bearing polymer by an immunoreaction using a Pz quartz crystal, has been developed for determining complement III (C_3). Instead of using latex as in the LPEIA method, carboxymethyl cellulose (CMC), a macromolecular polymer that has many hydroxy groups and ether bonds, was used. CMC forms a CMC-antibody complex with the specific antiserum through the hydrogen bonds between the hydroxy groups or ether bonds of CMC and the carboxy groups of the antibody. The agglutination of antibody-bearing polymer by immunoreaction results in a viscosity change in the solution, which can be monitored by a Pz quartz crystal. According to Eq. (2), the frequency change through this process, ΔF_{S1-S2}, is described by

$$\Delta F_{S1-S2} = -k[(\rho_{S2}\eta_{S2})^{1/2} - (\rho_{S1}\eta_{S1})^{1/2}], \quad (13)$$

where ρ_{S1} and η_{S1} are the respective density and viscosity of the CMC-antibody complex solution and ρ_{S2} and η_{S2} are the respective density and viscosity of the solution after immunoreaction. Because the latex is replaced by CMC polymer in the newly developed technique, the method is not limited by the need for commercially available latex reagents because the antibody- or antigen-bearing CMC polymer is easily prepared for determining the corresponding antigen or antibody. The effect of experimental conditions, such as polymer concentration, antibody dilution ratio, and the reaction temperature, on the frequency response were investigated. Two methods were applied to determine C_3, the end-point method and the initial rate method, both of which were modified according to the practical system involved. These two modified determination methods effectively eliminate interference caused by nonspecific adsorption. The linear ranges for C_3 concentration determined by the end-point method and the initial rate method cover 22.0–43.2 μg/mL and 22.0–49.1 μg/mL, respectively. Most other possible antigens in serum do not interfere with determining C_3. Analytical results of 10 clinical specimens obtained by the technique developed agreed satisfactorily with those given by rate diffusion turbidimetry. With the simple regeneration method devised, the crystal can be used repeatedly with acceptable reproducibility.

Polymer agglutination-based piezoelectric immunoassay for determining human serum albumin [59]. A similar immunoassay technique, based on the agglutination of antibody- or antigen-bearing polymer by an immunoreaction, has also been developed for determining human serum albumin (HSA). The linear range for HSA concentration determined by the end-point method covers 112–878 μg/mL. Most other possible antigens in serum do not interfere with the determination of HSA. Analytical results of five clinical specimens obtained by using the technique developed agreed satisfactorily with those given by rate diffusion turbidimetry.

4. CONCLUSION

Plant and animal tissues substitute well for isolated enzymes in preparing biosensors that have relatively high enzyme loadings and long lifetimes. Without significant loss of selectivity, simple and inexpensive sensor devices can be constructed for determining many inorganic and organic species.

The basis of the specificity of immunoassay is that a given antibody structure has specific binding sites for corresponding antigenic determinants formed on the surface of an antigen. Simple, direct potentiometric detection by binding the antibodies covalently on a polymeric membrane complex with corresponding antigens is limited by relatively low sensitivity. The use of piezoelectric devices in immunochemical sensing seems more promising, and more experimental work is needed to explore their analytical possibilities as an acceptable alternative to radioimmunoassay.

ACKNOWLEDGMENTS

This work was supported by the Foundation for Ph.D. Thesis Research of National Education Commission of China, the Natural Science Foundation of Hunan Province, and the Foundation for the Technological Development of Machinery Industry.

REFERENCES

[1] Rechnitz, G.A., Arnold, M.A., Meyerhoff, M.E. (1979). Bioselective membrane electrode using tissue slices. *Nature*, **278**, 466–467.

[2] Arnold, M.A., Rechnitz, G.A. (1980). Comparison of bacterial, mitochondrial, tissue and enzyme biocatalysts for glutamine selective membrane electrode. *Anal. Chem.*, **52**, 1170–1174.

[3] Arnold, M.A., Rechnitz, G.A. (1981). Selectivity enhancement of a tissue-based adenosine-sensing membrane electrode. *Anal. Chem.*, **53**, 515–518.

[4] Arnold, M.A., Rechnitz, G.A. (1981). Tissue-based membrane electrode with high biocatalytic activity for measurement of adenosine 5′-monophosphate. *Anal. Chem.*, **53**, 1837–1842.

[5] Kuriyama, S., Rechnitz, G.A. (1981). Plant tissue-based bioselective membrane electrode for glutamate. *Anal. Chim. Acta*, **131**, 91–96.

[6] Arnold, M.A., Rechnitz, G.A. (1982). Optimization of a tissue-based membrane electrode for guanine. *Anal. Chem.*, **54**, 777–782.

[7] Mascini, M., Jannelle, M., Palleschi, G. (1982). A liver tissue-based electrochemical sensor for hydrogen peroxide. *Anal. Chim. Acta*, **138**, 65–69.

[8] Fiocchi, J.A., Arnold, M.A. (1984). Rabbit muscle acetone powder as biocatalyst for adenosine 5′-monophosphate biosensor. *Anal. Lett.*, **17**, 2091–2109.

[9] Ma, Y.L., Rechnitz, G.A. (1985). Porcine kidney tissue-based membrane electrode for glucosamine-6-phosphate. *Anal. Lett.*, **18**(B13), 1635–1646.

[10] Ma, Y.L., Rechnitz, G.A. (1988). A sheep kidney tissue-based membrane electrode for D-amino acids. *Anal. Chem.*, **16**, 481–484 (Chinese).

[11] Shen, G.L., Li, Y.Z., Zeng, G.M., Yu, R.Q. (1988). Studies on biological tissue electrode sensitive to cytochrome C. *Acta Pharm. Sin.*, **23**, 762–766 (Chinese).

[12] Rechnitz, G.A. (1978). Bio-selective membrane electrode. *Chem. Eng. News*, **56**, 16.

[13] Mascini, M., Pallesehi, G. (1983). A tissue-based electrode for peroxidase assay: Preliminary results in hormone determination by EIA. *Anal. Lett.*, **16**, 1053–1066.

[14] Nikolskaya, E.B., Yagodina, O.V. (1987). Possibilities of using mitochondrial monoamine oxidase in drug analysis. *Khim.-farmzh. Zh.*, **21**, 1413–1417.

[15] Xu, F., Shen, G.L., Yu, R.Q. (1990). Studies on electrode for hydrogen peroxide using biological tissue slices. *Chem. Sensors*, **10**(1), 21–25 (Chinese).

[16] Li, J., Shen, G.L., Yu, R.Q. (1991). Study on enzyme reaction dynamics with KF. *Chem. Sensors*, **11**(3), 18–22 (Chinese).

[17] Xu, F., Shen, G.L., Yu, R.Q. (1990). Studies on sensor for L-ascorbic acid using biotissue slice. *Acta Pharm. Sin.*, **25**, 368–372 (Chinese).

[18] Xu, F., Shen, G.L., Yu, R.Q. (1991). Studies on electrode for uric acid using biological tissue slice. *J. Trans. Technol.*, **4**(4), 44–48 (Chinese).

[19] Sidwell, J.S., Rechnitz, G.A. (1985). An electrochemical biosensor for dopanime. *Biotechnol. Lett.*, **7**, 419–422.

[20] Fonong, T. (1986). Comparative study of potentiometric and amperometric tissue-based electrodes for oxalate. *Anal. Chim. Acta*, **186**, 301–305.

[21] Song, Y.Q., Shen, G.L. (1992). Banana tissue electrode sensitized for oxalate. *Anal. Chem.*, **20**, 186–189 (Chinese).

[22] Song, Y.Q., Shen, G.L. (1992). Preparation of *o*-dephenol sensor based on banana tissue membrane. *Chem. Sensors*, **12**(3), 41–45 (Chinese).

[23] Song, Y.Q., Shen, G.L., Yu, R.Q. (1993). Studies on flow type electrode for adreneine using banana tissue. *J. Hunan Univ.*, **20**(5), 41–45 (Chinese).

[24] Song, Y.Q., Shen, G.L. (1994). Studies on flow type electrode for the determination of methyldopa using banana tissue. *J. Pharm. Anal.*, **14**(1), 37–40 (Chinese).

[25] Smit, N., Rechnitz, G.A. (1984). Leaf based biocatalytic membrane electrodes. *Biotechnol. Lett.*, **6**, 209–214.

[26] Shen, G.L., Ma, M., Zeng, G.M., Yu, R.Q. (1989). A jack bean tissue electrode for urea determination. *Chem. J. Chin. Univ.*, **10**, 308–310 (Chinese).

[27] Schubert, F., Renneberg, R., Scheller, F.W., Kirstein, L. (1984). Plant tissue hybrid electrode for determination of phosphate and fluoride. *Anal. Chem.*, **56**, 1677–1682.

[28] Shen, G.L., Chen, G.Y., Shui, B., Yu, R.Q. (1990). Studies on biological tissue electrode sensitive for catechol. *Anal. Chem.*, **18**, 315–319 (Chinese).

[29] Janata, J. (1975). An immunoelectrode. *J. Am. Chem. Soc.*, **97**, 2914–2916.

[30] D'Orazlo, P., Rechnitz, G.A. (1977). Ion electrode measurement of complement and antibody levels using marker-loaded sheep red blood cell ghosts. *Anal. Chem.*, **49**, 2083–2086.

[31] Keating, M.Y., Rechnitz, G.A. (1984). Potentiometric digoxin antibody measurement with antigen-ionophore based membrane electrode. *Anal. Chem.*, **56**, 801–806.

[32] Lin, Z.H., Shen, G.L., Yu, R.Q. (1993). A new α-fetoprotein immunosensor. *Chem. Sensors*, **13**(4), 34–38 (Chinese).

[33] Lin, Z.H., Shen, G.L., Zhang, K., Yu, R.Q. (1994). Study on immunoglobulin M (IgM) immunoelectrode. *J. Trans. Technol.*, **7**(1), 37–41 (Chinese).

[34] Lin, Z.H., Shen, G.L., Zhan, C.F., Zhang, H.W. (1993). Study on C-reaction protein immunoelectrode. *Anal. Chem.*, **21**, 1013–1017 (Chinese).

[35] Lin, Z.H., Shen, G.L., Miao, Q., Yu, R.Q. (1996). A thyroid stimulating hormone immunoelectrode. *Anal. Chim. Acta*, **325**, 87–92.

[36] Lin, Z.H., Wang, H.M., Shen, G.L., Yu, R.Q. (1995). Study on thyroid (T_4) immunosensor. *J. Trans. Technol.*, **8**(3), 18–23 (Chinese).

[37] Sauerbrey, G.Z. (1959). Quartz oscillators for weighting thin layers and for microweighing. *Z. Phys.*, **155**, 206.

[38] Shons, A., Dorman, F., Najarian, J. (1972). Immunospecific microbalance. *J. Biomed. Mater. Res.*, **6**, 565–570.

[39] Kanazawa, K.K., Gordon, J.G.H. (1985). Frequency of a quartz microbalance in contact with liquid. *Anal. Chem.*, **57**, 1770–1771.

[40] Kanazawa, K.K., Gordon, J.G.H. (1985). The oscillation frequency of a quartz resonator in contact with a liquid. *Anal. Chim. Acta*, **175**, 99–105.

[41] Bruckenstein, S., Shay, M. (1985). Experimental aspects of use of the quartz crystal microbalance in solution. *Electrochim. Acta*, **30**, 1295–1300.

[42] Thompson, M., Arthur, C.L., Dhaliwal, G.K. (1986). Liquid-phase piezoelectric and acoustic transmission studies of interfacial immunochemistry. *Anal. Chem.*, **58**, 1206–1209.

[43] Muramatsu, H., Dicks, J.M., Tamiya, E., Karube, I. (1987). Piezoelectric crystal biosensor modified with protein A for determination of immunoglobulins. *Anal. Chem.*, **59**, 2760–2763.

[44] Davis, K.A., Leary, T.R. (1989). Continuous liquid-phase piezoelectric biosensor for kinetic immunoassay. *Anal. Chem.*, **61**, 1227–1230.

[45] Muramatsu, H., Kajiwara, K., Tamiya, E., Karube, I. (1986). Piezoelectric immuno sensor for the detection of *Candida albicans* microbes. *Anal. Chim. Acta*, **188**, 257–261.

[46] Ngeh-Ngwainbi, J., Foley, P.H., Kuan, S.S., Guilbault, G.G. (1986). Parathion antibodies on piezoelectric crystals. *J. Am. Chem. Soc.*, **108**, 5444–5447.

[47] Prusak-Sochaczewski, E., Luong, J.H.T. (1990). Detection of human transferrin by the piezoelectric crystal. *Anal. Lett.*, **23**, 401–409.

[48] Chu, X., Lin, Z.H., Shen, G.L., Yu, R.Q. (1995). A piezoelectric immunosensor for the determination of immunoglobulin M. *Analyst*, **120**, 2829–2832.

[49] Chu, X., Lin, Z.H., Shen, G.L., Yu, R.Q. (1996). Studies on the piezoelectric immunosensor for the determination of α-fetoprotein. *Chem. J. Chin. Univ.*, **17**, 870–873 (Chinese).

[50] Chu, X., Jiang, J.H., Shen, G.L., Yu, R.Q. (1996). Simultaneous immunoassay using piezoelectric immunosensor array and robust method. *Anal. Chim. Acta*, **336**, 118–128.

[51] Roederer, J.E., Bastiaans, G.J. (1983). Microgravimetric immunoassay with piezoelectric crystals. *Anal. Chem.*, **55**, 2333–2336.

[52] Konig, B., Gratzel, K. (1993). Human granulocytes detected with a piezoimmunosensor. *Anal. Lett.*, **26**, 2313–2328.

[53] Guilbault, G.G., Hock, B., Schmid, R. (1992). A piezoelectric immunobiosensor for atrazine in drinking water. *Biosensors and Bioelectronics*, **7**, 411.

[54] Kurosawa, S., Tawara, E., Kamo, N., Bastiaans, G., Ohta, F., Hosokawa, T. (1990). Oscillating frequency of piezoelectric quartz crystal in solution. *Chem. Pharm. Bull.*, **38**, 117–120 (Japanese).

[55] Muratsugu, M., Kurosawa, S., Kamo, N. (1992). Detection of antistreptolysin O antibody: Application of an initial rate method of latex piezoelectric immunoassay. *Anal. Chem.*, **64**, 2483–2487.

[56] Ghourchian, H.O., Kamo, N., Hosokawa, T., Akitaya, T. (1994). Improvement of latex piezoelectric immunoassay: Detection of rheumatoid factor. *Talanta*, **41**, 401–406.

[57] Chu, X., Lin, Z.H., Shen, G.L., Yu, R.Q. (1996). Kinetic studies of immunoglobulin M immunoreaction using quartz crystal microbalance methodology. *Chem. J. Chin. Univ.*, **17**, 1025–1029 (Chinese).

[58] Chu, X., Shen, G.L., Yu, R.Q. (1996). Polymer agglutination-based piezoelectric immunoassay for the determination of complement III. *Analyst*, **121**, 1689–1694.

[59] Chu, X., Shen, G.L., Xie, F.Y., Yu, R.Q. (1997). Polymer agglutination-based piezoelectric immunoassay for the determination of human serum albumin. *Anal. Lett.*, **30**(10), 1783–1796.

ELECTROCHEMICAL ENZYME, TISSUE, AND MICROBIAL BIOSENSORS

Jiaqi Deng

OUTLINE

Advances in Biosensors
Volume 4, pages 79–93.

ISBN: 0-7623-0073-6

ABSTRACT

A review of electrochemical enzyme, tissue and microbial biosensors is presented. Different enzymes, plant and animal tissues, and bacteria were used as the biocatalysts for constructing these biosensors.

1. INTRODUCTION

Clark and Lyons [1] were the first to suggest the use of an oxygen electrode as a transducer to measure glucose concentration based on the change of oxygen concentration in the glucose oxidase catalyzed reaction. Updike and Hicks [2] prepared quantitative biosensor for glucose. Since then, different kinds of biosensors have been rapidly developed. Some of these such as glucose, lactate, and BOD biosensors, have already been manufactured as commercial products.

Based on the different transducers used, biosensors can be divided into several classes, such as electrochemical, optical, thermal and piezoelectric types. In this chapter, only the electrochemical biosensors that we developed are discussed.

2. ENZYME BIOSENSORS

2.1 Glucose

A novel method for enzyme immobilization was developed based on natural silk fibroin obtained directly from living silkworms. The conformational transition of the fibroin occurred readily by physical means, resulting in forming a carrier which was insoluble in water. The glucose sensor formed in this way exhibited high biocatalytical activity, good membrane stability, and a sufficiently long lifetime. The linear range of determination of glucose was $1.5 \times 10^{-4} - 2.5 \times 10^{-3}$ mol/L [3–5].

Although natural silk fibroin is an ideal immobilization material, it is limited by the seasons for living silkworms. Regenerated silk fibroin (RSF), which can be obtained at any time, replicated the results of natural silk fibroin. Waste silk mill was treated with 0.5% aqueous $NaHCO_3$ at 100 °C for 30 min, washed with water, and then dissolved in 9.3 mol/L LiBr solution. After dialysis by water for three days, the solution was filtered and the aqueous solution of RSF was obtained. The glucose oxidase (GOD) membrane is prepared by casting the mixture of the RSF and glucose oxidase on a glass plate at room temperature in air. The structure of the blend membrane of RSF and glucose oxidase was studied [6,7].

Polyvinyl alcohol (PVA) is a nontoxic water soluble synthetic polymer that has good film-forming properties, impact strength, and water durability. With this in mind, a novel composite of RSF and PVA was prepared to obtain a material that maintains the compound's merits and improved the properties of RSF. IR and SEM were employed to give insights into the compound's structure. The water absorption property and maximum strength were also investigated. The linear range of glucose determination is $5 \times 10^{-5} - 1.6 \times 10^{-3}$ mol/L [8–10].

A high-performance amperometric glucose biosensor was developed based on immobilizing glucose oxidase in an electrochemically synthesized, nonconducting poly(*o*-aminophenol) film on a platinized glassy carbon electrode. The large microscopic surface area and porous morphology of the platinized glassy carbon electrode results in high enzyme loading, and the enzyme entrapped in the electrodeposited platinum microparticle matrix is more stable than on a platinum disk electrode surface. The response current of the sensor is 20-fold higher than that of the sensor prepared with a platinum disk electrode of the same geometric area. The experiments showed that the high sensitivity of the sensor is due to the large area and also to the high efficiency of transformation of H_2O_2 generated by enzymatic reaction to a current signal on the platinized glassy carbon electrode. The linear response of this enzyme sensor to glucose is from 1×10^{-6} to 1×10^{-3} mol/L and the detection limit is 5×10^{-7} mol/L. The lifetime is more than ten months [11].

In similar work, a sensitive and stable amperometric glucose sensor was constructed based on immobilizing GOD in electropolymerized poly(3,3-diaminobenzidine) film on a palladinized glassy carbon electrode [12].

A new amperometric biosensor was developed based for the first time on glucose oxidase automatically incorporated and immobilized in a modified Y zeolite substrate without using bovine serum albumin (BSA)-glutaraldehyde, as is usually the case. The large microscopic surface area and porous morphology of the zeolite matrix results in high enzyme loading. This zeolite-modified biosensor offers enhanced sensitivity, an extended linear range, and higher stability. Such improvements are attributed to the porous and hydrophilic characterization of zeolite that exposed the enzyme to the substrate solution and also to intermolecular interaction between zeolite and the enzyme, which was investigated by FT-IR. The resulting biosensors exhibit good reproducibility and excellent selectivity also because of the uniform pore structure and unique ion exchange property of zeolite. The sensor is favorably responsive to glucose and has a linear range from 2×10^{-6} to 3×10^{-3} mol/L. The detection limit is 5.0×10^{-7} mol/L. The calculated apparent Michaelis constant for the immobilized enzyme is 17.52 mmol/L [13].

The disadvantage of biosensors that use an oxygen electrode as the transducer is the variation of oxygen concentration. Mediators, such as ferrocene and its derivatives, tetrathiafulvalene (TTF), benzoquinone or methylene blue, can be used instead of oxygen to transfer electrons between the enzyme and the electrode. Turner used TTF as a mediator, which was modified simultaneously with GOD on the basic electrode, to construct a glucose sensor that has good reproducibility [14].

An amperometric mediated sensor for glucose was contrived by using BSA and glutaraldehyde to immobilize GOD on a Nafion-TTF modified glassy carbon electrode. The immobilized enzyme layer was further coated by Nafion. The inner Nafion membrane prevents leakage of TTF, and the outer Nafion film is a barrier to electroactive anionic interferents, such as ascorbate and urate, and it protected the biosensor from fouling agents. The experiment showed that TTF^{+} and TTF^{2+} oxidizes the reduced flavin adenine dinucleotide ($FADH_2$) of GOD. The biosensor responds to glucose in less than 50 s and its calibration curve is linear from 3.0×10^{-4} to 1.0×10^{-2} mol/L [15–17]. Following is the mechanism of the TTF-modified glucose sensor:

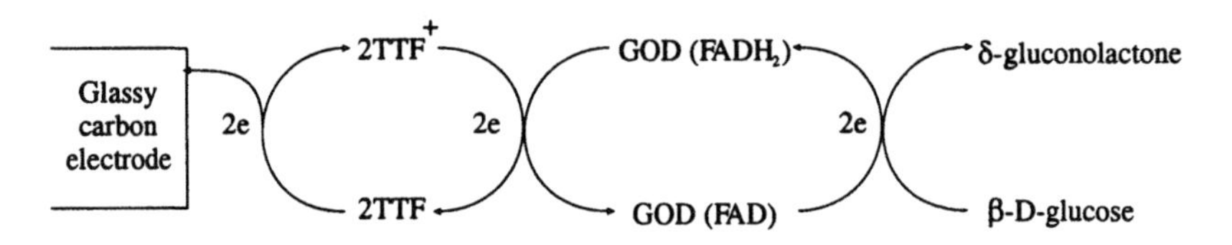

A similar glucose biosensor was prepared as follows. Tetrathiafulvalene was immobilized by a Eastman-AQ polymer, and GOD was immobilized on the TTF layer by bovine serum albumin and glutaraldehyde [18].

An amperometric glucose sensor was constructed by utilizing 1,1-dimethylferrocene as an electron transfer mediator. First, 1,1-dimethylferrocene was modified on a glassy carbon electrode (GCE) and then GOD was immobilized by RSF on the mediator-modified GCE to prepare the glucose sensor. We demonstrated the feasibility of 1,1-dimethylferrocene shuttling electrons between the glucose oxidase in the RSF membrane and a GCE. The experimental tests indicate good mechanical stability, no disruption of enzyme activity and no diffusion problems in incorporating glucose oxidase [8].

Another ferrocene derivative was also prepared as a mediator-modified biosensor for glucose. A GCE was used as the basic electrode. A thin layer of Eastman-AQ-55D polymer was modified on the surface of the GCE and then a small volume of dilute (dimethylaminomethyl)ferrocene benzene solution was applied to the polymer-modified layer. GOD was first dissolved in a RSF and polyvinylalcohol aqueous mixture and then spread on the ferrocene-modified electrode to prepare the glucose biosensor. Using the resulting biosensor, a typical reversible voltammogram of like this (dimethylaminomethyl)ferrocene attached to the polymer backbone is obtained in the absence of glucose. Addition of glucose to the test cell increases the oxidation current and decreases the reduction current. The reduction peak disappears completely and the oxidation peak appears as a plateau at slow scan rates (about 35 mV/s). No reduction current was observed in the reverse-potential scan, indicating that all ferricinium centers at the surface are reduced by the enzyme reaction for excess of substrate. Comparison of the voltammograms with and without glucose illustrated that (dimethylaminomethyl)ferrocene in Eastman-AQ film effectively shuttles electrons between glucose oxidase in the composite membrane and a glassy carbon electrode [19].

In addition, a glucose sensor based on GOD immobilized on a glassy carbon electrode by RSF was prepared when *p*-benzoquinone was added to the test solution as the effective mediator. The linear range for determining glucose is 1.0×10^{-4} – 5.0×10^{-3} mol/L [20]. Moreover, an amperometric biosensor for glucose was fabricated by immobilizing horseradish peroxidase and GOD on a Nafion-ferrocene modified electrode. The system was based on ferrocene as an electron shuttle between peroxidase and GCE. Because of the low working potential (0.0V vs. SCE), electroactive species, such as ascorbate and urate, do not interfere with the determination of glucose. The linear range of determination is from 5.0×10^{-4} – 2.5×10^{-2} mol/L [21].

2.2 Hydrogen Peroxide

Determining hydrogen peroxide is of great importance because it is the product of the reactions catalyzed by a large number of highly selective oxidase enzymes and it is essential in food, pharmaceutical, and environmental analyses. A new approach to constructing an amperometric biosensor is described. The classical dye methylene green and horseradish peroxidase as a probing-needle mediator and a base enzyme, respectively, were coimmobilized in the same montmorillonite-modified BSA-glutaraldehyde matrix to construct an H_2O_2 sensor. The immobilization matrix was formed from a pretreated sodium montmorillonite colloid in which the enzyme and the cross-linker were dissolved. Immobilization of methylene green from the dye mother solution was attributed to the adsorption function of the montmorillonite, whereas immobilization of horseradish peroxidase was attributed, as usual, to the cross-linking function of the BSA-glutaraldehyde. Cyclic voltammetry and potentiostatic measurements indicated that methylene green efficiently mediates electrons from the base electrode to the enzyme in the matrix. The sensor responds rapidly to low H_2O_2 concentration, achieves 95% of the steady-state current in less than 20 s. The detection limit is 4.0×10^{-7} mol/L H_2O_2 [22]. In addition, new methylene blue was successfully used instead of methylene green as an electron transfer mediator.

RSF was also used to immobilize horseradish peroxidase (HRP) to make an amperometric H_2O_2 sensor. The phenazine methosulphate in the test solution is used as an electron transfer mediator. The sensor is highly sensitive to H_2O_2 and has a detection limit of 1.0×10^{-7} mol/L H_2O_2 and a response time of less than 5 s. The effect of the applied potential and the mediator concentration on the Michaelis–Menten constant was calculated [23]. A similar experiment was accomplished using methylene blue as the mediator [24].

A nickelocene-mediating sensor for hydrogen peroxide was also fabricated. Nickelocene was coated on the GCE by pipetting a drop of diluted nickelocene-acetone solution on the GCE surface. Then after drying, a drop of HRP-RSF solution was deposited on the electrode and allowed to dry. The resulting sensor has a detection limit of 5.0×10^{-7} mol/L.

2.3 Lactate

There is a strong demand for lactate determination in blood, because of the association of lactate with several diseases (e.g., shock, heart disease, respiratory insufficiency), and in food analysis and biotechnological process control. Lactate and glucose sensors employing ferrocene derivatives meldola blue and TTF have been reported [25].

A lactate sensor was fabricated using glutaraldehyde and BSA to immobilize lactate oxidase onto an Eastman-AQ-TTF modified electrode. The TTF immobilized in Eastman-AQ functioned well to enhance electron shuttle between the redox center of the enzyme and the electrode. The sensor displays good stability and reproducibility and a fast response time for lactate [26]. The biocatalytic reaction is as follows:

$$\text{L-Lactate} + \text{LOD(FAD)} \longrightarrow \text{pyruvate} + \text{LOD(FADH}_2\text{)}$$

$$\text{L-LOD(FADH}_2\text{)} + 2\text{TTF}^+ \longrightarrow \text{LOD(FAD)} + 2\text{TTF} + 2\text{H}^+$$

$$2\text{TTF} \longrightarrow 2\text{TTF}^+ + 2\text{e (at the electrode)}$$

Another lactate biosensor was prepared by first modifying TTF on a GCE by Nafion and then immobilizing lactate oxidase on its surface by BSA and glutaraldehyde. The problem of mediator leaking in electrodes using TTF was overcome. Nafion membrane prevents leaching of positive TTF^+ ions by electrostatic attraction. Stability, interference, and factors, such as oxygen, applied potential, and pH that influence the performance of the sensor were examined and discussed [27].

2.4 Urea

Potentiometric urea electrodes were developed by Guilbault and Montalvo. The most useful approach is using an ion-selective electrode to track NH_4^+.

In our laboratory, urease was immobilized by fresh bombyx mori silk fibroin to form a membrane. The urea sensor was assembled by sandwiching the prepared membrane between a permeable Teflon membrane that was directly covered on the surface of an ammonia gas-sensing electrode and a nylon monofilament mesh. The sensor has a linear response over the range $1.8 \times 10^{-4} - 1.1 \times 10^{-2}$ mol/L urea. The slope is 56.0 mV per decade at 25 °C and the response time is about 2–3 minutes [28].

In particular, the urease was immobilized in a γ-aluminum oxide matrix without using BSA-glutaraldehyde, and it was first employed successfully for constructing a potentiometric urea biosensor. The intermolecular interaction between urease and the matrix was studied in a preliminary way by FT-IR [29].

2.5 Hypoxanthine

A mediated biosensor for hypoxanthine was fabricated by using BSA and glutaraldehyde, a cross-linker, to immobilize xanthine oxidase on a Nafion-ferrocene modified glassy carbon electrode. The response time to hypoxanthine is less than 70 s, the working potential is +0.2 V, and the linear response of the sensor is in the range of $5.0 \times 10^{-5} - 7.5 \times 10^{-4}$ mol/L [30]. The chemical reactions are as follows:

$$\text{Hypoxanthine} + \text{XOD(FAD)} \longrightarrow \text{Xanthine} + \text{XOD(FADH}_2\text{)}$$

$$\text{Xanthine} + \text{XOD(FAD)} \longrightarrow \text{Uric acid} + \text{XOD(FADH}_2\text{)}$$

$$2\text{Fc}^+ + \text{XOD(FADH}_2\text{)} \longrightarrow 2\text{Fc} + \text{XOD(FAD)} + 2\text{H}^+$$

$$2\text{Fc} - 2\text{e} \longrightarrow 2\text{Fc}^+ \text{ (at the electrode)}$$

3. TISSUE BIOSENSORS

Plant and animal tissues as biocatalytical materials provide advantages over some isolated enzymes for certain situations. The principal advantages are longer lifetime, lower cost, and higher catalytic activity. Arnold and Rechnitz were the first to use a slice of porcine kidney combined with an ammonia sensing electrode to construct a biosensor for glutamine [31].

3.1 Urea

A jack bean meal biosensor sensitive to urea has been reported, but it requires fresh jack bean. Fang and Deng found that pulverized soybean could be used to construct the urea sensor instead of jack bean. Several milligrams of the soybean meal was spread over the gas-permeable membrane and then immobilized by glutaraldehyde and covered on a NH_3 or CO_2 sensing electrode to form a urea sensor. The linear response range is from $1.3 \times 10^{-4} - 1.1 \times 10^{-2}$ mol/L urea, and the lifetime is 15 months [32].

3.2 Hydrogen Peroxide

A new plant tissue biosensor was fabricated by coupling an oxygen electrode with lettuce seed meal. Because lettuce seeds contain appreciable amounts of catalase, the resulting electrode has excellent stability and reproducibility when stored at room temperature and does not require special preservative reagents. It is possible that this biocatalytical material is a desirable replacement for catalase. The linear response range was $1.0 \times 10^{-4} - 2.0 \times 10^{-3}$ mol/L [33].

Another hydrogen peroxide tissue biosensor was prepared by coupling ginger powder as a biocatalyst with an oxygen electrode. The linear response range was $5.0 \times 10^{-5} - 2.0 \times 10^{-3}$ mol/L. Compared with the fresh ginger tissue sensor and the enzyme sensor, the ginger-powder sensor has better properties. Moreover, ginger powder is more easily obtained and stored [34]. Potato tissue can also be used as a biocatalyst for constructing a hydrogen peroxide biosensor [35].

3.3 Adenine Nucleotide

An animal tissue biosensor for adenine nucleotide was prepared. A slice of fresh porky liver, which contains adenine nucleotide deaminase, was covered on an ammonia sensing electrode to construct the sensor. The linear response range is $1.4 \times 10^{-4} - 1.0 \times 10^{-2}$ mol/L, and the lifetime is more than one month [36].

3.4 Guanine

A biosensor for determining guanine was prepared, which employed rat brain tissue as the biocatalyst and an ammonia gas sensing electrode as the base electrode. Its response characteristics, buffer condition and strategies of immobilization were studied. The slope of guanine response curve is 46.5 mV/decade and the linear range is $2.0 \times 10^{-5} - 6.3 \times 10^{-4}$ mol/L. The lifetime is more than 28 days [37].

3.5 Organic Acids

L-Lysine

A biosensor for L-lysine using chicken kidney as the biocatalyst was constructed. A slice of chicken kidney was covered on the Teflon membrane of an oxygen electrode and a cellulose acetate membrane was covered on the tissue layer to make the sensor. The linear response range is $5.0 \times 10^{-5} - 7.0 \times 10^{-3}$ mol/L, and the lifetime is 14 days [38].

Tyrosine

A tyrosine sensor constructed using sugar beet as the tyrosine oxidase source was studied. The enzyme was extracted from sugar beet by acetone and formed acetone-precipitated pulp. Then, it was immobilized by glutaraldehyde on an oxygen permeable membrane (Teflon film) which was coated on the surface of an oxygen electrode. The response of the sensor to the tyrosine has a linear range from 5.0×10^{-5} to 5×10^{-4} mol/L, and the lifetime is 30 days [39].

Serine

The construction and performance characteristics of a serine biosensor based on pork liver tissue with an ammonia gas-sensing electrode were studied. A slice of pork liver was set between a nylon cloth and an ammonia permeable membrane, then spread on the surface of an ammonia gas-sensing electrode to prepare the sensor. It has a linear response range to serine from 5.1×10^{-5} to 3.2×10^{-3} mol/L and a lifetime of about 30 days [40].

Ascorbic Acid

The determination of ascorbic acid is very important in medicine. An ascorbic acid biosensor was constructed using cauliflower tissue as the biocatalyst and an oxygen electrode as a transducer. Another sensor, a benzoquinone-modified graphite electrode, was used as the working electrode instead of an oxygen electrode. Here, the benzoquinone is a mediator to facilitate electron transfer from ascorbic acid oxidase within the cauliflower tissue to the graphite electrode. The linear response range of this sensor is $5.7 \times 10^{-4} - 5.7 \times 10^{-2}$ mol/L, and its lifetime is 30 days [41,42].

The analytical conditions of cauliflower tissue electrode for L-ascorbic acid and banana tissue electrode for catachols were optimized by using orthogonal experiments. According to the comparison with the method of monofactor analysis, more information is obtained by using orthogonal experimental design [43].

Glycollic Acid

Two biosensors for glycollic acid were prepared and studied. One used spinach, and another used aster indicus as the biocatalyst. An oxygen electrode was used as the base electrode. The linear range of the response curve for the spinach tissue sensor is $5.0 \times 10^{-6} - 1.0 \times 10^{-4}$ mol/L, and the lifetime is 10 days. The response linear range of the aster indicus sensor is $2.0 \times 10^{-6} - 1.0 \times 10^{-4}$ mol/L, and the lifetime is 30 days [44].

Uric Acid

Uric acid is the final catabolite of purine nucleotide in the human body, and the concentration of uric acid in serum provides information about some important diseases. Therefore, the determining uric acid clinically is of great importance. In our laboratory, fish liver was first used as a biocatalytical material to make a tissue biosensor for uric acid, because fish liver contains an enzyme complex that catalyzes the reaction sequence:

$$\text{Uric acid} + O_2 \xrightarrow{\text{uricase}} \text{Allantoin} + CO_2 + H_2O_2 \quad (1)$$

$$\text{Allantoin} + H_2O \xrightarrow{\text{allantoinase}} \text{Allanoic acid} \quad (2)$$

$$\text{Allanoin acid} + H_2O \xrightarrow{\text{allanyoinase}} \text{Urea} + \text{Glyoxylic acid} \quad (3)$$

In reaction 1, the consumption of oxygen is directly proportional to the uric acid content. Hence, an oxygen electrode was coupled to the fish liver to construct a biosensor. This tissue biosensor has a long lifetime of about three months. More than 20 compounds, such as some amino acids, were tested and showed no interference. When applied in the assay of serum, the results of this method agree with those of enzymatic methods. The recoveries in serum were 90 to 110% [45].

4. MICROBIAL BIOSENSORS

Since Clark established the first biosensor in 1962, a lot of biosensors have been developed for determining organic compounds. Enzyme sensors are highly specific to different substrates. But only a few biosensors have been manufactured as commercial products because the enzymes are first separated from plant and animal tissue and then purified. Thus they are generally expensive and unstable. Living microorganisms, such as bacteria, are a possible alternative to high cost isolated enzymes because of their good catalytic behavior, abundant source, and no limit on season. There are two kinds of microbial biosensors. One depends on the respiration of the bacteria, the biochemical oxygen demand (BOD) sensor, and the other uses the bacteria as an enzyme source, such as the lactate sensor.

4.1 BOD

BOD is an important index in measuring organic pollution. Usually, the BOD value takes five days to obtain. Therefore, a more rapid and reproducible method is needed for monitoring the BOD value. Karube was the first to use microorganisms in combination with an oxygen electrode to construct a BOD biosensor in 1977 [46].

A novel three-electrode-system amperometric BOD biosensor was designed. A large-area gold disk electrode was used as the working electrode, a titanium anode as the counter electrode, a silver-silver chloride electrode as the reference electrode, and 0.1 mol/L KCl as the inner electrolyte solution. The microorganism was immobilized by sodium alginate, then sandwiched between a piece of acetylcellulose and the oxygen permeable Teflon membrane of the oxygen electrode. Determination time is about 15 minutes, and the lifetime is 30 days. The results compare well with those of BOD_5 [47].

In addition a galvanic cell type BOD biosensor was fabricated. A large-area gold disk electrode was used as the cathode, a lead plate as the anode, and 0.1 mol/L KOH as the inner electrolyte solution. The sensor was convenient to use because it could be directly connected with the X-Y recorder without any other instrument. The results measured were close to those of BOD_5 [48]. Furthermore, the methods for immobilizing microorganisms to construct a rapid BOD biosensor were studied [49].

4.2 Organic Acids

Glutamic Acid

Two types of sensors for glutamic acid were prepared and studied. They employed glutamate decarboxylase and *E. coli* as the biocatalyst, respectively, and a carbon dioxide gas sensor as the base electrode. The results showed that each electrode has its own characteristics. The enzyme electrode has a linear range of $3.16 \times 10^{-4} - 1.0 \times 10^{-2}$ mol/L, a slope of 46 mV/decade, a response time of 6–8 minutes, and a lifetime of 5 days. The microbial sensor has a linear range of $4.2 \times 10^{-4} - 1.8 \times 10^{-2}$ mol/L, a slope of 48.5 mV/decade, a response time of 8–10 minutes, and a longer lifetime of 34 days. The two types of sensors have high selectivity for L-glutamic acid and enhanced activity in the presence of pyridoxal phosphate. The sensor was applied satisfactorily to determining sodium glutamate in a sample [50].

L-Asparagine

A bacterial sensor for L-asparagine was prepared by immobilizing *Corynebacterium glutamicum* on an ammonia gas sensor. The response of the sensor has a linear range between 7.1×10^{-5} and 1.1×10^{-2} mol/L L-asparagine and a slope of 51.4 mV/decade in a pH 8.5 boric acid-phosphate buffer at 30 °C. The response time is 4–7 minutes and the electrode is relatively stable for 30 days [51].

L-Aspartate

A potentiometric biosensor for L-aspartate was prepared. *Escherichia coli* was the biocatalyst and an ammonia gas electrode was the base electrode. The effects of the culture medium, immobilizing methods for the bacteria, and the response characteristics of the electrode were studied. The slope of the L-aspartate response curve E–lg C is 55 mV/decade, the linear range is $6.0 \times 19^{-4} - 8.0 \times 10^{-3}$ mol/L, and the lifetime is 10 days [52].

Uric Acid

A biosensor for determining uric acid was prepared. It employs a kind of *Bacillus* as the biocatalyst and an oxygen gas electrode as the base electrode. The effect of the culture medium, the response character of the electrode, and its kinetic mechanism were studied. At suitable pH, the uric acid concentration is linear in the range of $1.0 \times 10^{-5} - 4.0 \times 10^{-4}$ mol/L, the apparent Michaelis constant, K_m is 2.8×10^{-4} mol/L, the activation energy is 47,000 J/mol, and the temperature coefficient is 1.8 [53].

Lactic Acid

Mixed *Lactobacillus* were used as the source of lactate oxidase, coupled with an oxygen electrode to produce a microbial sensor for lactic acid. The conditions for bacteria culture and the different bacteria sources were studied. The linear range for determining lactic acid is $1.0 \times 10^{-5} - 5.0 \times 10^{-4}$ mol/L, the response time is 5–7 minutes, and the lifetime is about 30 days [54].

4.3 Urea

Two novel biosensors for urea based on immobilized *Corynebacterium glutamicum* 617 and *Corynebacterium glutamicum* ATCC13032 in calcium alginate gel, coupled with an ammonia gas-sensing electrode, were designed and constructed. The calibration plots of mV versus logarithmic urea concentration are linear in the range of $5.6 \times 10^{-5} - 1.4 \times 10^{-2}$ and $5.6 \times 10^{-5} - 1.1 \times 10^{-2}$ mol/L, and have slopes of 59.2 and 61.3 mV/decade, respectively, in pH 8.0, 0.1 mol/L phosphate buffer solution at 30 °C. The two urea biosensors were successfully applied to the actual measurement of urea in urine and they were relatively stable for 20 and 40 days [55,56].

4.4 DL-Phenylalanine

A biosensor for determining DL-phenylalanine based on the immobilized bacteria *P. Vulgaris* in calcium alginate gel coupled with an amperometric oxygen electrode was designed and constructed. The effect of the culture medium, the response character of the electrode, and its kinetic mechanism were studied, and the activity of phenylalanine deaminase was tested by biochemical methods. The biosensor exhibits a longer lifetime, higher sensitivity and faster response to phenylalanine in the linear range of $2.5 \times 10^{-5} - 2.5 \times 10^{-3}$ mol/L. The results indicate that the dynamic response process of the reaction catalyzed by bacteria is similar to that of isolated enzymes. The calculated apparent Michaelis constant K_m is 2.0×10^{-3} mol/L [54].

REFERENCES

[1] Clark, L. C., Lyon, C. (1962). Electrode systems for continuous monitoring in cardiovascular surgery. *Ann. NY Acad. Sci.*, **102**, 29–45.

[2] Updike, S. J., Hicks, G. P. (1967). The enzyme electrode, a miniature chemical transducer using immobilized enzyme activity. *Nature*, **214**, 986–988.

[3] Shao, Z., Fang, Y., Yu, T., Deng, J. (1991). A sensor of glucose oxidase immobilized in bombyx mori silk fibroin membrane. *Chem. J. Chin. Univ.*, **12**, 847–848.

[4] Fang, Y., Shao, Z., Deng, J., Yu, T. (1992). Immobilization of glucose oxidase with bombyx mori silk fibroin for the preparation of glucose sensor. *Chin. Sci. Bull. (English)*, **37**, 1437–1440.

[5] Fang, Y., Shao, Z., Deng, J., Yu, T. (1992). Bombyx mori silk fibroin-based immobilization of urea electrode. *Electroanalysis*, **4**, 669–672.

[6] Zhang, X., Liu, H., Liu, Y., Deng, J. (1995a). Structure of the blend membrane of regenerated silk fibroin and glucose oxidase and its application to glucose sensor. *Anal. Lett.*, **28**, 1593–1609.

[7] Qian, J., Liu, Y., Liu, H., Yu, T., Deng, J. (1996). Immobilization of glucose oxidase onto regenerated silk fibron and its application to glucose sensor. *Chem. J. Chin. Univ.*, **17**, 52–54.

[8] Liu, H., Liu, Y., Qian, J., Yu, T., Deng, J. (1996). Feature of entrapment of glucose oxidase in regenerated silk fibroin membrane and fabrication of 1,1-dimethylferrocene-mediating glucose sensor. *Microchem. J.*, **53**, 241–252.

[9] Liu, Y., Zhang, X., Liu, H., Yu, T., Deng, J. (1996). Immobilization of glucose oxidase onto the blend membrane of poly(vinyl alcohol) and regenerated silk fibroin: Morphology and application to glucose biosensor. *J. Biotechnol.*, **46**, 131–138.

[10] Liu, Y., Liu, H., Qian, J., Deng, J., Yu, T. (1996). Entrapment of both glucose oxidase and peroxidase in regenerated silk fibroin membrane. *Fresenius J. Anal. Chem.*, **355**, 78–82.

[11] Zhang, Z., Liu, H., Deng, J. (1996). A glucose biosensor based on immobilization of glucose oxidase in electropolymerized *o*-aminophenol film on platinized glassy carbon electrode. *Anal. Chem.*, **68**, 1632–1638.

[12] Zhang, Z., Lei, C., Deng, J. (1996). Electrochemical fabrication of amperometric glucose enzyme electrode by immobilizing glucose oxidase in electropolymerized poly (3,3-diaminobenzidine) film on palladinized glassy carbon electrode. *Analyst*, **121**, 971–976.

[13] Liu, B., Hu, R., Deng, J. (1997). Studies on the characterization of immobilization of enzyme in a modified Y zeolite membrane and its application to an amperometric glucose biosensor. *Anal. Chem.*, **69**, 2343–2348.

[14] Hendry, S.P., Cardosi, M.F., Turner, A.P.F. (1986). An amperometric enzyme electrode for glucose based on a tetrathiafulvalene-modified carbon electrode. *Biosensors*, (submitted).

[15] Liu, H., Lei, C., Deng, J. (1994). Amperometric glucose sensor based on Estaman-AQ-tetrathiafulvalene modified glassy carbon electrode. *Chem. J. Chin. Univ.*, **12**, 1757–1760.

[16] Liu, H., Li, H., Deng, J. (1994). Amperometric mediated biosensor based on tetrathiafulvalene Nafion chemically modified graphite electrode. *Chin. J. Anal. Chem.*, **22**, 882–886.

[17] Liu, H., Deng, J. (1995). Amperometric glucose sensor using tetrathiafulvalene in Nafion gel as electron shuttle. *Anal. Chim. Acta.*, **300**, 65–70.

[18] Liu, H., Deng J. (1996). An amperometric glucose sensor based on Eastman-AQ-tetrathiafulvalene modified electrode. *Biosensors and Bioelectronics*, **11**, 103–110.

[19] Liu, H., Qian, J., Liu, Y., Yu, T., Deng, J. (1996). Immobilization of glucose oxidase in composite membrane of regenerated silk fobroin and poly(vinyl alcohol): application to an amperometric glucose sensor. *Bioelectrochemistry and Bioenergetics*, **39**, 303–308.

[20] Qian, J., Liu, Y., Liu, H., Yu, T., Deng, J. (1996). Characteristics of regenrated silk fibroin membrane in its application to the immobilization of glucose oxidase and preparation of a *p*-benzoquinone mediating sensor for glucose. *Fresenius J. Anal. Chem.*, **354**, 173–178.

[21] Liu, H., Deng, J. (1995). Bienzyme sensor for glucose based on Nafion-ferrocene modified electrode. *Chin. J. Anal. Chem.*, **23**, 154–158.

[22] Lei, C., Deng, J. (1996). Studies on hydrogen peroxide sensor based on coimmobilized methylene green and horseradish peroxidase in the same montmorillonite-modified bovine serum albumin glutaraldehyde matrix on a glassy carbon electrode surface. *Anal. Chem.*, **68**, 3344–3349.

[23] Qian, J., Liu, Y., Liu, H., Yu, T., Deng, J. (1995). Characterization of regenerated silk fibroin membrane for immobilizing peroxidase and construction of an amperometric hydrogen peroxide sensor employing phenazine methosulphate as electron shuttle. *J. Electroanal. Chem.*, **397**, 157–162.

[24] Liu, Y., Liu, H., Qian, J., Deng, J., Yu, T. (1995). Regnerated silk fibrin as immobilization matrix for peroxidase and fabrication of a sensor for hydrogen peroxide utilizing methylene blue as electron shuttle. *Anal. Chim. Acta.*, **316**, 65–72.

[25] Newman, J.D., Turner, A.P.F., Marrazza, G. (1992). Ink-jet printing for the fabrication of amperometric glucose biosensors. *Anal. Chim. Acta*, **262**, 13–17.

[26] Liu, H., Kong, J., Deng, J. (1995). An amperometric lactate sensor using tetrathiafulvalene in polymer ionomer film as electron shuttle. *Anal. Lett.*, **28**, 563–579.

[27] Liu, H., Deng, J. (1995). An amperometric lactate sensor employing tetrathiafulvalene in Nafion film as electron shuttle. *Electrochim. Acta*, **40**, 1845–1849.

[28] Fang, Y., Shao, Z., Deng, J., Yu, T. (1992). Bombyx mori silk fibroin-based immobilization of urea electrode. *Electroanalysis*, **4**, 669–672.

[29] Liu, B., Hu, R., Deng, J. (1997). Studies on a potentiometric urea biosensor based on a γ-aluminium oxide matrix. *Anal. Chim Acta.*, **341**, 161–169.

[30] Liu, H., Deng, J. (1994). Amperometric mediated hypoxanthine biosensor based on Nafion-ferrocene modified glassy carbon electrode. *J. Anal. Sci. (Chinese)*, **10**, 7–10.

[31] Arnold, M.A., Rechnitz, G.A. (1982). Substrate consumption by biocatalytic potentiometric membrane electrode. *Anal. Chem.*, **54**, 2315–2317.

[32] Deng, J., Fang, Y., Cai, R. (1991). Soybean meal tissue based membrane bioelectrode for urea. *Electroanalysis*, **3**, 767–771.

[33] Fang, Y., Cai, R., Deng, J., Deng, Z. (1992). Lettuce seed meal tissue-based membrane electrode with high biocatalytic activity for hydrogen peroxide. *Electroanalysis*, **4**, 819–822.

[34] Zhang, X., Wang, Q., Deng, J. (1995). Preparation of ginger tissue-based membrane electrode for L-ascorbic acid and its analytical application. *Chin. J. Anal. Chem.*, **23**, 336–339.

[35] Dilibel, Deng, J. (1994). Potato tissue biosensor for hydrogen peroxide. *J. Analytical Science (Chinese)*, **10**, 53–56.

[36] Di, P., Deng, J. (1992). Study on the tissue electrode for adenine nucleotide. *Adv. Biochem. Biophys.*, **19**, 294–297.

[37] Deng, J., Kong, J., He, H. (1992). A sensor for guanine using brain tissue of rat. *Chem. J. Chin. Univ.*, **13**, 583–585.

[38] Kong, J., He, H., Deng, J. (1992). A sensor for L-lysine using kidney tissue of chicken. *Chin. J. Anal. Chem.*, **20**, 1265–1268.

[39] Fang, Y., Cai, R., Zhang, R., Wu, J., Deng, J. (1991). Sugar beet acetone pulp as tyrosine electrode. *Chin. J. Anal. Chem.*, **19**, 891–894.

[40] Di, P., Deng, J. (1992). Tissue membrane electrode sensitized for serine. *Chin. J. Anal. Chem.*, **20**, 663–665.

[41] Deng, Z., Fang, Y., Cai, R., Deng, J. (1992). Construction of cauliflower tissue for ascorbic acid. *J. Fudan Univ. (Chinese)*, **31**, 27–30.

[42] Ma, Q., Deng, J. (1994). Cauliflower tissue based membrane-benzoquinone modified electrode for the determination of L-ascorbic acid. *Chin. J. Anal. Chem.*, **22**, 121–124.

[43] Wang, Q., Deng, J. (1994). Optimization of analytical conditions of tissue electrode using othogonal experimental design. *Chin. J. Anal. Chem.*, **22**, 615–618.

[44] Ma, Q., Zhang, X., Chen, L., Deng, J. (1994). Study on spinach and aster indicus tissue-based membrane sensor for glycollic acid. *Chin. J. Anal. Chem.*, **22**, 668–671.

[45] Deng, J., Wang, Q. (1994). Fish liver tissue based membrane electrode for uric acid in serum. *Electroanalysis*, **6**, 689–693.

[46] Karube, I., Mitsuda, S., Matsunaga, T., Susuki, S. (1977). A rapid method for estimation of BOD by using immobilized microbial cells. *J. Ferment. Technol.*, **55**, 243–248.

[47] Deng, J., Kong, J., Jiang, Z., Cai, W. (1991b). An amperometric BOD biosensor. *Shanghai Environ. Sci. (Chinese)*, **10**, 25–27.

[48] Deng, J., Jiang, Z., Kong, J., Cai, W. (1991c). A galvanic cell type BOD biosensor. *Chem. World (Chinese)*, **10**, 456–460.

[49] Wu, Y., Liu, B., Kong, J., Deng, J. (1994). Property optimization of biochemical oxygen demand microbial sensor. *Chin. J. Anal. Chem.*, **22**, 647–651.

[50] Kong, J., He, H., Deng, J. (1993). Comparative studies on enzyme and microbial sensors for glutamic acid. *Chin. J. Anal. Chem.*, **21**, 251–254.

[51] Bao, Y., Lei, C., Deng, Q. (1996). Studies on bacterial electrode for determination of L-asparagine. *Chin. J. Anal. Chem.*, **24**, 535–538.

[52] Wang, Q., Deng, J. (1994). A microbial electrode for L-aspartate. *Chem. J. Chin. Univ.*, **15**, 974–976.

[53] Liu, B., Deng, J. (1996a). Studies on uric acid microbial biosensor and its dynamic response. *Chem. J. Chin. Univ.*, **17**, 702–706.

[54] Wu, Y., Kong, J., Deng, J. (1996). Preparation of microbial sensor using mixed *Lactobacillus* for determination of lactic acid. *J. Fudan Univ. (Chinese)*, **35**, 349–353.

[55] Lei, C., Bao, Y., Deng, J. (1995). Studies on urea biosensor based on immobilized corynebacterium glutamicum. *Progress in Biochemistry and Biophysics (Chinese)*, **22**, 555–559.

[56] Lei, C., Bao, Y., Deng, J., Lei, C. X. (1995). Studies on urea biosensor based on immobilized corynebacterium glutamicum and their kinetic response processes. *Talanta*, **42**, 1561–1566.

[57] Liu, B., Cui, Y., Deng, J. (1996). Studies on microbial biosensor for DL-phenylalanine and its dynamic response process. *Anal. Lett.*, **29**, 1497–1515.

OPTICAL CHEMICAL SENSORS AND BIOSENSORS

Zhujun Zhang and Manliang Feng

OUTLINE

ABSTRACT

In this chapter our research on optical sensors is described. These sensors are based on fluorescence, phosphorescence, chemiluminescence, reflectance, and absorbance. Fiber optics and flow-injection techniques were applied in these designs. Several materials were used as the sensing element. Sensors for immunoassay, clinical analysis, pharmaceutical analysis, and for electrolytes in body fluids were constructed.

Advances in Biosensors
Volume 4, pages 95–122.

ISBN: 0-7623-0073-6

1. INTRODUCTION

The increasing desirability of *in vivo*, *in situ*, and continuous monitoring of biomedically important substances, such as blood gases, metabolites, enzymes, coenzymes, immunological proteins, bacteria, inhibitors, electrolytes, and other analytes has led to the development of various devices to meet these demands. Thus sensors for different purposes have been developed, and research on this subject is becoming increasingly attractive and popular.

In evaluating the performance of a sensor, the importance of its reversibility, selectivity, stability (lifetime) and capability for continuous use have been emphasized. The special requirements of the specific application for which the sensor is designed tended to be ignored. Sensors have been confined to such devices that possess the following features: direct measurement, reversibility, continuous operation, accuracy, real-time and *in situ* response, long-term stability, and wide dynamic range. Because sensors are normally designed to operate under well-defined conditions for specific analytes in certain sample types [1] and for fixed purposes, therefore, emphasis should be focused on the suitability of the sensor for the specific analytical purpose. High selectivity, reversibility, and long lifetime are not always necessary. Fortunately, this concept has been accepted by more and more analysts and the perception of sensors is changing. Certain analytical devices, which are not fully reversible but are suitable for single use or can be easily and quickly regenerated, are considered sensors [2,3], e.g., sensors based on irreversible immunological reactions and sensing systems with the aid of flow injection analysis (FIA). In reality, in bioanalysis, especially clinical analysis, sensors that are disposable and suitable for single use are preferable if they are cheap enough or easy to formulate because these sensors can prevent patients from interinfection. In this situation, reversibility and lifetime are not key parameters. What is more, by coupling a sensor with specially designed reagent delivery systems, e.g. a reagent reservoir or a reagent-loaded column fitted in a FIA system, many previously irreversible sensing systems can be adapted as sensors for continuous use.

Sensors consist basically of two functional units [1]: a receptor and a transducer. The receptor is designed to recognize target species based on the chemical or biochemical interaction between an analyte and the sensing reagent, and it transforms the information on the target species (e.g., quantity, activity, class) into a species or a form of energy (e.g., pH, light, acoustic wave) which may be measured by the transducer. In some cases, a separator (e.g., a membrane or a polymer) is included in the receptor to improve the selectivity. In the transducer the energy or the chemical characteristic of a substance that carries the information about the target species is changed into an analytically useful signal. Some transducers act as sensors, but most transducers lack sufficient selectivity.

Sensors fall roughly into two broad types [4] according to their operational mode. First are the "batch" or probe type sensors which offer potential for *in situ*, *in vivo* use. However, in most cases these sensors are forced to operate under less than optimal conditions [5]. The second type is continuous or flow type sensors which have all the advantages of the first type, but they are easy to build, and the working conditions of the sensing zone can be controlled.

Biosensors can be subdivided into two groups. One group is bioreceptor-based sensors. Biosensors that use antibodies and lectins as sensing reagents are included in this group. Another group includes sensors that use biocatalysts as sensing reagents [6,7]. One representative of these sensors is an optic glucose sensor that couples glucose oxidase at the end of the sensor.

Because of great commercial and academic interest in this field, the development of biosensors is becoming one of the most important areas of bioscience, and tremendous effort has been devoted to research and applications in this field in recent years. Various kinds of biosensors based on different principles of receptors and transducers have been reported. Among them, the optical sensor is an attractive vehicle. Many optical sensors have been made and used for continuous, *in vivo* and *in situ* monitoring of biochemically and diagnostically important materials and also for testing blood samples *in vitro*. Optical sensors are based on the detection of optical changes encoded with information about the analyte. In most sensors of this type optical changes result from a chemical or biochemical interaction similar to that which occurs in solution chemistry. Thus traditional optical methods in solution form an important basis of optical sensors.

Chemical and biochemical analytical methods based on the optical properties of the analyte or analyte-related material, such as colorimetry, photometry, and spot tests, have always played a dominant role in the various fields of analytical science [2]. Absorptiometry, emission flame photometry, and radioactivity were the most commonly used techniques in chemical, biochemical and clinical analysis. In recent decades, the virtues of high sensitivity, wide linear range, low cost per test, and the relatively simple and inexpensive instrumentation of luminescence analysis have made an important impact in biochemical and clinical analysis [8]. Methods based on fluorescence, phosphorescence, and chemiluminescence are regarded as attractive replacements of colorimetry, spectrophotometry, and radioactivity in assays for biologically and clinically important species. These achievements have greatly accelerated the development of optical sensors.

It could be argued that optical sensors began with the advent of the pH strip. However, there was almost no advance in this area until fiber optic sensors appeared in the 1980s. Sensors based on fiber optics marked major progress in optical sensors [2], and they are considered a major breakthrough in analytical science.

Fiber optic sensors exhibit various attractive features relative to other kinds of sensors: (1) They are immune to electrical interference by static electricity of the body, radio frequency, and noise; no reference electrode is required, and they do not carry a risk of electric shock in clinical uses. (2) In principle, the diameter of the fiber optic can be on the order of the wavelength of transmitted light. The advent of miniaturized fiber has made available a small sensing tip and leads to the construction of a very small optic sensor which can be used for minute sample volumes and for invasive *in vivo* monitoring in biomedical analysis. (3) Because each fiber in a fiber bundle independently transmits optical information without interference, a bundle of small sensors for different analytes allows simultaneously monitoring various analytes at the same or different sites. (4) Flexibility and low attenuation permit sensing samples remotely located or in previously inaccessible sites. (5) The high information density of a fiber provides potential for simultaneous multianalyte detection, if the optical signals of different analytes vary in wavelength, lifetime, and polarization, etc. Fiber optic sensors also provide the benefits of real-time monitoring, no analyte consumption, no sample damage, freedom from inner filter effects, ease of use and, low cost make, and relative temperature independence. Because of these attractive features, fiber optic sensors are becoming increasingly attractive, and their use has become widespread.

Despite these attractive features, fiber sensors are subject to several limitations, such as ambient light inferences, relatively short-term stability of reagent phase, and limited dynamic ranges.

Considering the impressive amount of human investment in developing sensors, the practical results are rather limited. This lack of apparent success mainly results from inconsistent understanding of the connotation of sensors and the concepts of sensor evaluation. These concepts have limited research in three major ways. First is the desire for a reversible molecular recognition mechanism [9]. Many irreversible interactions cannot be employed in designing a sensor. The second limitation results from the requirement for selective sensors for every conceivable analyte. Major attention has focused on producing appropriate sensing layers or membranes by using a large selection of unique physical, chemical, or biochemical interactions. The sensing layer or membrane has to function as a barrier to exclude interfering components in a matrix from entering the sensing layer and interfering with the detected signal. In some cases, the membrane or layer is also used to manipulate the analyte concentration reaching and penetrating into the sensor. As a result, the performance of many sensors is diffusion-controlled. Thirdly, the need for a rod-shaped structure to facilitate *in situ*, simultaneous, and real-time application is restrictive. Consequently, a limited number of reagents and techniques can be used with a sensor, and therefore, sensors are often forced to operate under less than optimal conditions [5]. For instance, a sensor based on immobilized glucose oxidase for monitoring glucose in human blood must be covered with an appropriate diffusion impeding film or a membrane (barrier) to manipulate the glucose concentration that reaches the sensor so that first-order reaction kinetics for glucose are fulfilled. The reason for this is that the glucose concentration in a blood sample is higher than the Michaelis–Menten constant for glucose oxidase. In practice it is difficult to find an appropriate barrier, and when a barrier is used, the response of the sensor is delayed.

Operational compromises necessarily must be made and the concept of sensors broadened as a result. Techniques that can compensate for the shortcomings of conventional rod-shaped sensors and facilitate their applications have been employed in sensor construction. The flow-injection technique, which features strict and reproducible sampling and timing, has proven itself a complementary and an attractive tool to augment the performance of sensors [3,5]. Sensors using the flow-injection technique have been reported in recent years, and research into flow-type sensors has been initiated.

A typical flow-type sensor consists of two basic elements, a sensitive element, i.e. a flow-cell or a minireactor (column) immobilized with active materials, and a flow-feeding element. The sensitive element is incorporated in the flow line. A detectable signal is produced by a chemical or physical interaction between the analyte and sensing reagent. The interaction occurs in the flow line and/or in the sensitive element. Such configurations allow easy regulation of the amount and time that the analyte is exposed to the sensitive element or to the reagent in the flow stream by controlling the flow speed or by using appropriate manifolds. It is not subject to kinetic mass transfer through a membrane, which speeds up the response. Because separation, recognition, and transduction can occur at different sites in the flow system, they can be separately optimized, and they permit sensing under relatively optimal conditions. The ease of on-line regeneration or renewal of the sensitive element is one of the most salient features of this type of sensor, i.e., many previously irreversible sensors designed for single use are convertible to sensors suitable for continuous and repeated use.

According to the location of the sensitive element, flow-type sensors are divided into two main groups. In the first group, flow-through sensors, the sensitive element is integrated with the detector. Detection occurs directly in a column or in the flow-cell packed with sensing reagents. Obviously, sensors of this type always require that the sensitive element's support and the detection system be compatible. For example, a flow-type optical sensor must use a transparent support to prevent optical interference as a result of scattering or reflectance absorbance by the support. In the second group, the sensitive element and the detector are placed at different sites in the flow line, the detectable species is produced in or out of the sensitive element, and the optical signal is detected in a flow cell where no sensing reagent is packed. These two types of sensor designed by our team are schematically depicted in Figure 1.

We confine this chapter to our research on optical fiber biosensors and flow biosensors. These sensors are classified as fluorescence-based sensors, phosphorescence-based sensors, chemiluminescence-based sensors, reflectance-based sensors and absorbance-based sensors, according to the optical properties of detected signal.

2. FLUORESCENCE-BASED SENSORS

Fluorescent spectroscopy is a useful quantitative tool that combines excellent sensitivity and precision with high selectivity and has been extensively involved in

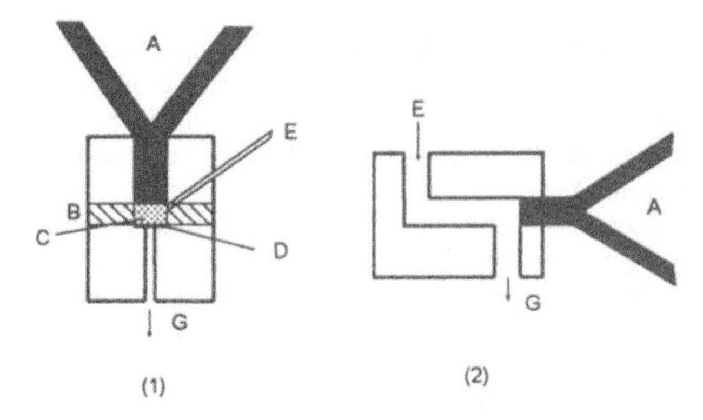

Figure 1. Schematics of flow-type sensor designs (**1**) flow-through sensor (**2**) flow-cell sensor A. fiber optic; B. rubber packing; C. substrate; D. filter disc; E. solvent entrance; G. exit.

sensor construction. Generally a fluorescent sensor is based on optically detecting a fluorescent change resulting from the analyte itself or a product formed in recognizing and transducing. The selectivity is dominated by the recognition reaction.

Given the virtues of selectivity of enzymes toward specific substrates, these catalytic reagents have gained increasing significance. Immobilized enzymes are used in numerous types of sensors to determine substrates by coupling with fluorescent indicators. Most of the biosensors of this type are based on light intensity change due to the interaction of the enzyme reaction product with the immobilized indicator. Thus such sensors suffer the disadvantage of irreversibility, long response times, and poor signal-to-noise (S/N) ratios.

We reported a simplified approach to develop a fiber optic sensor for hydrogen peroxide. To selectively determine hydrogen peroxide [10], horseradish peroxidase (HRP) is immobilized on an inert bovine albumin matrix with glutaraldehyde. Indicator is added to the sample solution. Several traditionally used fluorimetric substrates, such as tyrosine, alkali-activated diacetyldichlorofluorescein, and homovanillic acid have been tested as fluorimetric indicators for the sensor. The response of the sensors using tyrosine and homovanillic acid as an indicator is much smaller than expected. This is caused by the poor transmissibility of ultraviolet radiation in the glass optic fiber. The excitation wavelength of the oxidation products of tyrosine and homovanillic acid are at 305 nm and 307 nm, respectively. Diacetyldichlorofluorescein autohydrolyzes into fluorescein, and its oxidation rate is slow, so that the enzyme-based fiber optic sensor using diacetyldichlorofluorescein as the fluorimetric substrate has a poor S/N ratio and poor sensitivity. The higher sensitivity and stability of thiamine (1 week in refrigerator) and its oxidation product (at least 24 h) make it an excellent fluorimetric substrate for the simplified enzyme-based fiber optic sensor.

The oxidation of thiamine by hydrogen peroxide to fluorescent thiochrome in alkaline medium is always accompanied by the simultaneous formation of nonfluorescent thiamine disulfide. However, under the catalysis of HRP at pH 8.5, the oxidation reaction leads to a yield of thiochrome greater than 95%, which has strong fluorescence at 440 nm and excitation at 375 nm. The rate of fluorescent change is directly proportional to hydrogen peroxide concentration. Under optimized conditions the measuring range of the sensor is up to 1×10^{-4} M hydrogen peroxide with a detection limit of 5×10^{-7} M in a 5 min response period. This new sensor design has several advantages: First, it has a higher sensitivity than previously reported in that the fluorescent intensity increases with time. The sensitivity is increased by waiting longer for more product to build up. Secondly, the indicator is added in solution instead of immobilizing it on the membrane, so that the response of the sensor is completely reversible. Thirdly, the excitation wavelength of thiochrome is at 375 nm, but a glass fiber optic and tungsten lamp source can be used to lower the cost of the required instrumentation. One of the most attractive features is that it can be adapted as a base sensor for constructing multienzyme sensors involving oxidases, which produce hydrogen peroxide during the oxidation of its substrates. Then, biochemical substances which produce hydrogen peroxide under catalytic oxidation by their oxidase are detectable. This possibility has been tested for determining uric acid, D-amino acid, L-amino acid, glucose, cholesterol, choline, and acetylcholine using a membrane that has coimmobilized oxidase and HRP. Finally, the sensor constructed by binding the enzyme to bovine albumin is sensitive and has a stable response and a long lifetime relative to that prepared by immobilizing HRP on cellulose membrane (sensitivity is lower because of the smaller amounts of the immobilized enzymes) and entrapping enzyme in polyacrylamide gel (which leaks after a period of use).

Using this hydrogen peroxide sensor as a transducer, a simplified enzyme-based fiber optic biosensor for monitoring uric acid in serum and urine samples [11] was constructed by coimmobilizing uricase and horseradish peroxidase (HRP) to bovine albumin via glutaraldehyde, which demonstrates well the potential of this design. The immobilized enzyme membrane was prepared as follows: 50 μL of 0.1 M pH 8.5 phosphate buffer, 5 mg of uricase, 1 mg of HRP, 25 μL of bovine albumin solution (17.5%), and 30 μL of glutaraldehyde solution (2.5%) were added to a small test tube and mixed rapidly; then 50 μL of the mixed solution was transferred to the surface of the common end of the fiber optic bundle and allowed to stand at room temperature. When the membrane was nearly dry, it was immersed in a phosphate buffer solution (pH 8.5) for approximately 5 min with the aid of a nylon net and an O-ring.

When the membrane with the immobilized enzymes was immersed in a solution containing a certain amount of uric acid and excess thiamine, two enzyme-catalyzed reactions occur.

$$\text{Uric acid} + 2H_2O + O_2 \xrightarrow{\text{uricase}} \text{allantoin} + CO_2 + H_2O_2$$

$$\text{thiamine} + H_2O_2 \xrightarrow{\text{HRP}} \begin{cases} \text{thiochrome} + H_2O \text{ (fluorescent)} \\ \text{thiamine disulfide (nonfluorescent)} \end{cases}$$

The relationship between the reaction rate v and the concentrations of reactant thiochrome, H_2O_2 and uric acid can be represented as

$$v = \frac{dc_{\text{thio}}}{dt} = \frac{-dc_{\text{H2O2}}}{dt} = \frac{dc_{\text{uric}}}{dt} \tag{1}$$

where c_{thio}, c_{uric} and c_{H2O2} are the concentrations, respectively, of the reactants thiochrome, uric acid, and H_2O_2 at the surface of the membrane.

The reaction rate also obeys the Michaelis–Menten equation.

$$v = \frac{V\,[\text{uric acid}]}{(K_m + [\text{uric acid}])} \tag{2}$$

where [uric acid] is the concentration of the reactant uric acid at the surface of the membrane, V is related to the amount of immobilized enzyme $[E]$ and the rate constant k_0 of the conversion reaction from enzyme-substrate complex to product by $k_0[E]$. These two expressions of the reaction rate should be equal:

$$\frac{dc_{\text{thio}}}{dt} = \frac{V\,[\text{uric acid}]}{(K_{\text{m}} + [\text{uric acid}])} \tag{3}$$

Assuming $K_{\text{m}} >>$ [uric acid], then

$$dc_{\text{thio}} = \frac{V\,[\text{uric acid}]}{K_{\text{m}}} \tag{4}$$

The reaction occurs in a static solution, and the amount of thiamine in solution is much greater than that of uric acid. As the reaction proceeds, a diffusion layer is established at and near the membrane. According to the concept of a diffusion layer, the number of moles of uric acid reaching the surface of the membrane per time interval is given by

$$N = \frac{AD(C_{\text{b}} - C_{\text{s}})}{\delta} \tag{5}$$

where A is the area of the membrane, D is the diffusion coefficient, C_{b} and C_{s} are the concentrations of uric acid in solution and at the surface of the membrane, respectively, and δ is related to the thickness of the diffusion layer. Then, the concentration of the reactant uric acid can be represented as

$$[\text{uric acid}] = kN = \frac{kAD\,(C_b - C_s)}{\delta} \tag{6}$$

where k is a constant. In fact, the enzymatic reaction is very quick. It is assumed that $C_b >> C_s$. Thus, the expression can be simplified as:

$$[\text{uric acid}] = \frac{kADC_b}{\delta} \tag{7}$$

where δ follows the formula $\delta = (\pi Dt)^{1/2}$. Submitting δ and (7) into (4) and rearranging

$$\frac{dc_{\text{thio}}}{dt} = \frac{kADC_b}{\delta V} K_m = \frac{kADV}{(K_m\delta)\,C_b} \tag{8}$$

With the progression of the reaction, more and more thiochrome accumulates at the surface of the membrane. Assuming that all the produced thiochrome is detected by the sensor, then the amounts of the thiochrome product in the time range from 0 to t is represented as

$$c_{\text{thio}} = \int^{t} dc_{\text{thio}} = \int^{t} \frac{kAVDC_b}{K_m} (\pi D)^{1/2}\, dt$$

$$= \frac{2kAVDC_b}{K_m(\pi D)^{1/2} t^{1/2}} \tag{9}$$

Thus, the observed fluorescence intensity is given by

$$I_F = 2.3 Y_F I_0 \varepsilon bc = \frac{4.6 Y_F I_0 \varepsilon bkVADC_b t^{1/2}}{K_m(\pi D)^{1/2}} \tag{10}$$

where b is the thickness of the detectable solution.

Assuming $4.6Y_F I_0 \varepsilon bkVADC_b/K_m(\pi Dt)^{1/2} = K$ yields the final expression for the intensity as a function of the concentration of uric acid in solution:

$$I_F = KC_b t^{1/2} \tag{11}$$

According to this equation,

(1) The observed fluorescence signal is controlled by diffusion. The intensity is proportional to the concentration of uric acid in a fixed time.
(2) The response of the sensor is proportional to $t^{1/2}$ at a fixed concentration of uric acid, so that the sensitivity of the sensor is improved by waiting for a longer reaction time.

Following the same concept, a fiber optic biosensor that responds to D-amino acid is constructed by coimmobilizing D-amino acid oxidase and HRP on the inert matrix bovine albumin via glutaraldehyde [12].

The coimmobilization concept was introduced into a flow-injection fluorescence sensor for glucose in conjunction with the stopped-flow technique. In the sensing system, glucose oxidase and HRP were coimmobilized on silica gel to make a sensing column, and the column was fitted in the flow line. While glucose and fluorogenic substrate pass through the sensing column, thiochrome, which relates to the amount of glucose, is produced and detected in a post column flow cell by a fiber optic [13].

In addition to enzymes, other biomaterial, such as bacterial cells and intact mammalian and plant tissue, are available for preparing biosensors.

The specific interaction between a natural killer (NK) cell and a target cell, for instance, provides an approach to build up sensors for the cytotoxicity of NK cells. By labeling the target cell with a new fluorescent marker, the cytotoxicity of the NK cell can be quantified based on time-resolved fluorescent spectroscopy [14,15].

Electrolytes, O_2, CO_2, and pH in blood are critical parameters that warn of life-threatening trends, and what is more, the effects of trace elements, e.g., Zn, Fe, Se, and some rare earth elements, on human health have gained much more interest in recent years. Various types of sensors for these species have appeared.

Selectivity for alkali metal ions is achieved by using neutral ionophores, such as crown ethers, which selectively bind alkali metal ions. The problem is to couple this selectivity to optical detection [16].

One approach is to design an ionophore that undergoes a color change upon metal ion binding. Therefore, developing chromogenic ionophores has been an active area of research. This approach has been implemented for potassium ion sensing. However, it is subject to several difficulties. The first is that the structural constraints associated with a chromogenic ionophore make it difficult to achieve a high degree of selectivity. The second is that, in water, equilibrium constants for alkali metal ion binding by ionophores are too small to be suitable for sensing in the physiological response range because water bonds so strongly to alkali metal ions. Moreover, water undergoes deprotonation upon metal ion binding, resulting in a response which is pH-dependent.

An alternative approach is to base the indicator on ion-pair extraction. Ion-pairing reactions can be used to develop a reversible indicator system that selectively responds to both cations and anions, and we have succeeded in developing indicators that are reversible for cation and anion ions [17–19]. Cation detection systems include an uncharged immobilized ionophore, an anionic dye, and a cationic polyelectrolyte. Our starting point is an ionophore that selectively binds a particular ion. For the cations the ionophore is usually a microcyclic ligand, e.g., a crown ester. The metal ion reacts with a neutral ionophore to form a hydrophobic cation, which in turn associates with a hydrophobic anionic chromophore, caused by a combination of hydrophobic and electrostatic interaction, and results in an ion pair.

The ion pair can be extracted into an organic phase or onto a solid surface. The amount of chromophore extracted is related to the amount of metal ion in the original sample.

Using this approach we have achieved a selective and reversible response to sodium [20] using *N*,*N*″,*N*″-triheptyl-*N*,*N*′,*N*″-trimethyl-4,4′,4″-propylidynetris(3-oxabutyramide) as the ionophore and the sodium salt of 8-anilino-1-naphthalene-sulfonic acid (ANS) as the fluorophore. Response times for this system are on the order of 2–3 min depending on both the amount of ionophore and the sodium concentration.

Generally, several problems must be solved before constructing such an alkali metal ion sensor. The first problem encountered in doing ion pair extraction reversibly is deciding on an "organic phase." The second is to incorporate a neutral ionophore, which offers selectivity for alkali metal ions, into the "organic phase" of the system. Because most ionophores are poorly soluble in water, adsorption is a major means of achieving this. Next is the choice of chromophore. In conventional extraction the concentration of the chromophore is measured in the organic phase without any contribution from the chromophore remaining in the aqueous phase. However, when doing ion pair extraction continuously in a sensor, this may not be possible. The ammonium salt of 8-anilino-1-naphthalene sulfonic acid (ANS) was chosen in most cases because it is fluorescent in the organic phase but does not fluoresce in water.

The final, most challenging problem is to find a way of "immobilizing" the chromophore so that it can be used on a continuously reversible basis. To be available for ion-pair extraction formation, the chromophore must be in solution. Therefore, the only viable approach is increase the effective size of the chromophore so that it is retained by a size-selective membrane that is permeable to alkali metal ions. We chose a polyelectrolyte, poly(ethylenimine) (PEI), as the third component. The cationic complex of Cu(II)-PEI and ANS forms an ion pair by electrostatic and hydrophobic interactions and increases the effective size.

These problems have been satisfactorily solved in a series of sensors, whose performance is illustrated in an indicator system that responds reversibly to alkali metal ion concentration. The system consists of the Cu(II) complex of aqueous polyethylenimine, an anionic fluorophore, and a neutral ionophore immobilized by adsorption on silica particles. In the absence of metal ion, the anionic fluorophore binds to the Cu(II)-polyethylenimine complex where the presence of paramagnetic Cu(II) causes quenching. When alkali metal ion is present, it complexes with the ionophore to form a hydrophobic cation which binds to some of the anionic fluorophore rendering it fluorescent. The fluorescence of the indicator is measured using an optical fiber that results in a reversible sensor for alkali metal ions.

An important feature of the ion-pair approach is the ability to independently vary the anionic fluorophore and the ionophore which makes selective determination of other alkali metal ions possible.

Selectivity for anions is more frequently based on ligand exchange of anions by a metal complex. In an anion ion sensor for fluoride [19], the "ionophore" is a neutral Al(III) complex. Selectivity depends on the high affinity of Al(III) for fluoride. The ligand of the ionophore is a modified EDTA that couples to a C18 chain(C18-EDTA) and contacts the modified EDTA with an aqueous Al(III) solution in the presence of the immobilization substrate, either silica or cellulose, to form an Al(III)-C18-EDTA complex. Because the complex is uncharged and includes a C18 chain, it is minimally soluble in water and precipitates on the surface of the substrate. The cationic fluorophore used was pyronin, and the anionic polyelectrolyte was polyacrylic acid (PAA). Both were chosen because they yield a rapid response.

In principle, the ion-pairing approach could be used with any neutral ionophores. However it was found that with some ionophores, notably valinomycin, response times are unacceptably long due to slow chemical kinetics, and the reaction is subject to instability due to leaching of both the ionophore and the fluorophore.

The third approach is based on potential-sensitive dyes. Ions tend to adsorb strongly onto lipophilic surfaces. In the absence of potential across a biological membrane, there is equal surface coverage on the inside and outside surface of the membrane. When potential develops across the membrane, the relative coverage on either side of the membrane changes. Because potential-sensitive dyes form weakly fluorescent aggregates and are subject to concentration quenching, their fluorescence depends highly on concentration. Therefore, changes in cell potential are usually accompanied by a change in fluorescent intensity associated with the changes in the surface coverage on either side of the membrane. The direction and magnitude of the change depends on the ratio of dye to membrane and on the specific fluorescent characteristics of the dye.

The possibility of optical ion sensing based on this scheme has been recognized by other researchers. The specific system involves a cationic ester of rhodamine B as the potential-sensitive dye, valinomycin as the ionophore, and a lipid bilayer supported on the substrate [21].

We have reported an indicator system consisting of liposomes in the presence of valinomycin as the ionophore and merocyaine 540 as the potential-sensitive dye [16]. This sensor is more sensitive than the reported system of using rhodamine B as the potential-sensitive dye. The results produced by this sensor confirm the possibility of reversible, selective sensing of alkali metal ions using potential-sensitive dyes. Furthermore, the magnitude of the response that we have observed using liposomes is much greater than that previously reported using a solid-supported lipid bilayer. An interesting feature of this approach is the fact that the response depends on the ability of neutral ionophores to transport ions across the membrane rather than involving them in chemical equilibrium. The response property is exhibited in an indicator system consisting of a potential-sensitive fluorescent dye in the presence of liposomes and valinomycin. This sensor responds reversibly and selectively to changes in potassium concentration (Figure 2). It also allowed

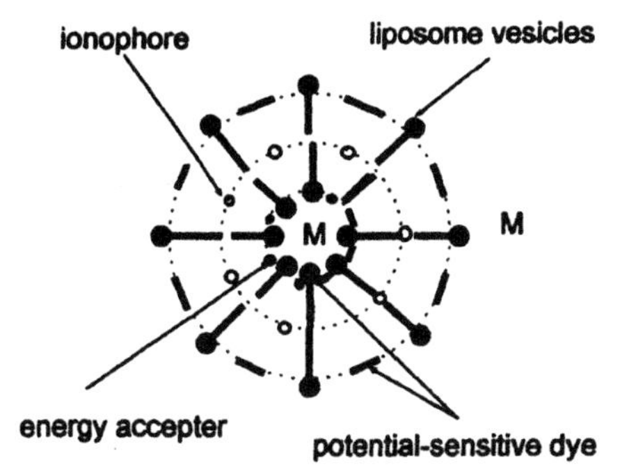

Figure 2. Response mechanism of liposomes and a potential-sensitive, dye-based sensor.

measurement at long wavelengths by incorporating an energy acceptor into the interior of the liposome.

Because some metal ions associate with fluorogenic ligands and change fluorescent characteristics, sensors for these metal ions may be prepared by immobilizing a fluorogenic ligand on the common end of a bifurcated fiber optic.

We have successfully immobilized several ligands including morin, quercetin, calcein, hydroxynaphthol blue (HNB), 8-hydroxyquinoline-5-sulfonate (8HQ5S), 2,2′,4-trihydroxyazobenzene and *p*-tosyl-8-aminoquinoline on silica gel, cellulose, and anion exchangers to formulate fiber optic ion sensors [22]. Fluorescence is excited through one arm of the fiber optic and observed through the other. When the immobilized ligand is placed in a solution of metal ions, assuming that a 1:1 ligand-metal ion complex is formed in the sensing layer, the equilibrium constant K_e of complex formation is given by

$$K_e = \frac{X_{MF}}{X_F [M]} \tag{12}$$

where X_F is the number of moles of immobilized ligand which do not associate with metal ions and X_{MF} is the number of moles of immobilized ligand associated with metal ions. $[M]$ is the concentration of metal ions in solution.

The fluorescent signal depends on the relative amounts of M and MF:

$$I_F = K_F X_F + K_{MF} X_{MF} \tag{13}$$

K_F and K_{MF} are proportionality constants that relate fluorescent intensity to the amounts of free and associated ligand, respectively. Assuming that the reagent layer does not absorb source radiation to a significant extent and that analyte depletion in the sample is negligible, Eq. (14) applies.

$$I_F = \frac{K_F C_F}{(1 + [M]K_e)} + \frac{K_{MF} C_F K_e [M]}{(1 + [M]K_e)} \tag{14}$$

C_F is the total number of immobilized ligands.

If the metal complex fluoresces and the ligand does not, the first term drops out. Fluorescence is proportional to low metal concentration. At higher concentrations, the response becomes nonlinear. If the ligand fluoresces by itself but the complex does not, the second term drops out. In this case fluorescence decreases with metal ion concentration in term of K_e. This kind of system is less desirable for analytical purposes because it is less sensitive and has a limited dynamic range.

Because the reversibility of the sensor is dominated by the binding constant, a compromise has to be made between reversibility and sensitivity, and the potential of several fluorogenic ligands has to be evaluated. Morin is immobilized to cellulose via cyanuric chloride. The immobilized ligand fluoresces weakly but forms strongly fluorescent complexes with both Al^{3+} and Be^{2+}. Quercetin is immobilized by the same procedure as morin. It responds to Al^{3+} similarly to morin, but is less sensitive. Hydroxynaphthol blue (HNB) belongs to the *o,o′*-dihydroxyazobenzene class of ligands which are nonfluorescent. They form fluorescent complexes with several metals. The three sulfonate groups allow for easy immobilization onto an anion exchange resin. HNB is suitable for Al^{3+}. 8HQ5S, like HNB, is immobilized on an anion exchanger. Because there is only one sulfonate group, it is more easily displaced from the ion exchanger. Sensors based on immobilized 8HQ5S have potential for Zn^{2+} and Cd^{2+} monitoring.

Calcein is immobilized to cellulose by the same procedure as morin. In solution it is nonfluorescent at higher pH but forms a fluorescent complex with calcium. This is not possible with calcein immobilized via cyanuric chloride because calcein hydrolyzes from the substrate at higher pH. However, at neutral pH it may be used to detect a variety of metals based on fluorescence quenching, e.g. Ni^{2+}, Cu^{2+} and Fe^{3+}. Notwithstanding its immobilized form, calcein is not suitable as a sensor, but it can be used for other purposes, such as detecting end points of complex titrations.

2,2′,4-tryhydroxyazobenzene is immobilized to silica gel through an additional noncomplexing hydroxy group, with which Al^{3+} forms a fluorescent complex. Like calcein, it binds the metal ion so tightly that it is not well suited for use in a sensor, but it is useful for preconcentration. *p*-Tosyl-8-aminoquinoline forms fluorescent complexes with Zn^{2+} and Cd^{2+}. Its potential is evaluated by immobilizing to silica gel.

A representative optical sensor based on complexation or chelating, responding to Al(III), Mg(II), Zn(II) and Cd(II), is prepared by immobilizing quinolin-8-ol-5-sulfonate (QS) on an ion exchange resin and attaching the resin to the end of a trifurcated fiber optic bundle [23]. Immobilization leads to weak fluorescence from QS and causes a shift in the fluorescent spectra of the QS/metal complexes. Detection limits for the metal ions studied are all below 1×10^{-6} M. The shape of the response curve fits a model that assumes a 1:1 metal/QS chelate is formed. Formation constants for immobilized QS complexes calculated from the model are similar to those observed for dissolved QS. Immobilized and dissolved QS behaves similarly with respect to pH and interferences. Electrostatically immobilizing QS on an ion-exchange resin is convenient and makes it possible to vary the relative amounts of ligand and substrate.

Recently, an optical fiber fluorosensor for cadmium ions was prepared by immobilizing 8-hydroxyl-7-iodoquinoline-5-sulfonic acid on diethyaminoethy(DEAE)-Sephadex [24]. Because the lowest excited singlet changes from $n - \pi^*$ to $\pi - \pi^*$, complexation between cadmium ion and the immobilized reagent causes a weakly fluorescent ligand to become a strongly fluorescent complex. Cadmium ions in solution were determined on the basis of the fluorescent intensity change of the immobilized phase at 505 nm. Because DEAE-Sephadex is a colorless transparent gel which contains diethylaminoethylglucan that has a low transfer resistance in water, the fluorosensor reaches 90% of the steady-state response in 10 s. The relationship between the response and the logarithm of cadmium ion concentration is linear in the range of 2.0×10^{-7}–7.0×10^{-5} mol/L. Using KI solution, the sensing layer saturated by cadmium ions is regenerated, and the immobilized 8-hydroxy-7-iodo-quinoline-5-sulfonic acid is not destroyed.

This sensing concept has also been extended to anion detection. A sensitive method for determining trace cyanide has been presented with ascorbic acid in a calcein-Cu^{2+}-CN^- system. When the pH is between 9 and 10, the formation of a Cu^+-calcein complex quenches fluorescence. CN^- displaces Cu^{2+} from the complex leading to fluorescence. Ascorbic acid enhances the fluorescent signal of this system [25].

The key to devising a good sensor is to shorten the response time and obtain good reversibility. Response time is controlled by the mass transfer process in the substrate. Polymer powder, porous glass and polymer membranes have been employed in constructing the sensing layer. An ideal substrate should possess low mass transfer resistance and be amenable to variation of the amount of indicator immobilized on it. In addition it should not absorb excitative and emissive light. Exploring new substrates has been an important area of our research.

Cross-linked poly(vinyl alcohol) (PVOH) has been evaluated as a substrate for immobilizing indicators used in fiber optic chemical sensors [26]. The OH group made it possible to covalently immobilize indicators via cyanuric chloride. Cross-linking can be implemented by adding glutaraldehyde and HCl to an aqueous PVOH solution. The resulting gel is clear and transparent in the visible and ultraviolet regions down to about 230 nm. Swelling properties depend on the amounts of glutaraldehyde and HCl. One attractive feature of this procedure is that it controls both the amount of indicator and the amount of substrate. This makes it possible to evaluate the effects of the amounts of indicator and the thickness of the indicator layer in fabricating a sensor. This was not previously possible with other methods that combine covalently immobilized indicators with fiber optics. Another interesting feature of this procedure is that it allows *in situ* formation of the sensing layer. The feasibility of this substrate has been illustrated in a fiber optic pH sensor [27] by covalently coupling fluoresceinamine to poly(vinyl alcohol) (PVOH) via cyanuric chloride. A drop of aqueous PVOH/indicator conjugate is cross-linked with glutaraldehyde *in situ* on the common end of a bifurcated fiber optic bundle. Fluorescent intensity excited at 480 nm and observed at 520 nm increases as the pH is varied from 3 to 8. Fluorescent intensity excited at 420 nm and observed at 520 nm is constant over the same range and is a reference to compensate for instrumental fluctuation.

PVOH also has potential in metal ions sensors. A Mg(II) sensor can be prepared by coimmobilizing eriochrome black T (EBT) and fluoresceinamine on PVOH. At pH 9.6, increasing the concentration of Mg(II) from 0 to 20 micromolar leads to a decrease of fluoresceinamine emissive intensity. This occurs because the absorption spectrum of Mg(II)-EBT overlaps the fluoresceinamine emission, causing quenching via fluorescent energy transfer.

Taking advantage of relatively low mass transfer resistance, a sensor using PVOH as the substrate for ferric ion based fluorescent quenching has been developed [28]. The fluorescent indicator, *o*-aminobenzoic acid, is covalently immobilized on cross-linked polyvinyl alcohol. The fluorescence of *o*-aminobenzoic acid is quenched by the ferric ion, which acts as a collisional quencher, at pH 4. The response time of the sensor is on the order of seconds, and reversibility is excellent. An optical sensor responding to Al^{3+} was also prepared by covalently coupling morin to poly(vinyl alcohol). The steady-state response is reached in 5 to 10 s [29].

The pH value of human blood is among the critical physiological parameters. Sensors for quantifying the pH value in physiology are of clinical and biomedical interest. A fluorescent indicator which has a pK_a in the middle physiological pH range can be exploited to design a sensor for this purpose. The trisodium salt of 8-hydroxyl-1,3,6-pyrene trisulfonic acid (HOPSA) which has a pK_a of 7.3 is an alternative. Because HOPSA has three sulfonate groups on an otherwise hydrophobic structure, it can be immobilized conveniently and essentially irreversibly on an anion exchanger. Another attractive feature of HOPSA is that it is not subject to concentration quenching in the immobilized state. The excited state of HOPSA ionizes more rapidly than it returns to the ground state. As a consequence, below pH 7.3, the observed fluorescence is characteristic of the excited state of $OPSA^-$ ($OPSA^{-*}$), even though the HOPSA is the major ground-state species. However, the excitation spectra for HOPSA and $OPSA^-$ differ considerably (Figure 3). Consequently. A pH-dependent response can be observed by exciting at a wavelength which is selectively absorbed by either HOPSA (405 nm) or $OPSA^-$ (470 nm). A better approach is to measure the ratio of intensities for fluorescence excited at two different wavelengths, one selective for HOPSA and the other for $OPSA^-$. The intensity ratio is insensitive to factors, such as variations of source intensity, fluorescent quenching, and slow loss of HOPSA which would affect the absolute intensity values [30].

The pK_a^* for excited state ionization is 1.4, indicating that this sensor can also be used to sense pH values between 0.5 and 2.5.

This pH sensor can be adapted to make a CO_2 sensor [31]. We fabricated the first fiber optical fluorescent sensor for CO_2, another critical physiological parameter,

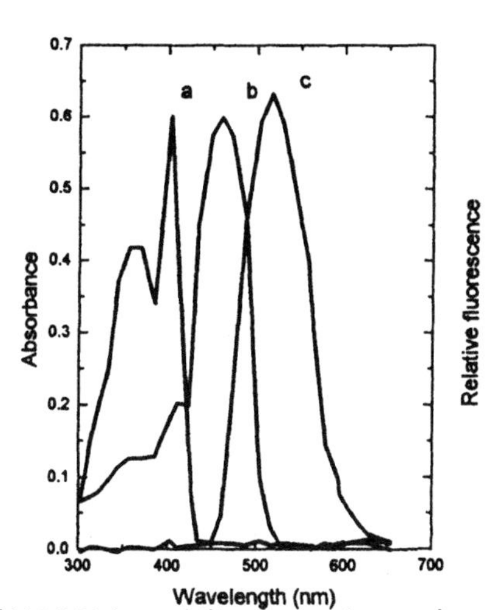

Figure 3. Spectra of HOPSA immobilized on anion-exchange resin: (**a**) absorption spectrum in 0.1 M HCl, (**b**) absorption spectrum in 2 M KOH, (**c**) fluorescence spectrum.

by covering the pH sensor with a CO_2-permeable membrane and contacting the pH-sensitive membrane with a reservoir of hydrogen carbonate. The pH-sensitive membrane and the CO_2-permeable silicone rubber membrane are held over the end of a piece of glass tube by an O-ring. The glass tube is filled with hydrogen carbonate solution of known concentration, inside of which a bifurcated fiber optic fits. As carbon dioxide diffuses across the membrane, it causes a change in pH which is detected by measuring the change in fluorescence from the base form of the pH-sensitive fluorescent dye. The usable range of response depends on the concentration of hydrogen carbonate in contact with the membrane. Unlike the electrode, the optical CO_2 sensor does not require a thin layer of solution between the CO_2-permeable membrane and the pH sensor, although it does require diffusion in the pH-sensitive membrane. Thus, in principle, it should be possible to formulate an optical sensor that would respond more rapidly than the electrode if the membrane of the optical sensor could be made thin enough (or if a single membrane could be devised that would serve both the functions of CO_2 permeability and pH response). It takes about 3 min to reach a steady-state response with a bilayer sensing unit. This same principle can also be used in sensors responding to sulfide and sulfite.

Fiber optics are also an attractive vehicle in kinetic spectrofluorimetry with FIA and make these measurements simple. A flow sensing system has been developed using 8-hydroxyquinoline-5-sulfonate aluminum as the kinetic fluorescent indicator [32,33]. Mn(II), which catalyzes the oxidation of Al-8HQ5S complex by $NaIO_4$ and causes the quenching of fluorescence, can be determined. Another flow sensing system, Mn(II), is based on the fact that Mn(II) catalyzes the oxidation of thiamine by sodium periodate and forms a fluorescent compound.

Although selectivity is the most important feature of a sensor, a useful sensor is not necessarily required to respond to an analyte specifically with respect to the possible interference parameters in a sample. Therefore, receptors that exhibit adequate selectivity can be used in sensors for certain purposes.

C18 silica gel is the most widely used stationary phase for reversed-phase liquid chromatography. For a given mobile system, many substances have their own characteristic retention which can be used to separate them chromatographically. We found that tryptophan is easily retained on the C18 column and shows strong fluorescence, about seven times that in aqueous solution when a 0.1 M NaH_2PO_4-0.1 M NaOH buffer solution (pH 8.0) is selected as the carrier stream [34]. The retained tryptophan is also quickly and completely eluted out of the C18 column by a mixture of methanol and water (30:70,V/V). However, under the same conditions, many other substances are not retained on the C18 column, or they did not show fluorescence, although they are retained. Based on this property, an optosensor for tryptophan was developed with C18 silica gel as the molecule recognition agent. The retention of the cinchona alkaloids under appropriate pH on the C18 column and the enhancement of their fluorescence has also been used to formulate a flow-through optosensor for cinchona alkaloids in pharmaceutical preparations and soft drinks [35]. The detection limits for quinine, cinchonine, quinidine and cinchonidine are 2.3 ng mL^{-1}, 31.6 ng mL^{-1}, 2.3 ng mL^{-1} and 31.6 ng mL^{-1}, respectively. Most of the common species do not interfere. Another naturally fluorescent species that is well retained on a C18 column is riboflavin [36]. It shows about ten times stronger fluorescence than that in aqueous solution. These proposed sensors proved simple, rapid, and sensitive, and required only conventional instrumentation.

In contrast with the spectra in aqueous solution, the fluorescence spectrum of a fluorescent analyte on a C18 column in the spectrum on C18 silica gel generally has a blue or red shift in its emission and excitation wavelength. For example, the spectrum of riboflavin on C18 silica gel has a 12 nm red shift in its excitation wavelength and a 8 nm blue shift in its emission wavelength. Fixation on C18 silica gel also results in an enhanced fluorescence signal that is about ten times greater than that in aqueous solution for riboflavin and seven times greater for quinine. This is the result of accumulation of the analyte on C18 silica gel and its associated change of the environment. When the sample solution is passed through the flow cell, it fixes on the packed C18 silica gel, and the effect of water molecules on the fluorescent emission is reduced. Thus, a more favorable environment results in stronger fluorescence.

One of the main advantages of the sensor is the potential increase in sensitivity with an increase in the sample volume taken for analysis. This effect can be assessed by measuring the fluorescent intensity on C18 silica gel with different volumes of solution containing the same concentration of quinine passed through the flow cell.

A CH_3OH-H_2O mixture was used to elute analytes fixed on C18 silica gel and to regenerate the sensor.

A similar retention mechanism has to be used in a flow optosensor for Al^{3+}, Ga^{3+}, In^{3+}, Zr^{4+} and Hf^{4+} based on their ternary complexes with quinolin-8-ol-5-sulfonate and the chelating resin of Chelex 100 [37,38]. This sensing system can greatly improve the selectivity and sensitivity for determining these metal ions. Most of the common metal ions do not interfere, except Fe(III), which causes serious interference and should be masked with 1,10-phenanthroline.

Cyclodextrins (CD) are sugar molecules whose structure is a hollow truncated cone with a hydrophobic cavity. Their complexation ability has been attributed to four factors [39]: (1) Van der Waals interaction, (2) hydrogen bonding, (3) displacement of high-energy water molecules from the cavity, and (4) release of the strain energy of the CD on inclusion of the guest molecule. The hydrophobicity of the guest molecule also plays a key role in the stability of the complex. The stoichiometry of the complexes is not necessarily 1:1. Fluorescence is enhanced by inclusion complexation of many fluorophores with β-cyclodextrin (β-CD). Inclusion of the analyte (inclusate) in the cavity of β-CD places it in a more hydrophobic environment where it becomes protected from bulk solution quenchers (e.g., water). The extent of complexation (i.e., selectivity and sensitivity) strongly depends on analyte geometry and functional group orientation. Thus, β-CD is another potential sensing reagent for certain species.

Combining the selectivity of the β-CD immobilized on silica gel with the enhanced fluorescence of the complexed molecule, we have reported a new type of flow-through optosensor for riboflavin [40] in aqueous media. Immobilized β-cyclodextrins have also proved to be a potential sensing agent for very low levels of tryptophan, phenylalanine, tyrosine, and quinine under appropriate conditions [41–43].

3. PHOSPHORESCENCE-BASED SENSORS

Room temperature phosphorimetry (RTP) has added new dimensions to the area of luminescence analysis. Phosphorescence is emission by an excited triplet to ground singlet state transition. Because of the spin-forbidden nature of this process it has a relatively long life and is quite susceptible to radiationless decay by molecular collision. Consequently a rigorous support or ordered medium that eliminates molecular collision is preferred in yielding RTP. Cellulose, silica gel, and sodium acetate have been used for this purpose. Because of hydrolysis of the hydrogen bond between the phosphor and the matrix, RTP is usually sensitive to moisture.

The phenomenon of a phosphor producing room temperature phosphorescence (RTP) in an aqueous medium, when the complex formed by 8HQ5S with hafnium is adsorbed on a strongly basic anion exchange resin, has been employed for developing a RTP sensor. Ferron complexes of Al^{3+}, Ga^{3+}, In^{3+}, Zr^{4+}, and Hf^{4+} show

RTP behavior, but it is impossible to selectively determine one of them without separation because they have similar spectral characteristics. However, when 8HQ5S is immobilized on an ion exchange resin, hafnium can be selectively determined in samples containing Zr^{4+}. The new optosensing method [44,45] has several advantages. First, the RTP stabilized by an exchange resin is not sensitive to moisture, so an RTP measurement can be made in a flowing aqueous solution and an "ordered medium" is not needed. This provides a method with a simple procedure and high sampling speed. Secondly, the method has relatively high selectivity. In contrast to their ferron complex, the complexes formed by 8HQ5S with Al^{3+}, Ga^{3+}, In^{3+} and Zr^{4+} have no RTP behavior when adsorbed on an exchange resin.

When excited at 393 nm, the complex of hafnium-8HQ5S produces RTP that has a maximum wavelength at 585 nm and fluorescence at 500 nm. A delay time of 0.04 ms ensures negligible background fluorescent interference because the lifetime of the triplet state is 0.402 ms. Chelating resins, such as Chelex 100, are another substrate used in RTP sensors. Based on the RTP behavior of Eu(III)-thenoyltrifluoroacetone (TTA) immobilized on this resin, an optosensing system with FIA was established for Eu(III) [46]. This sensor is a potential approach to a time-resolved fluoroimmunosensor. Based on the phosphorescence quench of the TTA-Eu(III) complex by picric acid, a reversible sensor for picric acid has been also established [47]. The phosphorescent property of the binary complex of terbium with 1,4-bis (1′-phenyl-3′-methyl-5′-pyrazolone-4′-) butanedione-(1,4) (BPMPBD) on a strongly acid cation exchange resin has been used to design a flow optical sensor for terbium. Gd(III) with BPMPBD on Chelex 100 has also been employed in a sensor for Ga(III) in conjunction with time-resolved RTP [48,49].

4. CHEMILUMINESCENCE-BASED BIOSENSORS

Because of its high sensitivity and relatively simplified instrumentation, chemiluminescence (CL) analysis has gained much more attention than other methods and numerous CL sensors have been reported. Many CL reactions involve an oxidizing agent, such as H_2O_2, so they are easily coupled to an oxidizing agent for the related enzymatic reaction in designing sensors. In a typical design [50] we reported that lactate oxidase, which is used as a receptor, and horseradish peroxidase, which is used as a transducing reagent, were covalently coimmobilized on 3-aminopropyltriethoxysilane modified porous silica. Then, the silica gel coimmobilized with the enzyme is attached to the end of a fiber optic as a biocatalytical sensing layer. In the sensitive microzone, the enzyme-catalyzed oxidation reaction of lactate acid is coupled to the HRP catalyzed CL reaction of luminol through H_2O_2:

$$\text{Lactic acid} + 2O_2 + 2H_2O \xrightarrow{\text{LOD}} CH_3COCOOH + 3\,H_2O_2$$

$$\text{Luminol} + 2H_2O_2 + 2OH^- \xrightarrow{HRP} \text{3-Aminophthalate} + N_2 + h\nu$$

The CL signal that carries the information of lactic acid concentration is conveyed through a single fiber optic to the detector. No pretreatment of the sample is needed. Because the kinetics of the CL reaction are rapid, the response time is controlled by the molecular recognition reaction. Coimmobilization of the two enzymes eliminates the mass transfer of hydrogen peroxide and results in a rapid response.

Another typical example of CL is a fiber optic biosensor for determining both cholesterol esters and cholesterol [51]. It is prepared by covalently coupling cholesterol oxidase and horseradish peroxidase to bovine serum albumin via glutaraldehyde and making a membrane on the end of a fiber optic bundle. In the sensing layer, cholesterol esterase catalyzes the hydrolysis of cholesterol esters to free cholesterol. Then the cholesterol oxidase in the presence of oxygen oxidizes the free cholesterol to 4-cholestene-3-one and hydrogen peroxide, which reacts with luminol under the catalysis of horseradish peroxidase to produce CL [52]. The sensing layer is 20 μm thick, and therefore, the response is relatively quick (2 min.) (Figure 4).

A similar principle is implemented in a fiber optic biosensor that responds to uric acid. It is prepared by covalently coupling uricase and horseradish peroxidase to 3-aminopropyl porous silica via glutaraldehyde and attaching the silica to the end of fiber optic bundle. In the microenvironment of the membrane, the uricase catalyzes the oxidation of uric acid to produce hydrogen peroxide, which then reacts with luminol under the catalysis of HRP to produce CL. We have also designed a fiber optic CL sensor for glucose in serum. The sensor exploits the $K_3Fe(CN)_6$-

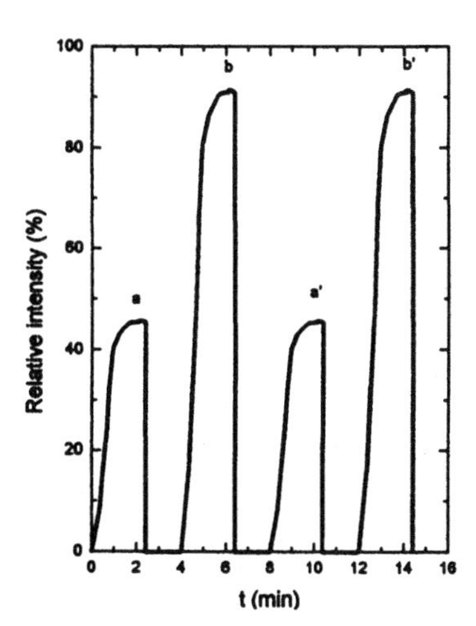

Figure 4. Response curves for cholesterol; **a, a′**: 10 μg/mL cholesterol; **b, b′**: 16 μg/mL cholesterol.

luminol CL reaction to detect hydrogen peroxide produced by the glucose oxidase catalyzed oxidate of glucose. The sensing reagent is confined to the tip of the fiber optic by a size-exclusion dialytic membrane [53].

The specific affinity between antigens and antibodies is another major tool for biosensor design. These sensors are usually irreversible and disposable because of the high binding constant of antigen-antibody complex formation. Following the basic theory of affinity chromatography and flow injection technology, we have established a reusable sandwich-type immunoassay system for HBsAg in serum. HBsAb, the recognition reagent, is immobilized on controlled pore glass, and the signal is produced by an indophenol enhanced CL reaction relating to the amount of HBsAb-HRP conjugate retained in the sensing zone or by an I_2-luminol CL reaction coupled to the HRP catalyzed oxidation of KI [54,55].

Because CL is measured on or near a solid surface in most designs, scattering and absorbance by the support is mainly responsible for the poor accuracy (precision) and low sensitivity, in contrast to the corresponding liquid phase CL. We have reported a new configuration for designing a CL sensor. The starting point of this configuration is to locate the receptor and transducer of a sensor separately to prevent interference by the substrate (support). A simple, rapid, economical and sensitive method for determining acetylcholine (ACh) and choline (Ch) in neuronal tissue using a FIA system with an immobilized enzyme and chemiluminescence detection has been devised (Figure 5) [56].

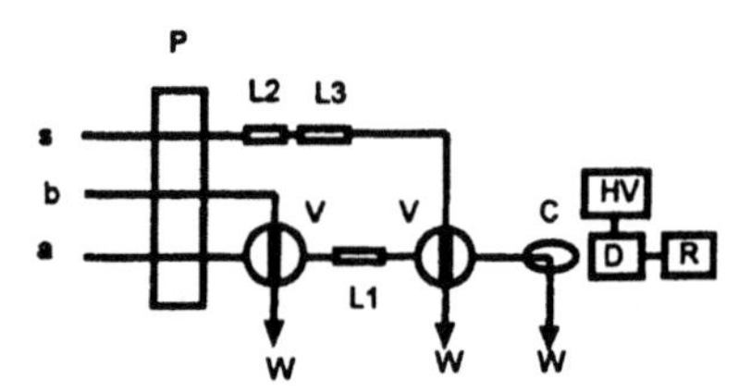

Figure 5. Schematic of sensor for choline and acetylcholine; **a**: water; **b**: KOH; **s**: sample; **v**: six-way valve; **L1**: immobilized cobalt and luminol; **L2**: immobilized AChE; **L3**: immobilized ChO; **c**: flow-cell **D**: detector; **HV**: negative high voltage supply; **R**: recorder; **P**: pump; **W**: waste.

The receptor unit of the sensor consists of an immobilized acetylcholinesterase (AChE) column and an immobilized choline oxidase (ChO) column, both of which use silanized controlled pore glass as matrices [57]. The transducer unit of the sensor is a column that has electrostatically coimmobilized cobalt (II) and luminol on the ion exchange resin. When a stream of a sample flowing through a sequence of enzyme reactors contains acetylcholinesterase and choline oxidase, hydrogen peroxide is generated:

$$ACh + H_2O \xrightarrow{AChE} Ch + CH_3COOH$$

$$\text{Ch} + H_2O \xrightarrow{\text{ChO}} \text{Betaine} + H_2O_2$$

The hydrogen peroxide generated, which is encoded with the information of analyte concentration, is monitored by a downstream chemiluminescent reaction with luminol and Co(II) that elutes from the transducer. Because CL is in the liquid phase, this sensor does not suffer from the disadvantages stated previously.

This idea has also been applied to an amygdalin sensor [58]. Glucosidase immobilized on chitosan is the receptor, luminol embedded on ion exchange resin acts as the transducer, and the amount of Amygdain is sensed by the liquid phase CL of luminol with CN^-, which is generated in the receptor column:

$$\text{Amygdalin} + H_2O \xrightarrow{\beta\text{-Glucoside}} CN^- + \text{Phenylformic acid} + \text{Glucose}$$

$$CN^- + O_2 + \text{Luminol} \longrightarrow \text{Phthalate} + h\nu$$

This configuration, a consumed CL-based sensor, has also served as a way for making sensors for pharmaceuticals or metal ions.

Consumed chemiluminescence sensors for ascorbic acid [59], cyanide [60], Mn^{2+}, Co^{2+} and H_2O_2 [61], chlorine [62], V(V) [63], and Cr(VI) [64] have been developed. All of the reagents involved in the chemiluminescent reactions are immobilized electrostatically on Amberlyst A-27 or D151 ion exchange resin. The analytes of interest are sensed directly by reaction with the chemiluminescent reagents, which are eluted by Na_3PO_4 or NaCl from the immobilization column before the chemiluminescence reaction. These sensors have a wide linear range, high sensitivity and simplicity for instrumentation, and the immobilization methods of the chemiluminescent reagents are also simple. They have been applied successfully to determining analytes in various samples.

The analytical characteristics of these systems can be illustrated in a prototype CL-based sensor for cyanide, which is based on the CL reaction of luminol immobilized on Amberlyst A-27 anion exchange resin and copper ion immobilized on D151 large-pore cation exchange resin with CN^- in alkaline solution. The amount of luminol taken up is 0.4 mmol/g resin and the amount of Cu^{2+} bound per gram resin is 0.8 mmol, in contrast with the amount of luminol and Cu^{2+} released per injection of 0.10 μmol and 0.12 μmol, respectively. Therefore, the sensor could, in principle, be used more than 200 times. In another sensor for H_2O_2, an anion exchanger is connected to the flow line before the sensing column, resulting in higher selectivity.

5. REFLECTANCE-BASED SENSORS

Another potential tool for optical sensors is reflected spectroscopy. The hemoglobin-based O_2 sensor [65] represents our early research. Although hemoglobin is associated with oxygen, it exhibits a shift in the Soret absorption band and a change in color which in turn affects the reflected spectrum of hemoglobin. The oxygen sensor consists of a 0.5 mm thick layer of deoxyhemoglobin immobilized by electrostatic attraction to preswollen CM-Sephadex C-50-120 ion exchange resin and positioned on the common end of a bifurcated fiber optic bundle. An O_2-permeable TFE Teflon membrane separates the immobilized reagents from the sample. The specific parameter measured is the ratio of reflected intensity at 435 and 405 nm. O_2 partial pressure from 20 to 100 torr is also measured. The reflected intensity and intensity ratio decreases with the amount of immobilized hemoglobin, and it takes approximately 3 min to reach steady state when an O_2 sample is introduced.

One attractive feature of this approach is that it involves a true equilibrium. In contrast, the oxygen electrode requires steady-state mass transfer of O_2 to the electrode surface and thus is subject to error if factors that affect O_2 mass transfer are not properly controlled. The same is true for an O_2 sensor based on CL. Optical O_2 sensors based on fluorescent quenching avoid this problem but are not true equilibrium sensors because the response depends on the relative rates of fluorescence and nonradiative return to the ground state via interaction with O_2. Thus, this type of sensor is sensitive to a slight change in the medium, which affects fluorescence, and to the presence of other quenchers.

A more fundamental advantage of this approach is that it is based on a spectral shift. Because the oxygenated and deoxygenated forms each have distinct spectra, there is inherently more information available than from a fluorescent quenching sensor where the measured parameter results in a decrease in intensity.

Using DEAE-Sephadex as a substrate, and pyrocatecholdisulfonic acid (PDA) as an indicator, an optical Fe(III) sensor [66] based on reflected intensity was developed in conjunction with the FIA technique. When the sample passes through the reagent phase, the immobilized PDA reacts with iron(III) to form the complex which causes a large change in reflectance at 460 nm. The sensor reaches a 90% steady-state response in 10 s because of the low mass transfer resistance of the substrate. Similarly, copper(II) has been measured using a reflectance-based optic fiber sensor using flow injection analysis. The sensing layer consists of a nitroso-R salt immobilized on Amberlyst A-27 resin [67].

As discussed previously, quite a number of sensor schemes are based on change in luminescence or absorbance of indicators. However, these systems are unstable because they require photoexcitation of dyes. Therefore, a dye-free sensor is desirable because it is stable, less toxic, and also suitable for *in vivo* use.

Recently, the variation in reflectance of some polymers resulting from polymer swelling has been employed as a new tool for sensor construction. The sensing element is a polymer that changes size as a function of analyte concentration. According to Flory's theory of polymer swelling, the five-thirds power of equilibrium swelling ratio of polymer should increase with the square of the fixed charge on the polymer. This principle is used for the first time in a new pH sensor [68] that consists of a porous polymer membrane of polystyrene aminated by diethanolamine. As the pH is lowered, the proportion of the amine introduces charges to the polymer, which causes it to swell due to electrostatic repulsion. Swelling is accompanied by a decreased intensity of light reflected by the polymer. The reflected intensity is measured using an LED and a photodiode connected to a two-arm fiber optic bundle.

Another new application of this principle is constructing sensors by utilizing the ion-pair forming reaction. This type of sensor is the fourth way of sensing alkali metal ions. The sensor consists of a porous polymer membrane of sulfonated polystyrene and a neutral ionophore that has a cavity sized to accommodate potassium, sodium, or lithium ions. When alkali metal ions are bound with the selective ionophore, they can migrate into the bulk of polymer matrix. The hydrophobic cation combines with the sulfonate anion to form the ion pair. In this case, the reflected intensity of the polymer membrane is increased by polymer shrinking and becomes cloudy because the charge is removed. These sensors have high selectivity and sensitivity and are also tough and reversible. Because they do not require a specific wavelength, they can use near infrared light emitting diode as the source and a photodiode as the detector. This lowers the cost of the required instrumentation and also reduces the possibility of optical interference from the sample matrix because the absorbance of a biological sample in this spectral region is minimal.

Similar to fluorescence, reflectance, and chemiluminescence, the absorption spectra of some dyes are another alternative route to optical sensors. The changes in absorbance of some dyes, including bromophenol blue, bromocresol green, nitrazine yellow, bromocresol purple, *o*-cresol red and thymol blue, which result from pH change, can be used for pH monitoring. A series of pH sensors has been developed by using this property. Such sensors serve in turn as transducers in some gas sensors, e.g., CO_2, SO_2 sensors [69,70].

6. CONCLUSION

There is substantial current interest in developing sensors for biomedical and biochemical applications. Optical sensors provide important advantages in performing these tasks. In this paper, our research on optical biosensors is described. Fluorescence, phosphorescence, chemiluminescence, reflectance, and absorbance were exploited in designing these sensors. Sensors for immunoassays, clinical use, pharmaceutical analysis and for measuring electrolytes in body fluid are presented.

Fiber optics and flow-injection techniques are two useful tools in designing sensors for biomedical use. Sensors based on fiber optics are suitable for *in situ* and *in vivo* analysis. They possess such advantages as immunity to electromagnetic interference, and have potential for remote and multicomponent sensing. With the aid of FIA, sensors can work under optimized conditions. FIA can also make many chemical and biochemical reactions, which have not been used in designing sensors previously, useful, and consequently broaden the research field. Our present attention is focused on adapting these flow-type sensors for *in vivo* and *in vitro* use.

REFERENCES

[1] Hulanicki, A., Glab, S., Ingman, F. (1991). Chemical sensors: Definitions and classification. *Pure Appl. Chem.*, **63**, 1247.

[2] Wolfbeis, O.S. (1991). *Fiber Optic Chemical Sensors and Biosensors*, Volume I, Wiley, New York, pp. 2.

[3] Valcarcel, M., Luque de Castro, M.D. (1993). Flow-through (bio)Chemical sensors. *Analyst*, **118**, 593.

[4] Valcarcel, M., Luque de Castro, M.D. (1991). Integration of reaction(Retention) and spectroscopic detection in continuous-flow systems. *Analyst*, **115**, 699.

[5] Hansen, E.H. (1994). Flow injection analysis: A complementary or alternative concept to biosensors. *Talanta*, **41**, 939.

[6] Arnold, M.A. Fiber-optic-based biocatalytic biosensors. In *Chemical Sensors and Microinstrumentation* (ACS Syposium Series 403). Murray, R.W. et al. (Eds.), American Chemical Society, Washington, D.C., 1989, pp. 303.

[7] Narayanaswamy, R. (1991). Current developments in optical biochemical sensors. *Biosensors & Bioelectronics*, **6**, 467.

[8] Whitehead, T.P., Kricka, L.J., Carter, T.J.N., Thorpe, G.H.G. (1979). Analytical luminescence: Its potential in the clinical laboratory. *Clin. Chem.*, **25**, 1531.

[9] Egorov, C.C., Ruzicka, J. (1995). Flow-injection renewable fiber optic sensor system. Principle and validation on spectrophotometry of chromium(VI). *Analyst*, **120**, 1959.

[10] Li, J., Zhang, Z. (1994). A simplified enzyme-based fiber optic sensor for hydrogen peroxide and oxidase substrates. *Talanta*, **41**, 1999.

[11] Gong, Z., Zhang, Z. (1996). A fiber optic biosensor for uric acid based on immobilized enzymes. *Anal. Lett.*, **29**(5), 695.

[12] Zhang, Z., Gong, Z., Ma, W. (1995). An enzyme-based fiber optic biosensor for determination of D-amino acid in serum. *Microchem. J.*, **52**, 131.

[13] Gong, Z., Zhang, Z. (1996). Flow-injection fluorometry for determination of glucose in serum based on immobilized enzyme. *Fenxi Huaxue*, **24**, 998.

[14] Li, J., Zhang, Z. (1996). Natural killer cell cytotoxicity assay with time-resolved fluorimetry. *Sci. China* (series C), **39**(2), 217.

[15] Li, J., Zhang, Z., Jin, B. (1995). Assay of NK cell cytotoxicity with time-resolved fluorimetry. *Zhongguo Kexue* (series B), **25**(4), 393.

[16] Zhang, Z., Seitz, W.R. (1988). Ion-selective sensing based on potential sensitive dyes. *Proc. SPIE*, **906**, 74.

[17] Zhang, Z., Mullin, J.L., Tang, Y., Seitz, W.R. (1987). Optical ion detection via ion pairing. *Biosensors*, **10**, 229.

[18] Seitz, W.R., Zhang, Z., Mullin, J.L. (1986). Reversible indicators for alkali metal ion optical sensors. *Proc. SPIE*, **713**, 126.

[19] Zhang, Z., Mullin, J.L., Seitz, W.R. (1986). Optical sensor for sodium based on ion pair extraction and fluorescence. *Anal. Chim. Acta*, **184**, 251.

[20] Seitz, W.R., Zhang, Z. (1986). Fiber optic sensor determines sodium. *C & EN*, **5**, 38.

[21] Wolfbeis, O.S., Schaffar, B.P.H. (1987). An ion-selective optode for potassium. *Anal. Chim. Acta*, **198**, 1.

[22] Seitz, W.R., Saari, L.A., Zhang, Z. (1985). Metal ion sensors based on immobilized fluorogenic ligands. *ASTM Special Technical Publication* 863, pp. 63.

[23] Zhang, Z., Seitz, W.R. (1985). A fluorescent sensor for aluminum(III), magnesium(II), zinc(II) and cadmium(II) based on electrostatically immobilized quinolin-8-ol-sulfonate. *Anal. Chim. Acta*, **171**, 251.

[24] Lu, J., Zhang, Z. (1995). Optical fiber fluorosensor for cadmium with diethylaminoethyl-sephadex as a substrate. *Analyst*, **120**(2), 453.

[25] Zhang, Z., Zhang, Y. (1986). Determination of trace cyanide in water by fiber optic fluorometry. *Fenxi Haxue*, **14**, 415.

[26] Zhang, Z., Zhang, Y., Ma, W., Seitz, W.R. (1989). Poly(vinyl alcohol) as a substrate for indicator immobilization for fiber optic chemical sensors. *Anal. Chem.*, **61**, 202.

[27] Ma, W., Zhang, Z., Seitz, W.R. (1989). Poly(vinly alcohol)-based indicators for optical pH and Mg(II) sensing. In *Chemical Sensors and Microinstrumentation (ACS Symposium Series 403)*, R.W. Murray et al. (Eds.), American Chemical Society, Washington, D.C., pp. 273.

[28] Zhang, Z., Tie, J., Li, J. (1994). A fiber-optic ferric sensor with fast response characteristic. *Huaxue Xuebao*, **52**, 492.

[29] Ma, W., Zhang, Z. (1991). The investigation of a fiber optical aluminum sensor using poly(vinyl alcohol) gel as a substrate. *Gaodeng Xuexiao, Huaxue Xuebao*, **12**(10), 1304.

[30] Zhang, Z., Seitz, W.R. (1984). An optic sensor for quantifying pH in the range from 6.5 to 8.5. *Anal. Chim. Acta*, **160**, 47.

[31] Zhang, Z., Seitz, W.R. (1984). A carbon dioxide sensor based on fluorescence. *Anal. Chim. Acta*, **160**, 305.

[32] Wang, J., Zhang, Z. (1995). Determination of manganese(II) by a fiber optic kinetic method with a FIA and binary fluorescent complex indicator: Manganese-binary fluorescent complex-sodium periodate. *Gaodeng Xuexiao, Huaxue Xuebao*, **16**(2), 188.

[33] Wang, J., Zhang, Z. (1995). Fluorescence spectrophotometric determination of trace manganese(II) by optical fiber kinetic method: Flow injection analysis with the manganese-thiamine-sodium periodate system. *Fenxi Huaxue*, **23**(3), 299.

[34] Gong, Z., Zhang, Z. (1997). An optosensor for tryptophan with C18 silica gel as a substrate. *Fresenius J. Anal. Chem.*, **357**, 1097.

[35] Gong, Z., Zhang, Z. (1997). An optosensor for cinchona alkaloids with C18 silica gel as a substrate. *Analyst*, **122**, 283.

[36] Gong, Z., Zhang, Z. (1997). An optosensor for riboflavin with C18 silica gel as a substrate. *Anal. Chim. Acta*, **339**, 161.

[37] Gong, Z., Zhang, Z. (1996). An optosensor based on the fluorescence on metal complexes adsorbed on Chelex 100. *Anal. Chim. Acta*, **325**, 201.

[38] Gong, Z., Zhang, Z. (1996). Luminescence characteristics based on chelating resin Chelex 100 and its analytical applications. *Fenxi Huaxue*, **24**, 1007.

[39] Szejtli, J. (1982). *Cyclodextrins and Their Inclusion Complexes*. Alademiai Kiado, Budapest.

[40] Gong, Z., Zhang, Z. (1996). A cyclodextrin-based optosensor for determination of riboflavin in pharmaceutical preparations. *Analyst*, **121**(8), 1119.

[41] Gong, Z., Zhang, Z. A cyclodextrin-based optosensor for determination of tryptophan. *Mikrochim. Acta*, in press.

[42] Gong, Z., Zhang, Z. Cyclodextrin-based optosensor for the determination of quinine. *Fresenius J. Anal. Chem.*, in press.

[43] Gong, Z., Zhang, Z. (1996). A fluorosensor with immobilized cyclodextrin as a substrate. *Anal. Lett.* **29**, 2441.

[44] Li, J., Zhang, Z. (1994). Determination of hafnium with SS-RTP optosensing. *Anal. Lett.*, **27**, 2769.

[45] Li, J., Zhang, Z. (1993). Determination of zirconium and hafnium with SS-RTP optosensing. *Chemical Sensor II*, Butler, M. et al. (Eds.), The Electrochemical Society, Pennington, NJ, p. 103.

[46] Lu, J., Zhang, Z. (1995). Determination of europium with solid-surface room-temperature phosphorescence optosensing. *Analyst*, **120**(10), 2585.

[47] Lu, J., Zhang, Z. (1996). A reversible optical sensing layer for picric acid based on the luminescence quenching of the euthenoyltriflouacetone. *Anal. Chim. Acta*, **318**(2), 175.

[48] Gong, Z., Zhang, Z. (1997). Room temperature phosphorescence optosensing for gadolinium. *Mikrochim. Acta*, **126**, 117.

[49] Gong, Z., Zhang, Z. (1996). Room temperature phosphorescence optosensing for *Terbium. Anal. Lett.*, **29**(4), 515.

[50] Zhang, Z., Ma, W. (1993). Studies on fiber optic chemiluminescence sensor for lactic acid. *Kexue Tongbao*, **38**(3), 230.

[51] Zhang, Z., Ma, W. (1993). A fiber optic biosensor for the determination of cholesterol. *Gaodeng Xuexiao, Huaxue Xuebao*, **14**(10), 1366.

[52] Zhang, Z., Ma, W., Yang, M. (1992). A fiber optic biosensor based on chemiluminescence for the determination of uric acid. *Fenxi Huaxue*, **20**, 1048.

[53] Zhang, Z., Li, J. (1987). A chemiluminescence fiber optic sensor for determination of glucose in blood. *Shaanxi Shida Xuebao*, (4), 57.

[54] Zhang, X., Feng, M., Lu, J., Zhang, Z. (1994). Studies on flow injection chemiluminescence immunoassay I determination of HBsAg in serum. *Huaxue Xuebao*, **52**(2), 83.

[55] Shao, Q., Zhang, Z., Ma, W., Feng, F. Studies on flow injection chemiluminescence immunoassay II determination of HRP and its conjugates by coupled reaction. *Huaxue Xuebao*, in press.

[56] Song, Z., Zhang, Z. (1998). A chemiluminescence biosensor for determination of acetylcholine and choline. *Acta Chim. Sin.*, **56**(12), 1207.

[57] Fan, W., Zhang, Z. (1996). Determination of acetylcholine and choline in rat brain tissue by FIA with immobilized enzymes and chemiluminescence detection. *Microchem. J.* **53**, 290.

[58] Song, Z., Zhang, Z. A chemiluminescence biosensor for determination of amygdalin. *Chin. Sci. Bull.*, in press.

[59] Zhang, Z., Qin, W. (1996). Chemiluminescence flow sensor for the determination of ascorbic acid with immobilized reagents. *Talanta*, **43**, 119.

[60] Lu, J., Qin, W., Zhang, Z. (1995). A flow-injection type chemiluminescence-based sensor for cyanide. *Anal. Chim. Acta*, **304**, 369.

[61] Qin, W., Zhang, Z., Chen, H. Chemiluminescence flow sensor for monitoring of hydrogen peroxide in rain water. *Int. J. Environ. Anal.*, in press.

[62] Qin, W., Zhang, Z., Liu, S. (1997). Flow-injection chemiluminescence sensor for the determination of free chlorine in tap water. *Anal. Lett.*, **30**, 11.

[63] Qin, W., Zhang, Z. (1997). Chemiluminescence flow system for vanadium(V) with immobilized reagents. *Analyst*, **122**, 685.

[64] Zhang, Z., Qin, W., Liu, S. (1995). Chemiluminescence flow system for the monitoring of chromium(VI) in water. *Anal. Chim. Acta*, **318**, 71.

[65] Zhang, Z., Seitz, W.R. (1986). Optical sensor for oxygen based on immobilized hemoglobin. *Anal. Chem.*, **58**(1), 220.

[66] Lu, J., Zhang, Z. (1994). A fiber optic iron sensor with DEAE-Sephadex as a substrate. *Anal. Lett.*, **27**(13), 2431.

[67] Lu, J., Zhang, Z. (1995). A fiber optic copper sensor with Sephadex as a substrate. *Huaxue Xuebao*, **53**, 895.

[68] Zhang, Z., Shakhsher, Z., Seitz, W.R. (1995). Aminated polystyrene membrane for a fiber optic pH sensor based on reflectance changes accompanying polymer swelling. *Mikrochim. Acta*, **121**, 41.

[69] Zhang, Z.Q., Zhang, Z. (1987). pH optic sensors based on absorption spectra of immobilized dyes. *Huaxue Xuebao*, **45**(3), 239.

[70] Zhang, Z.Q., Zhang, Z. (1987). Sulfur dioxide optic sensor based on absorption of immobilized dye. *Huaxue Xuebao*, **45**(5), 497.

SURFACE PLASMON RESONANCE BIOSENSORS BASED ON BIOFUNCTIONALIZED INTERFACES

Sen-fang Sui, Xiao Caide, Yue Zhou,
Wenzhang Xie, and Junfeng Liang

OUTLINE

ABSTRACT

Surface plasmon resonance biosensors have become more powerful in molecular recognition detection involved in biochemical reactions. A surface plasmon resonance

Advances in Biosensors
Volume 4, pages 123–137.

ISBN: 0-7623-0073-6

biosensor developed in our laboratory combined spectroscopy with microscopy. First we examine the characteristics of the spectrum of surface plasmon resonance, analyzed by a numerical approach. It shows that in an aqueous environment the shift of the resonance angle is approximately proportional to the thickness of the covering layer and the difference between the value of the refractive index of the covering layer and that of the buffer, when the thickness of the covering biomolecular layer is less than 12 nm. The variation of the thickness of the gold film substrate may change the shift of the resonance angle but does not affect the linear correlation between the shift of the resonance angle and the thickness of the covering layer.

In the second part, a functionalized interface based on biotin/avidin binding is presented. As an application of such a sensitive chip, we detected the specific binding of anti-DNA antibody extracted from the serum of a patient who had systemic lupus erythematosus by using DNA which was biotin-labeled and immobilized on the top of the avidin layer.

In the third part, we present surface plasmon resonance imaging analysis of the covering lipid layer, which is not homogenous when viewed in two dimensions.

1. INTRODUCTION

Functionalized interfaces formed by organization of biomolecular assemblies on a solid support are fundamentally important in biomembrane research [1–7], the design of biomimetic materials, and the development of biosensors and biochips [8–10]. The characterization of the structure and function of organized molecular assemblies at interfaces has stimulated the development of novel techniques for sensing biochemical reactions at interfaces. Among the numerous experimental techniques, surface plasmon resonance (SPR), an optical method, has become particularly useful because of its high surface sensitivity and specificity [11]. Based on optical detection, the SPR technique allows monitoring molecular interactions at interfaces without interference from the bulk solution and the need for a special label. In our laboratory, we are investigating the fundamental processes involved in lipid–protein and ligand–receptor interactions of biomembranes [5,12,13]. In such studies we investigated the buildup of functionalized, highly organized monolayers and bilayers supported on substrates. This article first briefly reviews the physics of surface plasmon resonance in general, and then the characteristics of the SPR spectrum are analyzed. We focus in particular on the physical and geometrical factors in arranging the apparatus. Using SPR apparatus developed in our laboratory, the specific recognition reactions between biotin-lipid containing monolayers and the protein avidin are presented as an example of receptor-based biosensors. Surface plasmon microscopy is also presented for imaging the nonhomogeneous covering layer organized on the support. By varying the resonance angles for the regions in gel domains or for those in fluid domains, the image of the supported lipid layer in a phase separation state was inverted. It is hoped that SPR biosensors will become more powerful by combining spectroscopy and microscopy of surface plasmon resonance.

2. SPR BIOSENSORS BASED ON SURFACE PLASMON SPECTROSCOPY

In the Kretschmann configuration [14], the incident light beam falls on a prism under total reflective conditions and onto a metal (gold or silver) film evaporated on the base of the prism. In our case for monitoring biomolecular reactions, the receptor that contains molecular assemblies is immobilized on the surface of the thin metal film prepared on a cover glass, as shown in the schematic drawing of the optical system of the SPR biosensor (Figure 1). The optical system is composed of a prism, a cover slide, a gold film, a sample layer, and a buffer, which are labeled with the Arabic numerals 0, 1, 2, 3, and 4 in the sequence from prism to buffer,

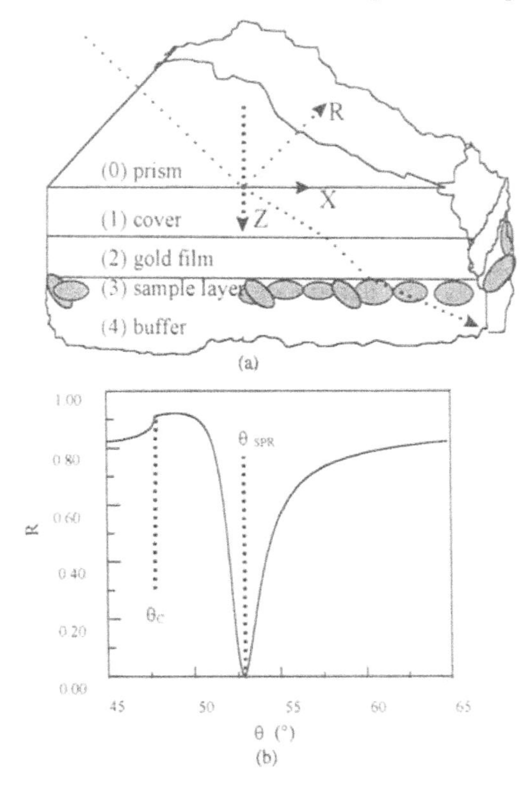

Figure 1. Modified Kretschmann configuration of (**a**) the surface plasmon resonance biosensor and (**b**) a typical SPR spectrum, where θ_C is the critical incident angle and θ_{SPR} is the resonance angle.

respectively. The cover glass is attached to the base of the prism with index-matched oil. The capital letters X and Z indicate the directions of the coordinate axes for the optical system, and R represents the reflectivity of light intensity. The reflectivity R is a function of the light wavelength λ, the dielectric constant ε_i ($i = 0, 1, \ldots, 4$), the medium thickness d_i ($i = 1, 2, 3$) and the angle of incidence θ. Above the critical incident angle, θ_c, the light intensity does not fall to zero across the interface. Instead there is an evanescent wave whose amplitude decays exponentially normal to the interface. Below a certain angle of incidence, θ_{SPR}, the surface plasmons at the interface of the metal and dielectric are excited by the evanescent wave. The resonance angle θ_{SPR} corresponds to the peak of the SPR spectrum as shown in Figure 1(b). According to the Kretschmann geometry shown in Figure 1, we describe the relationship of R versus θ by

$$R(\theta) = \left| r_{0,4}(\theta) \right|^2 \tag{1}$$

$$r_{i,4}(\theta) = \frac{r_{i,i+1}(\theta) + r_{i+1,4}(\theta) \cdot \exp[2 \cdot j \cdot d_{i+1} k_{zi+1}(\theta)]}{1 + r_{i,i+1}(\theta) \cdot r_{i+1,4}(\theta) \cdot \exp[2 \cdot j \cdot d_{i+1} k_{zi+1}(\theta)]} \qquad (i = 2, 1, 0 \quad j = \sqrt{-1})$$

$$r_{i,i+1}(\theta) = \frac{\zeta_{i+1}(\theta) - \zeta_i(\theta)}{\zeta_{i+1}(\theta) + \zeta_i(\theta)} \qquad (i = 0, 1, 2, 3)$$

$$k_{zi}(\theta) = \frac{2\pi}{\lambda} \sqrt{\varepsilon_i - \varepsilon_0 \sin^2(\theta)}$$

$$\zeta_i(\theta) = \frac{\varepsilon_i}{k_{zi}(\theta)} \qquad (i = 0, 1, 2, 3, 4)$$

where k_{zi} represents the wave vector of the transmission light in the Z direction in the medium i ($i = 0, 1, \ldots, 4$), and $r_{i,i+1}$ is the reflectivity of the interface between two adjacent media. In terms of Eq. (1) the SPR spectrum can to be analyzed by numerical calculation and the resonance angle θ_{SPR} can be obtained from $dR/d\theta = 0$.

Figure 2 shows the relationship of resonance angle θ_{SPR} to sample thickness d_3 and refractive index n_3. From Figure 2 we demonstrate that when $d_3 \leq 12$ nm and $1.3 < n_3 < 1.55$, $\Delta\theta_{SPR}$ has a linear relationship with the thickness d_3 and the difference of the refractive indexes between the sample layer and buffer. The correlation factor is greater than 0.9999. Such a linear relationship is given by

$$\Delta\theta_{SPR} = 52.887 + 0.570\,(n_3 - n_4)d_3 \tag{2}$$

Equation (2) was developed on the basis of the following parameters: $\lambda = 670$ nm, $\varepsilon_0 = 1.795^2$, $\varepsilon_1 = 1.516^2$, $\varepsilon_2 = \varepsilon_{Au}$, and $\varepsilon_4 = \varepsilon_{water}$. The parameters chosen are representative of an aqueous environment. Therefore, the relationship of Eq. (2) is useful for a number of biochemical reactions. Under such conditions, a shift in the

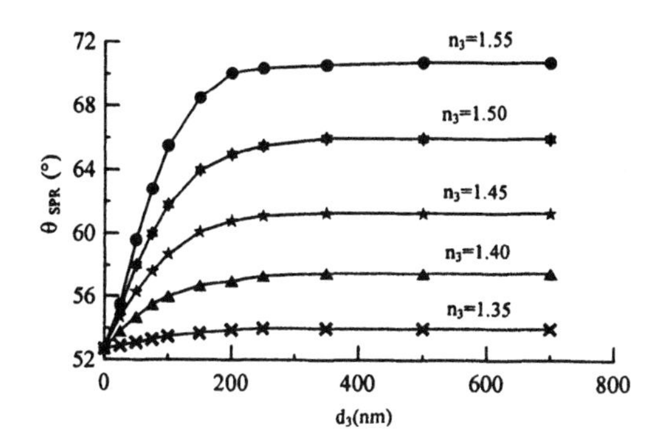

Figure 2. Effects of the thickness d_3 and the refractive index n_3 of the sample layer on the resonance angle θ_{SPR}.

θ_{SPR} value is a measure of d_3. When $d_3 > 200$ nm, however, θ_{SPR} reaches a constant value which no longer depends on $(n_3 - n_4)d_3$. This phenomenon is reasonable because the depth of penetration of the evanescent wave is about 205 nm when $\varepsilon_2 = 13.4 + 1.4j$ (gold) and $\varepsilon_4 = 1.33$ (water). The gold film for SPR measurement is usually evaporated to a thickness of about 50 nm. However, it is difficult to prepare gold films of exactly the same thickness. Why is a thickness of 50 nm chosen for evaporated gold films? Results shown in Figure 3(a) indicate that SPR spectra are remarkably sensitive to the thickness of the gold film. The SPR spectrum has the lowest reflectivity at θ_{SPR} when the gold film is about 50 nm, which provides the most accurate measurement. An intriguing result is shown in Figure 3(b) where a change in thickness d_2 of the gold film causes a change in the value of θ_{SPR}, but with little effect on the linear correlation between θ_{SPR} shift and d_3. This finding implies that the thickness of gold film does not have to be controlled at 50 nm, providing great convenience for applications of the SPR technique.

A laboratory-built computer-controlled surface plasmon resonance biosensor is schematically shown in Figure 4. A semiconductor laser beam whose wavelength is 670 nm is used as the incident light source. The measurement is performed by varying the angle of incidence θ. To characterize the interaction between ligands and the receptor-containing monolayer, the measurements are made as follows. After transferring the functionalized lipid monolayer onto the surface of the gold-coated cover glass, the slide is carefully stuck with matching fluid onto the bottom surface of the triangular prism. Then the sample cell is installed and the buffer is pumped slowly into it. The initial peak position of the SPR curve with the buffer is recorded. Then, the protein solution is pumped into the cell, and the second SPR curve is measured after equilibrium. Because the protein layer is very thin (less than 10 nm), the displacement of the SPR peak position is linearly related to the layer thickness, as mentioned previously. As a direct comparison, the preparations

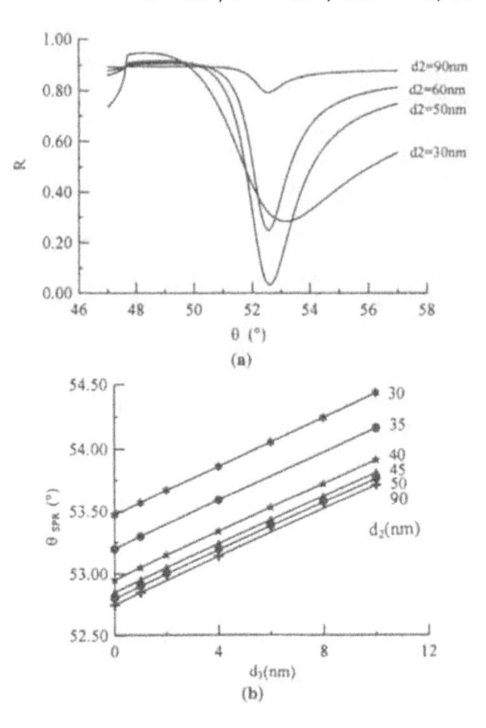

Figure 3. Effects of the thickness of gold film (**a**) on the shape of SPR spectra and (**b**) on the relationship between θ_{SPR} and d_3.

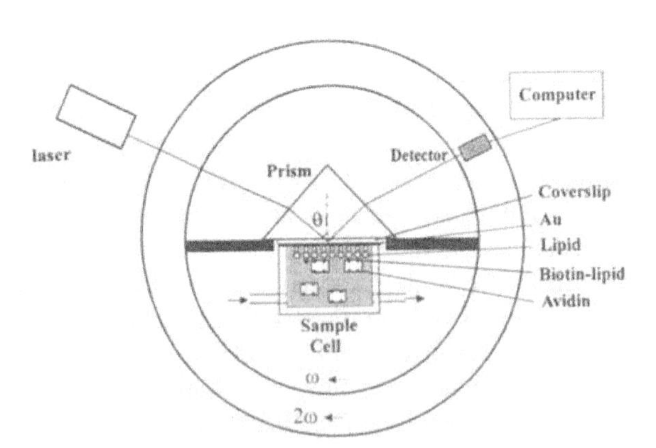

Figure 4. Schematic illustration of the arrangement of the SPR setup. The triangular prism (refractive index 1.795) is fitted onto the small rotating stage with an angular velocity of ω (0.5 °/s). The photodiode fitted on the big rotating stage moves at 2ω (1.0 °/s). The SPR curves are recorded as the photodiode moves in response to the reflected light beam.

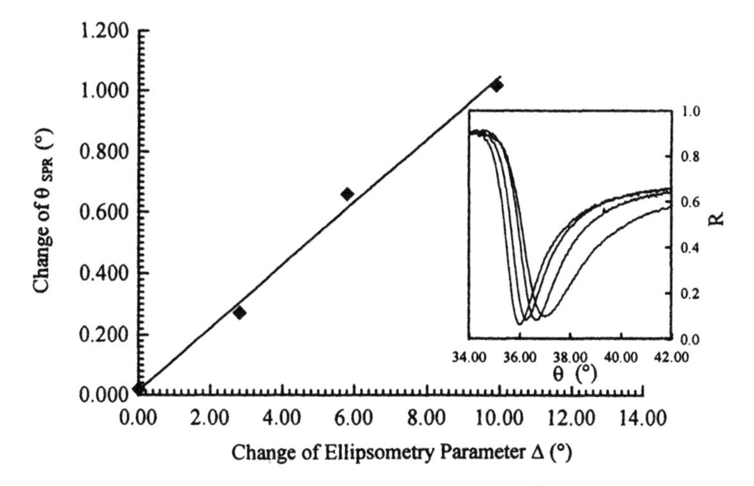

Figure 5. Direct comparison of SPR biosensor and ellipsometry measurements under identical conditions. The SPR spectra are shown in the insert.

are also measured by ellipsometry under identical conditions [15]. Figure 5 shows the results obtained for multilayers of stearic acid measured by two techniques, ellipsometry in parameter Δ and SPR in θ. Figure 5 demonstrates that a satisfactory linear correlation is obtained by the two techniques, evidence that the two techniques are comparable. In Figure 6 we show the time dependence of θ_{SPR} for an antigen/antibody interaction. BSA as an antigen is first immobilized on the surface of the gold film by adsorption. A remarkable change in θ_{SPR} is observed following

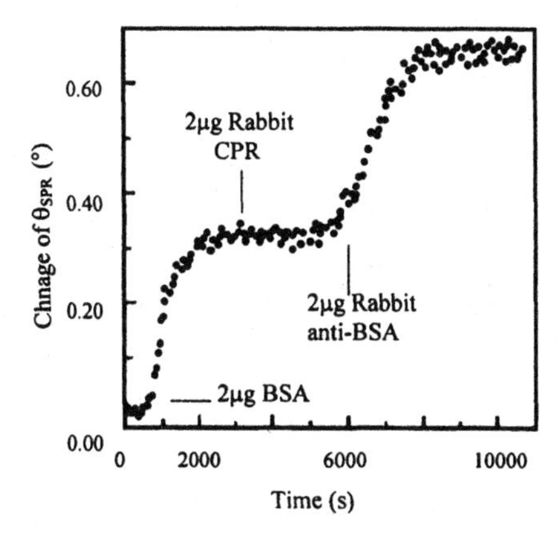

Figure 6. Time dependence of surface plasmon resonance angle θ_{SPR} in antigen/antibody interaction.

BSA adsorption, which is displayed as a rapid increase in the value of θ_{SPR} (the rapid rise in the first part of the plot in Figure 6). After adding a certain amount of rabbit C-reactive protein (rCRP), however, the isotherm retains its smooth curve because there is no interaction between BSA and rCRP. Then, as rabbit anti-BSA is added, the isotherm starts to increase sharply again (the rapid rise in the second part of the plot in Figure 6).

3. BIOFUNCTIONALIZED INTERFACES BASED ON BIOTIN/AVIDIN BINDING

The high-affinity avidin/biotin system was used in our laboratory for several reasons to develop the receptor-based SPR biosensor. The affinity of avidin for biotin (15) ($10^{15}M^{-1}$) under neutral conditions is one of the highest known. It is just one order of magnitude less than chemical bonds but is much stronger than noncovalent bonds. Another characteristic of the avidin-biotin complex is its stability over a broad range of pH values and in the presence of high concentrations of denaturants. Although biotin can be covalently linked to many chemical compounds by its carboxyl group, this does not affect the affinity of the cyclic ureido group for avidin [16]. In our case we attached the biotin covalently to the head group of appropriate phospholipids and then proceeded to assemble artificial membrane systems containing the biotin group as a membrane-bound receptor [17,18]. We used avidin rather than streptavidin as a ligand because avidin from egg white is

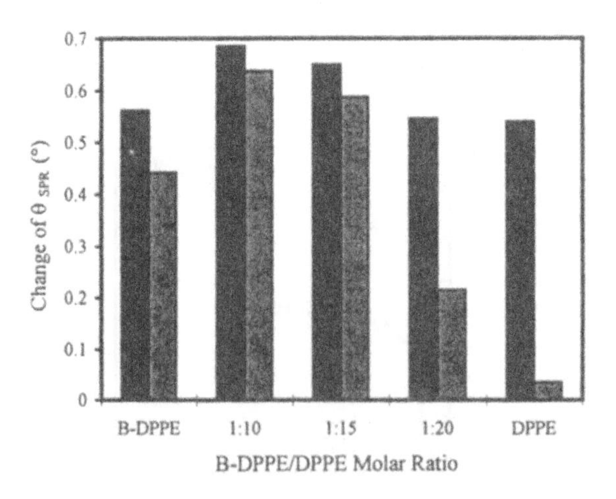

Figure 7. Comparison of the amounts (measured in the change of θ_{SPR}) of avidin binding onto supported planar monolayers that contain various molar fractions of biotin-DPPE lipids was studied with the SPR biosensor at 20 °C, before and after desorption.

readily obtainable, whereas the isolation of streptavidin from *Streptomyces avidinii* is much more complicated and expensive.

For receptor-based biosensor research, the nonspecific adsorption of proteins is an interference which must be resolved. In response to the problem of nonspecific binding of avidin to a lipid monolayer, we found that nonspecific adsorption is eliminated by adding a certain amount of bivalent alkaline-earth metal cations, such as Ca^{2+}, Mg^{2+}, and Ba^{2+} [5,17]. After solving the problem of nonspecific adsorption, the specific binding of avidin to supported membranes was investigated. To analyze the influence of receptor density on ligand binding, the molar ratios of biotin-DPPE/DPPE mixtures were varied carefully by the LB technique. The specific binding characteristics of avidin were precisely analyzed by surface plasmon spectroscopy (as shown in Figure 7), which displays a significant dependence on membrane composition. By SPR (shown in Figure 7) and ellipsometry (results not displayed), we found an optimum molar ratio of biotin-DPPE/DPPE at 1:12.

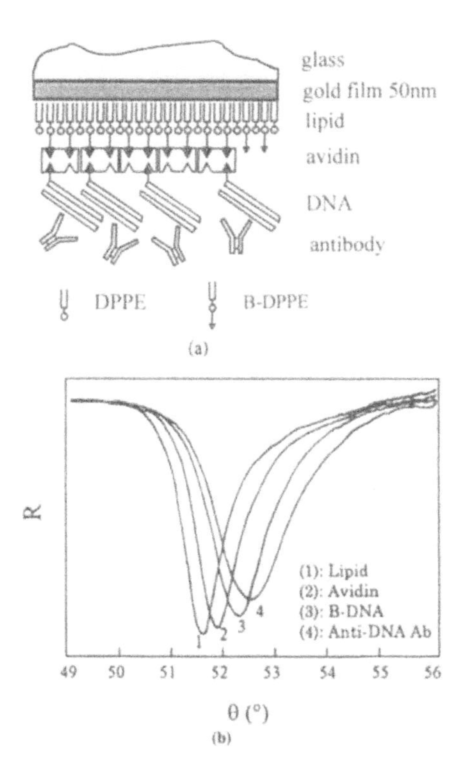

Figure 8. Schematic illustration of the multilayer assembly: (**a**) the first inner monolayer containing biotin-lipid, the second layer of avidin, the third layer of biotin-DNA, and the outer layer of anti-DNA antibody. (**b**) The SPR spectra corresponding to the formation of the four layers.

Around this ratio the amount of protein specifically bound is the highest, and the nonspecific adsorption is negligible. Above or below the ratio of 1:12, the amount of the specific binding of avidin decreases. An intriguing suggestion from the results of Figure 7 is that nonspecific adsorption is almost eliminated if the monolayer composition is around the optimum molar ratio. Thus, if one organizes a functional lipid layer so that the layer composition is chosen only as the optimum ratio, the nonspecific binding is of minor importance. This implies a possible approach to overcoming the problem of nonspecific adsorption, which would help improve the molecular engineering applications of biotin-functionalized lipid monolayers.

Human anti-DNA antibody, the subject of intense clinical interest recently, is present in trace amounts in human serum under normal conditions but is greatly elevated in patients who have systemic lupus erythematosus. Analysis of anti-DNA antibody is essential in clinical serum diagnosis of systemic lupus erythematosis. Using the SPR biosensor in our laboratory, we studied the interaction between immobilized DNA and anti-DNA antibody extracted from patients' serum [19]. The principal goal of this research was to develop a simple method to detect systemic lupus erythematosus in real-time analysis. The core technology of such a biosensor system is a DNA probe modified layer structure on top of the surface layer of the gold substrate. In terms of biotin/avidin binding, the biotin-labeled DNA probes were immobilized onto the two-dimensional, organized avidins at the biotin-functionalized surface as shown in Figure 8(a). Figure 8(b) exhibits the measured SPR spectra for monitoring the multilayer formation based on molecular recognition.

4. SPR BIOSENSORS BASED ON SURFACE PLASMON MICROSCOPY

Surface plasmon microscopy (SPM) as reported by Yeatman and Ash [20] and Rothenhausier and Knoll [21] uses surface plasmon resonance as the contrast mechanism. A typical experimental arrangement of SPM in the Kretschmann configuration is also used, and the light reflected from the interface passes through a lens to form a virtual image in real space, as shown in Figure 9. Any small change in the dielectric constant and/or thickness of the overlayer changes the local conditions for resonance and thus generates high contrast available for imaging. To obtain a higher contrast picture, a He-Ne laser light source, a BK7 glass prism, and a CCD camera were employed, and two polarizers are used to control the light density and make it match the camera sensitivity.

The SPM was used in our laboratory to survey the formation of organized multilayers on a solid surface. Multilayers of fatty acid that have different compositions and thicknesses were first transferred onto different regions of a substrate marked A, B, C, and D and then observed by SPM as shown in Figure 10. The SPM images shown in Figure 10 (a) to (d) were obtained by changing the incident angle. When the incident angle of the light beam matches the resonance angle of a certain

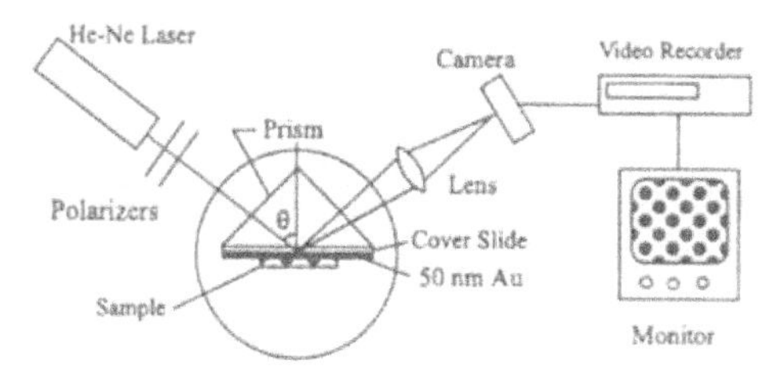

Figure 9. Schematic representation of the arrangement of the surface plasmon resonance microscope (SPM).

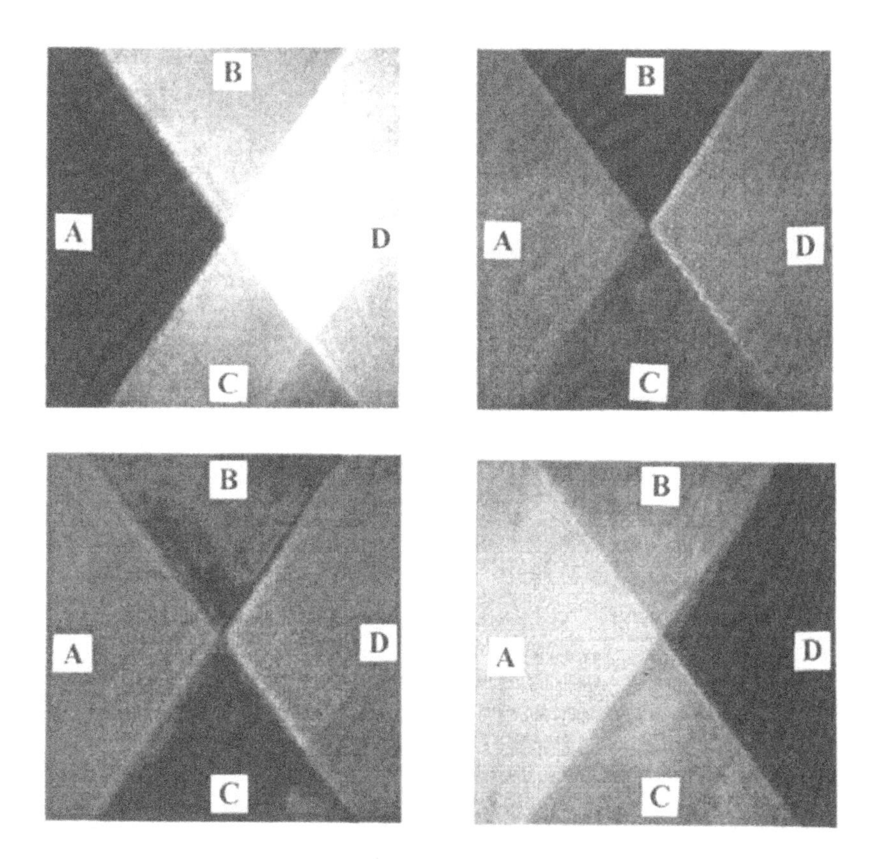

Figure 10. SPM images of organized multilayers of fatty acid LB films under different incident angles are shown in (**A**)–(**D**), vision field: 400 μm × 300 μm. **A**: one layer of stearic acid, **B**: one layer of stearic acid and two layers of palmitic acid, **C**: three layers of stearic acid, **D**: three layers of stearic acid and two layers of palmitic acid.

region of the multilayers, this region appears the darkest in the view whereas other regions of different thickness have different reflectivities. The multilayer compositions of the four regions are presented schematically in Figure 11(a). Using the method described previously we obtained the resonance angles of the other three regions. Those were identical to those obtained from the SPR curves corresponding to each region and free gold film as shown in Figure 11(b). The shift of resonance angles in the regions A, B, C, and D are 0.40°, 0.93°, 1.22°, and 1.75°, respectively.

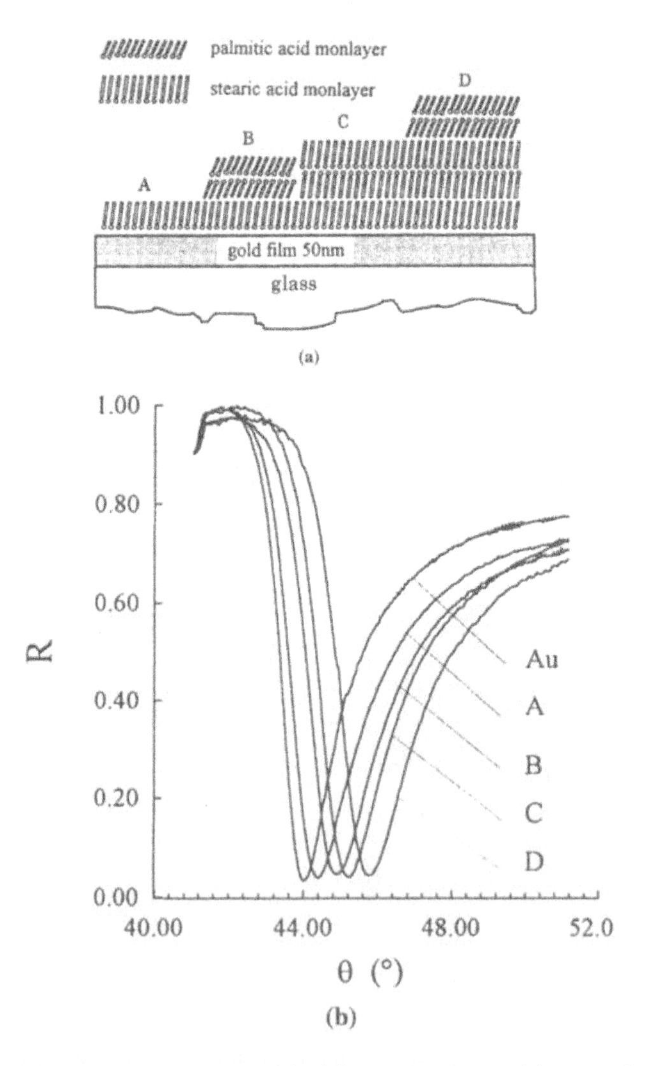

Figure 11. Schematic representation **(a)** of the composition of the organized multilayers imaged in Figure 10 and **(b)** the corresponding SPR spectra.

Assuming that the fatty acids used have the same refractive index, the calculated thicknesses of the four regions in alphabetical sequence are 2.3 nm, 5.2 nm, 6.9 nm, and 10.0 nm.

The phase separation of the organized lipid monolayer was observed by SPM in real time. Figure 12(a) shows a schematic representation of the DMPE monolayer in a phase separation state on a modified support. The substrate was first modified so that a DSPC monolayer in the solid phase was transferred onto the hydrophobically treated surface of the gold film on the cover glass, and then a DMPE monolayer compressed to phase separation state was deposited on top of the DSPC-covered substrate by lowering the subphase level. Figures 12(b) and (c) display two SPM images of the same view obtained with a difference of 0.20° in incident angles. The former is at θ_{SPR} corresponding to the layer region in liquid state [dark region in Figure 12(b)] and the latter is that in the solid state [dark region in Figure 12(c)]. Based on the difference in the θ_{SPR}, one can further calculate the thickness difference between the two phases.

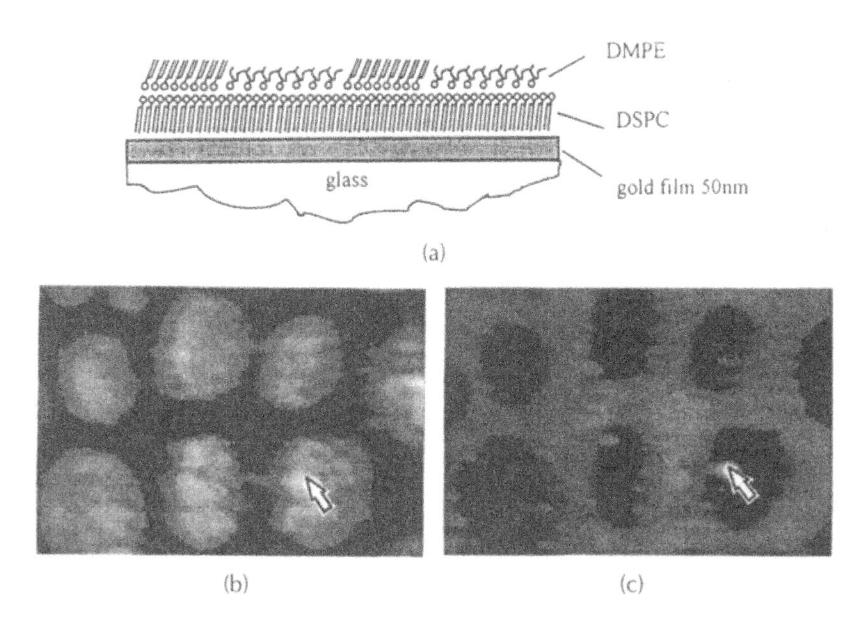

Figure 12. SPM images of the organized two layer film. The outer monolayer of DMPE is in a phase separation state, and the inner layer of DSPC is in a homogeneous phase. (**a**) schematic representation of the organized film; (**b**) and (**c**) images taken at different incident angles.

5. SUMMARY

A SPR biosensor was developed in our laboratory for molecular recognition analysis. According to the theoretical analysis of the SPR optical structure, we found that the thickness of the gold film does not affect the linear relationship between the shift of the resonance angle and the mass of the biomolecule adsorbed. We demonstrated that the shift of the resonance angle in an aqueous environment is proportional to the thickness of the adsorption layer and the difference between the refractive indexes of the sample and the buffer. The interaction of avidin with a biotin-containing membrane has been studied by SPR biosensor. The results show that the problem of nonspecific adsorption of avidin is eliminated if the surface density of the membrane-bound receptor was chosen as optimum. Based on the controlled assembly of biotin-lipid/avidin/biotin-DNA, a functionalized surface structure was organized on a gold substrate. With such a sensing surface, the specific adsorption of an anti-DNA antibody was examined. Because analysis of anti-DNA antibody is essential in clinical serum diagnosis of systemic lupus erythematosus, this research is significant for developing a simple method to detect systemic lupus erythematosis in real time. SPR microscopy was employed together with SPR spectroscopy in the present work. It is expected that SPR biosensors will become more powerful by combining spectroscopy and microscopy.

ABBREVIATIONS

B-DPPE: Biotin-L-α-phosphatidylethanolamine dipalmitoyl
BSA: Bovine serum albumin
DMPE: L-α-Phosphatidylethanolamine dimyristoyl
DPPE: L-α-Phosphatidylethanolamine dipalmitoyl
DSPC: L-α-Phosphatidylcholine distearoyl
LB: Langmuir–Blodgett
SPR: Surface plasmon resonance
SPM: Surface plasmon microscopy

ACKNOWLEDGMENTS

This work is supported by grants from the National Natural Science Foundation of China and the National Climbing Program of China.

REFERENCES

[1] Sackmann, E. (1996). Supported membranes: Scientific and practiced applications. *Science*, **271**, 43–44.

[2] McConnell, H.M., Tamm, L.K., Weiss, R.M. (1984). Periodic structures in lipid monolayer phase transitions. *Proc. Natl. Acad. Sci. USA*, **81**, 3249–3254.

[3] Sui, S.-F., Urumow, T., Sackmann, E. (1988). Interaction of insulin receptors with lipid bilayers and specific and nonspecific binding of insulin to supported membranes. *Biochemistry*, **27**, 7463–7469.

[4] Sui, S.-F., Wu, H., Sheng, J., Guo, Y.J. (1994). Conformational changes of protein at interface by supported planar phosphatidyl acid monolayer. *Biochemistry*, **115**, 1053–1057.

[5] Liu, Z., Qin, H., Xiao, C., Wen, C., Wang, S., Sui, S.-F. (1995). Specific binding of avidin to biotin containing lipid lamella surfaces studied with monolayers and liposomes. *Euro. Biophys.*, **24**, 31–38.

[6] Xiao, C., Liu, Z., Gao, Q., Zhou, Q., Sui, S.-F. (1996). Specific interaction of rabbit C-reactive protein with phospholipid membrane. *Thin Solid Films*, **284**, 793–796.

[7] Sui, S.-F. (1989). Evanescent field flourescence detection of adsorption of FITC-insulin to supported asymmetric bilayers. *Thin Solid Films*, **180**, 89–92.

[8] Fodor, S.P.A., Rava, R.P., Huang, X.C., Pease, A.C., Holmes, C.P., Adams, C. (1993). Multiplexed biochemical assays with biological chips. *Nature*, **364**, 555–556.

[9] Malmquist, M. (1993). Biospecific interaction analysis using biosensor technology. *Nature*, **361**, 186–187.

[10] Stelzle, M., Sackmann, E. (1989). Sensitive detection of protein adsorption to supported lipid bilayers by frequency-dependent capacitance measurements and microelectrophoresis. *Biochim. Biophys. Acta*, **981**, 135–142.

[11] Knoll, W. (1991). Optical characterization of organic thin films and interfaces with evanescent waves. *MRS Bulletin*, **XVI**, 28–39.

[12] Mi, L.Z., Wang, H.W., Sui, S.-F. (1997). Interaction of rabbit C-reactive protein with phospholipid monolayers studies by microflourescence film balance with an externally applied electric field. *Biophys. J.* **73**, 446–451.

[13] Sui, S.-F., Wu, H.Y., Chen, K. (1994). Conformational changes of melittin upon insertion studied by phospholipid monolayer and vesicle. *Biochemistry*, **116**, 482–487.

[14] Kretschmann, R.E. (1971). The determination of the optical constants of metals by the excitation of surface plasmons. *Physics*, **241**, 313–324.

[15] Wilchek, M., Bayer, E.A. (1990). Applications of avidin-biotin technology: Literature survey. *Methods Enzymol.* **184**, 14–45.

[16] Bayer, E.A., Rivany, B., Skutelsky, E. (1979). On the mode of liposome-cell interactions with biotin-conjugated lipids as ultra structural probes. *Biochim Biophys. Acta*, **550**, 464–473.

[17] Qin, H., Xie, W., Chen, H., Sui, S.-F. (1993). Study on interaction of avidin with biotin lipid monolayer. *Chin. Sci. Bull.*, **38**, 486–490.

[18] Qin, H., Liu, Z., Sui, S.-F. (1995). Two-dimensional crystallization of avidin on biotinylated lipid monolayers. *Biophys. J.* **68**, 2493–2496.

[19] Xie, W., Liang, J., Sui, S.-F. (1996). Molecular recognition induced formation of protein/nucleic acid multilayers. *Acta Biophys. Sin. (Chinese)*, **12**, 239–242.

[20] Yeatman, E., Ash, E. (1987). (1987). Surface plasmon microscopy. *Electron. Lett.* **23**, 1091–1092.

[21] Rothenhausier, B., Knoll, W. (1988). Surface plasmon microscopy. *Nature*, **332**, 615–616.

OLFACTORY AND TASTE BIOSENSOR SYSTEM

Wang Ping, Xie Jun, Tan Yi, Wang Yongqiang, Li Rong, and Zhang Qintao

OUTLINE

Advances in Biosensors
Volume 4, pages 139–155.

ISBN: 0-7623-0073-6

ABSTRACT

A review of an olfactory and taste biosensor system is presented in this chapter. The implementation and application of an artificial olfactory and taste sensor system based on the odor or gas and taste sensor array is described. Signal processing and pattern recognition of these systems are also discussed.

1. INTRODUCTION

As two of the basic senses of human beings, smell and taste play a very important role in daily life. Olfactory and taste sensor systems are very useful in the food industry and in environmental protection. At present, people have a considerable amount of basic knowledge about vision and hearing. A robot possessing these senses has already been put into practice. However, the study of olfactory and taste is still at an early stage. Only a few kinds of olfactory and taste sensor systems are in commercial use. Therefore an artificial olfactory and taste system is very important for future development.

Some research work has already been carried out in these fields. In our previous work, we used an odor or gas and taste sensor array combined with pattern recognition to implement the artificial olfactory and taste sensor system. Some integrated sensor arrays were produced. To optimize the system, a method that could adjust the structures of both the sensor array and the artificial neural network (ANN) was introduced into the system. Because pattern recognition of odor and taste is carried out by the brain using neural networks and fuzzy logic, a pattern recognition method that combines ANN and fuzzy recognition was applied to mimic human pattern recognition. Because the unusual characteristics of the Light-Addressable Potentiometric Sensor (LAPS), it is a very promising method for implementing a taste and olfactory sensor. Successful applications of LAPS can be found in some research work.

2. OLFACTORY INTEGRATED SENSOR ARRAY

The micromachine is the key technology for the rapid development of solid-state chemical sensors in recent years. It has emerged as an important and expanding extension of integrated circuit technology. Basically, micromachining is a combination of precise etching, deposition, insulator-to-silicon or silicon-to-silicon bonding, packaging, connection between layers, and standard integrated circuit methods used to fabricate precise three-dimensional silicon-based microstructures of great diversity, including thin film technique, microbridges, miniature cantilever beams, and so on. These micromachined structures, combined with special purpose thin films and high performance electronics, have been successfully employed to realize a large variety of solid-state sensors for gas composition, molecular concentration, and other chemical sensors [1]. We produced an integrated olfactory sensor array using microelectronic fabrication techniques. The recent advancement in micromachining techniques, such as chemical anisotropic etching and sacrificial layers, adds a new dimension to the advancement of chemical sensor research. Controlled temperature and operating conditions can be accomplished using these techniques, and it is possible to construct low mass, low power devices. A sensor array can be constructed for a multichannel sensing system to identify various gases and their concentrations. A thin oxide-based olfactory sensor and other chemical sensors are used to illustrate the advantages of this novel sensor structure. This work includes the design of a novel integrated olfactory sensor array with a heat insulation structure as shown in Fig. 1(a), microelectronic fabrication techniques, the design of an independent gas sensing element and its control of operating temperature, the recognition methods of various gases, and so on [2,3].

The experimental results show that the previously mentioned design methods for an integrated olfactory sensor array have obvious advantages compared to a separated gas sensor array. Secondly, the temperature distribution of the integrated

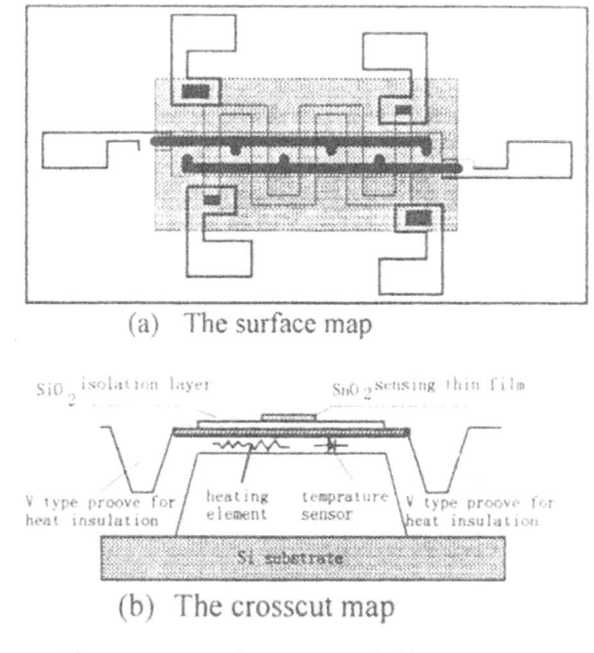

Figure 1. Map of an integrated olfactory sensor.

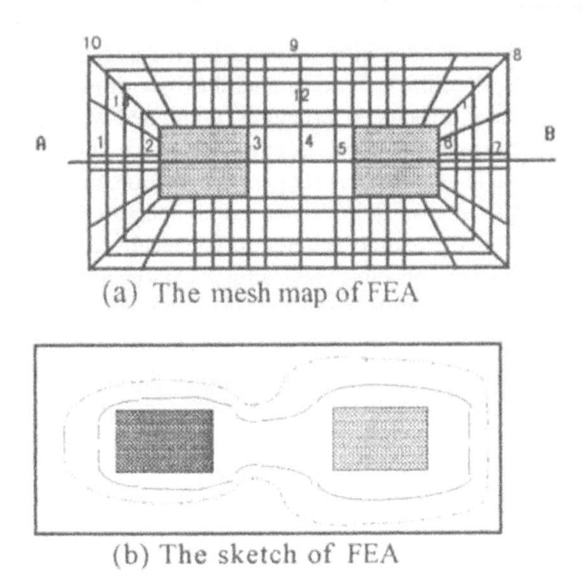

Figure 2. The mesh and sketch of FEA on the integrated gas sensor element.

gas sensor which should be taken into account in manufacturing design is discussed in the previous work. The thermal analysis model derived from the array's exact structure and physical features, the equation, and its boundary condition are obtained from this model. According to the complexity of the bounds, in the analysis of an array model, finite element analysis (FEA) software is often used. In this analysis, the result is obtained from special FEA software. We also suggest a gas detection method which adjusts the element temperature in an array. The detection result can be processed by a clustering method.

A map of an integrated gas sensor is shown in Figure 1. Its shape is not regular enough to analyze, so it is necessary to regularize its shape. As a simplified model shape, it can be expressed as a two-dimensional thermal analysis problem. The array model consist of only two sensor elements. In the equations of this model, two hypotheses are used: a. In a steady-state condition, the heat transfer ratio to air is fixed. b. The heat transfer ratio of the inside thermal source is also fixed. Figure 2 shows the process of the model's mesh and a sketch of FEA on the integrated gas sensor element [4].

3. THE FLEXIBLE STRUCTURE OF OLFACTORY SENSOR ARRAY

At present, most olfactory sensors are made of metal-oxide-semiconductor sensing material or biological and polymer organic sensing material. Among them, the metal-oxide-semiconductor sensing material is the most widely used because of its broad range of gases, its high sensitivity, its rapid response, and easy manufacture. With the development of microelectronic and microprocessing technology, sensors made of this material have potential for good unity, miniaturization, and integration of the sensor array. A general defect of odor sensing is the poor stability and sensitivity to selectivity. After in-depth research on the functional membrane component surface status and membrane technology of semiconductor sensing material based on SnO_2, we applied the advanced nanometer technique from material science to the manufacture of odor sensors. We achieved nanometer construction of odor sensing material based on the sol-gel technique. The sol-gel preparation technique makes it possible to make semiconductor sensing material on a molecular level. In the processing of a sol-gel, components are mixed with each other in the liquid phase at the molecular level to prepare a homogeneous pure material or a homogeneous mixed material of the required components which finally form nanometer crystalline grains. The technique clearly improves the unity and stability of sensors and offers a basis for the optimum odor sensor array and its stable function.

We propose a flexible design of artificial olfactory sensors as follows. Generally, olfactory sensors work at a constant temperature t and humidity h. Here we make gas sensing elements work at different temperatures according to the design of optimum feature space in pattern recognition, so that gases to be recognized take up different pattern areas in the feature space. It is known from theoretical analysis and experiments that semiconductor gas sensors differ in sensitivity and selectivity at different temperatures, as shown in Figure 3. The relationship can be expressed as $R = F(c, t, h)$, and this feature can be used to improve the similarity of the sensor array response mode. The basic method is described as follows:

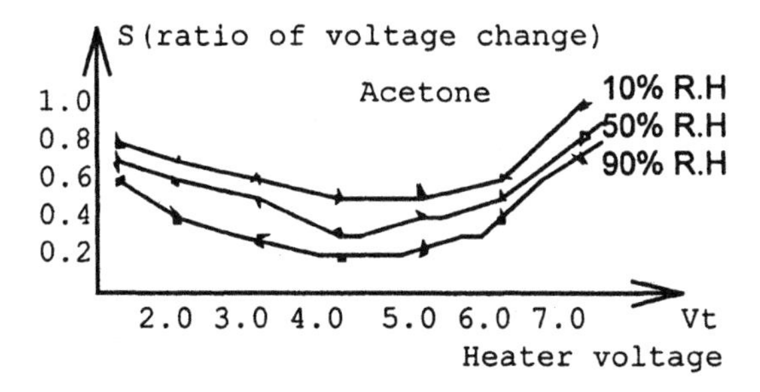

Figure 3. The response of an odor sensor to temperature and humidity.

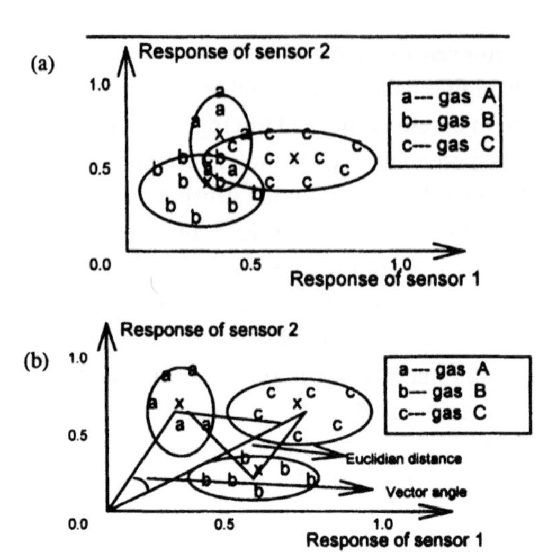

Figure 4. Responses of two elements to three gases (ratio of voltage change). (**a**) The response of two sensors to three kinds of gas at given temperatures t_1 and t_2 before optimal design. (**b**) The response of two sensors to three kinds of gas at given temperatures t_1, t_2 after optimal design.

1. Measure the relationship between working temperatures and the responses of each sensing element for measurable gases.
2. Measure the relationship between working humidities and responses of each sensing element for measurable gases.
3. Define the features of the sensor response to measurable gases, and complete the optimal feature extraction.
4. Compute the nonsimilarity between different gases of every sensing element of a given working temperature and humidity.
5. Determine the optimal working temperature and humidity of every sensing element according to the criterion of maximum nonsimilarity.

The design and comparative results of a flexible olfactory sensor array are shown in Figure 4 [5].

4. IDENTIFICATION METHODS FOR MIXED GASES BASED ON A FLEXIBLE NEURAL NETWORK

4.1 Existing Methods

Considerable interest has recently arisen in the use of arrays of gas sensors together with an associated artificial neural network technique to structure an artificial olfactory system or "electronic nose." But recent ANNs include the fact that these network structures had no flexible capability or adaptivity, that is, network structure including the hidden layer and neuron cannot be adaptively adjusted when measured gases are different.

We introduce a neural network with a flexible structure and its algorithm for an olfactory sensory system. This method included two steps: first, we built a larger network structure with more hidden nodes which is completed off-line based on tested gases. Second, by on-line learning and calculating the correlations between hidden neurons and the dispersity of the hidden neurons, useless or low-activity neurons were merged or deleted to obtain a smaller feasible network. In addition, tested gases can adaptively adjust the transfer characteristics of each neuron, i.e., activation functions to obtain optimum active characteristics. The experimental results show that the previously mentioned flexible design methods for a neural network can reduce die structure of the network considerably in the artificial olfactory system and can increase its flexible or adaptive characteristics. In addition, this design can reduce the training cycle of the neural network at the same time as it improves the recognition accuracy of mixed gases.

At present, many ANN paradigms are used in artificial olfactory systems or electronic noses, such as Back Propagation (BP), Self-Organizing Map (SOM) [6, 7] . But the disadvantages of the above mentioned ANNs include the fact that these network structures have no flexible capability. Therefore the training times are increased and the stability and sensitivity are also reduced. In fact, because measured gases and environmental influences often change, the neural networks in olfactory systems should respond adaptively to mixed gases, i.e., they should be flexible. In general neural networks, the input and output layers are determined on the basis of practical problems, but the selection of hidden layers is very difficult. Theories show that a network with one hidden layer can realize arbitrary continuous monotonic nonlinear maps, and with an additional hidden layer, it has an appropriate output representation of odors. Therefore, networks with one or two hidden layers are often adopted. On the other hand, the choice of hidden units of ANN is very random. Therefore the ANN construction is not optimal [8].

4.2 Olfactory Identification Algorithm

Section 4.1 discusses the adaptive structure of neural network and its algorithm for an olfactory sensory system—the electronic nose. The comparison of results before and after the addition of a flexible structure is shown in Figure 5. The basic methods are described as follows:

1. Build an ANN off-line.
2. Determine the hidden layer and its neural units.
3. Train the weights and thresholds.
4. Adjust the ANN on-line.

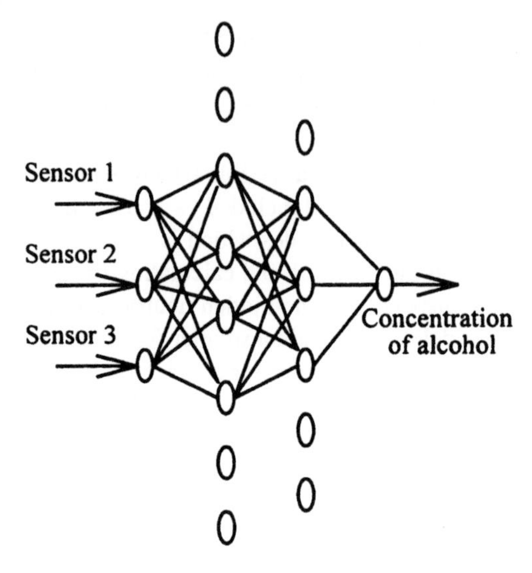

Figure 5. The structure after adding a flexible neural network.

5. Input samples of gases.
6. Calculate the correlation between units.
7. Calculate the dispersion of each neuron.
8. Compress redundant hidden nodes.

4.3 Experimental Results

The sensor array consists of three metal-oxide-semiconductor gas sensory elements or Surface Acoustic Wave (SAW) gas sensing elements [9]. The response of three gas sensors to a mixture of alcohol and acetone was obtained. In comparison, given twenty different concentrations of the mixture and the same learning accuracy of 1%, the experiment results show that conventional neural networks, such as the two-layer back propagation network, are trained in 5000 iterations, whereas the flexible network researched in this work needed only 3000 iterations to meet the same network output accuracy. Furthermore, the training speed is increased by two times, and its structure is reduced considerably. The structure of ANN after flexible design is shown in Figure 5. The test error for alcohols is reduced from 4.3% to 3.1% for learned samples and from 4.9% to 4.0% for unlearned ones, respectively, when acetone is changed from 0 to 20% (ppm) alcohol content. The experimental results show that the previously mentioned flexible design methods for a neural network reduce the network structure considerably in an artificial olfactory system and increase its flexible or adaptive characteristics. In addition, this design reduces the training cycle of the neural network, at the same time as it improves the recognition accuracy for mixed gases [10].

5. APPLICATION OF AN OLFACTORY SENSOR SYSTEM

5.1 A Noninvasive Method for Diabetes Diagnosis Based on an Electronic Nose

Diabetes is a chronic lifelong disease caused by carbohydrate metabolic block. The International Diabetes Association (IDA), based in Brussels, reported that there are about 1,000,000 diabetic sufferers in the world, accounting for 6% of the adult population. In industrial countries, diabetes is the third highest cause of death after heart diseases and cancer. In China, according to statistics, about 700,000 diabetes cases are reported every year among those over the age of 25, accounting for 0.13% of the population over 25 years old. With improvements in living conditions, the incidence of diabetes increases steadily. The conditions of diabetes are very complicated and vary frequently, so the disease is difficult to cure. If states of diabetes cannot be well controlled, the function and matter metabolism of some tissues and organs are disordered, which result in weakness, poor immunity, and complications. These complications can bring great pain to patients and can even endanger their lives.

Diabetes can be well controlled by such convenient means as diet adjustment, if it is diagnosed in time. Otherwise, once it reaches an advanced stage, it is likely to result in serious illnesses, such as heart and renal disease, blindness, and paraplegia. The president of IDA believes that early diagnosis is the only way to cure diabetes. Diagnosis, taking advantage of odors, which is called 'xiuzhen' in Chinese medicine, has been applied in clinics since ancient times in China. According to this method, we first proposed applying artificial olfactory recognition to diabetes diagnosis. The novel method has proved convenient, efficient, and simple to operate. It causes no pain and is noninvasive to patients, so it has a bright prospect and great practical value.

As mentioned before, diabetes is caused by the disturbance of carbohydrate metabolism. The main symptom of diabetes is high blood sugar concentration. Diabetics cannot make full use of glucose. Simultaneously fat resolution is accelerated to produce more fatty acid which is converted into ketone bodies. If only a few of the latter are produced, they can be completely absorbed by tissues, especially muscles. But when the production exceeds the capacity of application, these ketones are excreted and produce odor. Acetone is a volatile matter and is expired while breathing, forming acetone expired gases. Therefore a diabetic's breath often smells like decaying apples which give off acetone.

During diagnosis, the patient is required to breathe a certain volume of gases from his nose into a container. Diagnosis is realized by an artificial olfactory system, which decides whether there is acetone in the patient's breath. Figure 6 shows the

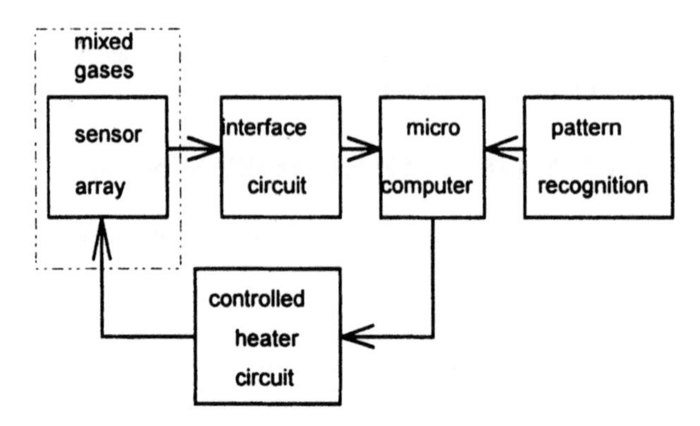

Figure 6. The structure of an artificial olfactory system.

schematic representation of an artificial olfactory system, which simulates the human olfactory system [11]. The human olfactory system is a process of gas molecule recognition. Gas molecules first get into the primary neuron (formed by a receptor and an olfactory neuron). Receptor signals pass to the olfactory bulb through an axon. The olfactory bulb is formed by about 10,000 secondary neurons, assembled together to adjust and refine signals from the primary neuron. The final signals are sent to the cerebrum for recognition.

In the artificial olfactory system, odor sensors are formed by an array of mental-oxide-semiconductor gas sensors. Because a single odor sensor lacks selectivity and sensitivity, we apply sensing elements that have different selectivity and sensitivity to form the sensor array and make every element work at different temperatures through a controlled heater circuit. Thus, a flexible olfactory sensor array and optimum recognition are obtained. The response signals to expired gases produced by the sensor array are fed into the computer via the preprocessing and interface circuits. Different kinds of odors are classified and recognized by algorithms, such as ANN pattern recognition and fuzzy cluster pattern recognition.

5.2 Experiment Results and Discussion

A preliminary experiment to verify this method has been done by examining diabetics and normal persons at Zhejiang Hospital. Thirty-two volunteers attended our experiments, 18 diabetics and 14 normal persons. All of the diabetics had idiopathy diabetes. We detected their expired gases with an odor sensor array consisting of five sensing elements before a meal, 0.5 hour after a meal and, 1 hour after a meal, respectively, and then recognized the responses for each detection. Simultaneously, blood sugar was detected for comparison, as shown in Figure 7 [13].

We designated all samples so that they would cluster into 2 classes, Class I and Class II. Recognition results are shown in Table 1–3. From the results, we find that

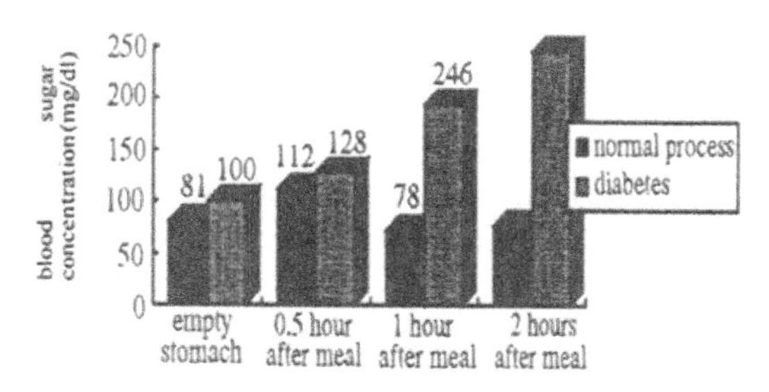

Figure 7. The test results of blood sugar concentration in diabetics and normal persons.

Table 1. Blood Sugar Level and Gas Analysis Results for Detection Before a Meal

	Number with Higher Blood Sugar	*Number of Class I among Column 1*	*Number with Normal Blood Sugar*	*Number of Class I among Column 2*	*Percent of Higher Blood Sugar*	*Percent of Class I*
Diabetics	12	11	6	3	12/18	14/18
Normal persons	2	1	12	4	2/14	5/14

Table 2. Blood Sugar Level and Gas Analysis Results for Detection 0.5 Hour after a Meal

	Number with Higher Blood Sugar	*Number of Class I among Column 1*	*Number with Normal Blood Sugar*	*Number of Class I among Column 2*	*Rate for Higher Blood Sugar*	*Total Acetone Positive Rate*
Diabetics	14	13	4	2	14/18	15/18
Normal persons	1	0	13	2	1/14	2/14

Table 3. Blood Sugar Level and Gas Analysis Results for Detection 1 Hour after a Meal

	Number with Higher Blood Sugar	*Number of Class I among Column 1*	*Number with Normal Blood Sugar*	*Number of Class I among Column 2*	*Rate for Higher Blood Sugar*	*Total Acetone Positive Rate*
Diabetics	18	18	0	0	100	100
Normal persons	2	0	12	0	2/14	0

Class I has a high correlation with diabetics, and so does Class II with normal persons. Noticeably, before a meal, most diabetics are diagnosed correctly, and some normal persons are mistaken for diabetics. We infer that the main reason for that is hunger, which can also lead to the rise in acetone concentration, as indicated by the fact that their acetone concentration is restored to normal 1 hour after a meal. On the other hand, the acetone concentration of diabetics remained high after a meal. This means that detection of expiration after a meal may give valuable diagnostic information.

Several gas analysis methods have been introduced into electronic noses [12,24]. Here, a nonsupervised fuzzy clustering algorithm, which is discussed in detail elsewhere [24], is applied to the analysis of the sensor responses. If in the feature space, samples of diabetics are clustered as a class and those of normal persons are clustered as another, then diabetics can be distinguished from normal persons [14, 15].

6. TASTE SENSOR ARRAY

Existing taste sensors detect only some physical-chemical taste properties of the taste substance, but they cannot simulate the taste sense of the real biological system. These measurements are often disturbed by tasteless substances. In addition, the acquired properties cannot indicate the relationship between the different taste substances, for example, the inhibition and harmony effect. On the other hand, these taste sensors are designed to improve their selectivity to get a better measurement of the taste substance, but it is very hard to manufacture a sensor that has good selectivity for several taste substances. An effective method for solving this problem is to use a material which is very similar to the material in the real biological system as the sensitive membrane. Japanese researchers applied the biosensor mechanism and biologically sensitive material to the realization of the taste sensor [13,14]. Although some results were valuable neither the sensitivity or stability of the sensitive material, nor the taste pattern recognition and quantitative detection gave significant results [15].

In previous research, we introduced an experiment which uses a multichannel electrode taste sensor made up of biolipid material to detect five basic types of taste sense (sourness, sweetness, saltiness, bitterness, and 'umami'). We adopted the dip method to make the lipid membrane. This approach is different from that adopted by the Japanese researchers. This method improved the characteristics of the membrane because the thickness of the membrane is decreased by the dip method and the uniformity of membrane is improved by it. When making the membrane, tetrahydrofuran (THF) was used as the solvent. Three lipids, oleic acid, lecithin, and cholesterol were mixed with the plasticizer and dissolved in THF. The silver electrodes were immersed into the solution and then lifted up at a given speed. In the given environment in which temperature and time were precisely controlled, the solution formed a uniform, thin membrane on the surface of the electrodes. Several experiments were carried out to find the optimal combination of these parameters. The Ag/AgCl electrode was chosen as the reference electrode. The finished electrodes were preserved in a 1 mM KCl solution. In another study we improved this design by adopting the light-addressable potentiometric sensor (LAPS) technique to integrate many sensor devices on a single semiconductor substrate [16].

In the experiment, five chemicals were used as basic taste substances: glucose for sweetness, HCl for sourness, NaCl for saltiness, quinine for bitterness, and monosodium glutamate for 'umami.' From the results of the experiment, we discovered that this type of taste sensor, using a biological lipid membrane, responds only to the substances which have taste and not to tasteless substances. The characteristics of the membrane, especially the responses to the saltiness, were improved greatly in the experiments compared to those of other researchers [13, 14]. These sensors had long-term drift and short-term drift, and both of them were also reduced. Short-term drift is related to the equilibrium time of the reaction, but long-term drift is related to the aging of the lipid membrane. The reason for this phenomenon is that the artificial lipid membrane does not have the activation of the biological membrane. The biological membrane has a metabolism and the ability to self-adjust, so it has a continuous reactivity and maintains the stability of the membrane potential. However, the artificial membrane can be used only a few times before it ceases to function. In spite of the limited number of taste substances, the taste substances in the experiment stand for the five basic tastes and the results are satisfactory [17].

7. ARTIFICIAL TASTE SENSOR SYSTEM

Taste includes mainly sourness, saltiness, bitterness, sweetness, and 'umami.' In a biological taste sensing system, taste substances are accepted on thin microhole taste cells. Then electrical response signals are produced, and the different regions of the tongue have different sensitivities and selectivities to the different taste substances. The electrical responses are sent to the brain through neural networks. Finally, taste discrimination is carried out by the brain which has already learned some sample taste patterns. In the real biological system, both the response pattern of taste sensors and the neural stimulation pattern of taste senses differ according to the different taste substances. These differences are recognized by the brain. So we find that the taste sense is based on multiparameter analysis and that the type of taste substance must be identified through multichannel pattern recognition. The structure of a taste sensor which is similar to the human taste sense is shown in Figure 8 [19].

Brain science tells us that the recognition procedure of the brain is fuzzy and that the brain uses fuzzy logic not fixed logic. For this reason, an improved self-organ-

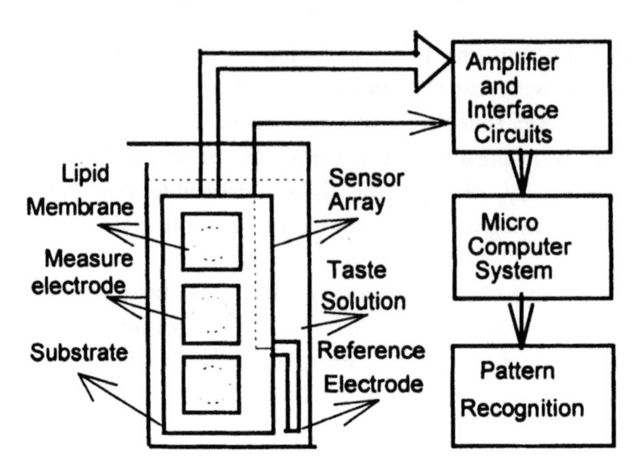

Figure 8. The structure of an artificial taste system.

izing map (SOM) which is an artificial neural network (ANN) that has a fuzzy learning algorithm was applied into the pattern recognition of the five basic taste senses. The ANN was successfully used in the artificial odor system (electronic nose) we made [11]. To realize fuzzy recognition, we proposed an improved learning algorithm basing on the fuzzy c-means algorithm (FCMA) [20]. Instead of the winner-take-all rule that SOM uses, the FCMA uses the Euclidian distance to calculate the membership between the sample and the class. Also, when the FCMA updates its weight, all of the samples are incorporated, whereas the SOM uses only one sample to update its weight at one time. For this reason, SOM spends more time than FCMA in the learning procedure and may cause local oscillation. From this point of view, FCMA is fuzzy and parallel and clusters more accurately and faster. Another important property of the improved method is that the classification result of this method uses fuzzy logic which can be found in the real biological system. Therefore, this result really simulates the taste sense of the actual biological system.

After the normalization procedure, only the sensor response characteristics relating to the type of taste substance are extracted as the input of taste patterns in the artificial neural network. The results of the experiments proved that the performance of the method in taste recognition was very good. It can solve some problems that were faced by the ordinary ANN method. The details of the method and the experiments are described in [21].

In general, the taste sensor using the biolipid material gets a different sensitivity and selectivity to the five basic taste sense types. If combined with the proper ANN information processing method, this type of multichannel taste sensor can achieve effective recognition of the basic taste types. At present, the stability and reproducibility of the designed sensor is unsatisfactory, and the sensitive membrane suffers from fatigue. Moreover, the response time is long, that is, it takes a long time for the response to stabilize. These problems appear in many biosensor systems. So, in further work, we must improve the manufacture of both the biolipid material and the lipid membrane. In addition to this, we should study the information extraction method of the instantaneous sensor response, the fast accuracy recognition, and quantitative analysis of the different taste types using ANN and other pattern recognition methods, so that the taste sensor developed meets the needs of real applications.

8. LIGHT-ADDRESSABLE POTENTIOMETRIC SENSOR (LAPS)

Because of the unusual characteristics of LAPS, it has been widely used to obtain chemical and biological imaging in some sensor system [22,23]. Recently, some research work has applied LAPS to construct a taste sensor by combining modified Langmuir-Blodgett (LB) methods to deposit different lipid membranes on a single semiconductor surface. The advantage of this method is the simple structure of the sensing system which enables integrating multiple sensing elements on a semiconductor surface. Modified LB methods are used to achieve integration of multiple sensing elements with different sensitivities and selectivities on a semiconductor surface. The basic structure of this taste sensor is the electrolyte/insulator/semiconductor (EIS) structure of LAPS. This type of taste sensor simulates the human tongue whereby different areas have different sensitivities and selectivities to different taste substances. Similarly, another researcher has used the scanning light-pulse technique to create a sensor that obtains a visual image of gas mixtures. The sensor simulates the human nose which can get the olfactory images. Now we

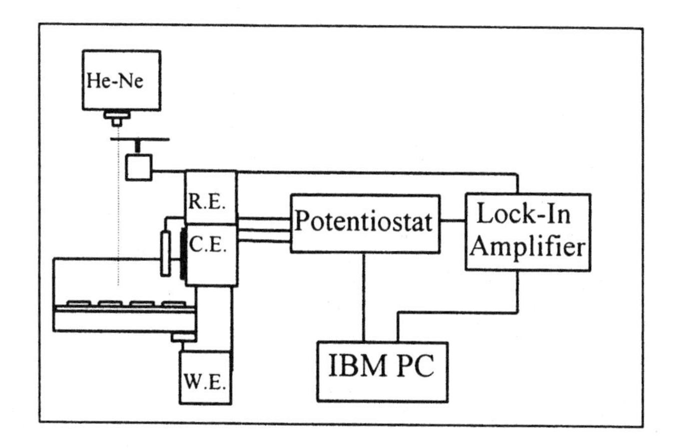

Figure 9. The structure of the LAPS taste sensor.

are developing a sensor system based on LAPS, and we will use this system to construct a new taste and olfactory sensor, as shown in Figure 9 [24,25].

9. CONCLUSION

Though the history of study of artificial olfactory and taste is not long, its progress has been rapid. Many methods have been applied in both the manufacture and signal processing of the sensor system, and some positive results have been obtained. On the basis of existing research, it can be anticipated that some kind of sensor system for practical commercial use will appear in the next few years.

ACKNOWLEDGMENTS

This research was supported by the National Science Fund of China (Grant No. :39400034, 69402005), the National Post-doctoral Science Fund, the Zhejiang Province National Science Fund of China (Grant No. 695041, 396082) and The Open Fund of Electronic Materials Research Laboratory, Xian Jiaotong University (Grant No. 9505).

REFERENCES

[1] Liu C.C. (1992). Development of chemical sensors using microfabrication and micromachining techniques. *Proc. Int. Meeting on Chemical Sensors*, pp. 2–5.

[2] Yongqiang, W., Ping, W., Yuquan, C. (1997). The anti-fatigue integrated olfactory sensor array based on micromachining technique. *The Conference on STC'97*, to be published (in Chinese).

[3] Rongqiang, W., Ping, W., Yuelin, W. A novel integrated microsystem for gas detection. *EACCS'97*, Seoul, Korea, Nov. 5–6, 1997, to be published.

[4] Rongqiang, W., Ping, W., Xiaoxiang, W. The thermal analysis of a novel integrated gas sensor array. *EACCS'97*, Seoul, Korea, Nov. 5–6, 1997, to be published.

[5] Ping, W., Rong, L., Yuquan, C. (1995). The implementation of flexible olfactory sensor array for electronic nose. *EACCS'95*, Xi'AN, China, pp. 71–73.

[6] Gardner, J.W., Hines, E.L., Wilkinson, M. (1990). Application of artificial neural network to an electronic olfactory system. *Meas. Sci. Tech.*, **1**, 446–451.

[7] Lemos, R.A., Nakamura, M., Sugimoto, I., Kuwano, H. (1993). A self-organizing map for chemical vapor classification. *Transducer'93*, Japan, pp. 1082–1085.

[8] Ping, W., Yuquan, C., Weixue, L. (1994). Development of gas sensor arrays for detection of mixed gases. *J. Zhejiang Univ.*, **28**, 172–176 (in Chinese).

[9] Ping, W., Yuquan, C., Weixue, L. An improved design of SAW gas sensors. *The 5th International Meeting on Chemical Sensors*, July 11–14, 1994, Rome, pp. 65–68.

[10] Ping, W., Rong, L., Weixue, L. (1995). Study on flexible neural network for olfactory sensory system. *Transducers'95*, Sweden, pp. 715–717.

[11] Ping, W., Yuquan, C., Jian, W., Yan, S. (1996). The study of artificial olfactory—electronic nose. *Chin. J. Biomed. Eng.* (in Chinese), **V15**, (4) 346–353.

[12] Ping, W., Yi, T., Rong, L. (1996). A novel method for diagnosing diabetes using an electronic nose. *Biosensors'96*, Thailand, May, 1996, p. 256.

[13] Ping, W., Yi, T., Haibao, X., Farong, S. (1997). A novel method for diagnosing diabetes using an electronic nose. *Biosensors & Bioelectronics* (revised).

[14] Liyma, S. (1986). Effect of bitter substance on a model membrane system of taste reception. *Agric. Biol. Chem.*, **50**, 2709–2714.

[15] Hayashi, K., Yamanaka, M. (1990). Multichannel taste sensor using lipid membrane. *Sensors & Actuators* B2, 205–213.

[16] Ping, W., Yuquan, C., Weixue, L. (1995). Progress of taste sensing technique. *BME Fascicle of Foreign Med. Sci.*, **N1**, 1–5 (in Chinese).

[17] Sasaki, Y., Kanai, Y. (1994). Highly sensitive taste sensor with a new differential LAPS method. *The Fifth International Meeting on Chemical Sensors*, Italy, July 11–14, 1994, pp. 1118–1121.

[18] Ping, W., Rong, L., Ying, Z. (1995). The study on taste sensor based on biological lipid membranes. *J. Functional Mater. Devices* (in Chinese), **1** (2), pp. 71–76.

[19] Ping, W., Jun, X., Rong, L. (1996). The study of artificial taste system. *Biosensors'96*, Thailand, May, 1996, p. 83.

[20] Ping, W., Jun, X. (1996). A novel method combined neural network with fuzzy logic for odour recognition. *Meas. Sci. Tech.* **7** (12), 1707–1712.

[21] Ping, W., Jun, X. (1997). A recognition method using neural network and fuzzy recognition for electronic nose, *Sensors Actuators B*, **37**, 169–174.

[22] Hafeman, D.G. (1988). Light-addressable potentiometric sensor for biochemical systems. *Science*, **240**, 1182–1185.

[23] Jun, X., Ping, W. A novel application of LAPS for determination of serotonin. *Transducers'97*, Chicago, USA, June 16–19, 1997, pp. 489–492.

[24] Jun, X., Ping, W. (1996). The study progress of olfactory and taste imaging sensor. *BME Fascicle Foreign Med. Sci.*, **19** (5), 1–8 (in Chinese).

[25] Jun, X., Ping, W. (1998). The study of light-addressable potentiometric imaging sensor. *J. Chinese Instrum.*, **19**(1), 50–55 (in Chinese).

KINETIC STUDIES OF LIGAND–DNA INTERACTIONS USING AN EVANESCENT-WAVE BIOSENSOR

Mengsu Yang, Hing Leung Chan,
Pui Yan Tsoi, and Li-Qiang Ren

OUTLINE

Advances in Biosensors
Volume 4, pages 157–178.

ISBN: 0-7623-0073-6

ABSTRACT

Recent advances in biosensor technology have contributed significantly to our understanding of the kinetic basis of molecular recognition processes. Biosensors enable real-time studies of biomolecular interactions without the need for labeling. Recently, we employed an optical biosensor based on evanescent-wave detection to study the affinity and kinetics of drug–DNA and protein–DNA interactions. Specifically, the optical biosensor has been used to study the DNA-binding characteristics of doxorubicin, an antitumor drug, conjugated with various polymeric dextran molecules; and the interactions between various DNA polymerases and primer–template DNA with perfectly matched or mismatched sequences. The observed binding profiles are discussed to highlight the uniqueness of the biosensor technique in obtaining kinetic information about ligand–DNA interactions.

1. INTRODUCTION

Real-time kinetic studies of biochemical processes using biosensor technology are contributing significantly to understanding the molecular basis of immune response, signal transduction, gene expression/regulation, and drug discovery [1–4]. The elegance of biosensors is that a biospecific interaction between a ligand in solution and a receptor immobilized on the surface of an electronic or optical device leads to enhanced concentration of the ligand at the device–solution interface and that it provides a detectable signal directly related to the biological species and the interaction. In our laboratory, we are interested in applying biosensor technology as one of the tools for examining the fundamental processes involved in ligand–DNA interactions. Recently, we employed an optical biosensor based on a resonant mirror mounted in a stirred cuvette [5–8] to study the affinity and kinetics of drug–DNA and protein–DNA interactions. Specifically, the optical biosensor was used to determine the binding characteristics between an immobilized double-stranded DNA and the antitumor drug doxorubicin conjugated with various polymeric dextran molecules and the interactions between various DNA polymerases and immobilized primer-template DNA with perfectly matched or mismatched sequences. This chapter includes brief descriptions of the basic principles of the optical biosensor, the strategies for DNA immobilization, and the kinetic analysis of real-time binding data. Selected results from the drug–DNA and protein–DNA binding studies are presented to highlight the uniqueness of using the biosensor technique to obtain kinetic information about the molecular basis of ligand–DNA interactions.

2. EVANESCENT-WAVE OPTICAL BIOSENSOR

The biosensor used in this study is the IAsys optical sensor from Affinity Sensors (Cambridge, UK). The evanescent-wave sensor device, also called a resonant mirror

[5], is comprised of dielectric resonant layer of a high refractive index material, such as titania or hafnia, deposited on the surface of a low-index coupling layer (silica) and coupled to a prism block (Figure 1). Laser light is directed at the prism over a range of angles and total internal reflection occurs from the prism to coupling layer interface. At one unique angle (resonant point), light tunnels through the coupling layer and propagates in the high-index (resonant) layer before coupling back into the prism, a phenomenon known as waveguiding. The light propagating in the resonant layer generates an evanescent field that decays exponentially beyond the resonant layer and has a decaying length of about 100 nanometers. A shift in the phase of the reflected light occurs on resonance, and the incident angle for resonance is very sensitive to changes of the refractive index/thickness of the interface. For biosensing, the surface of the resonant mirror may be modified with a layer of a polymeric matrix (dextran) or a functionalized organic monolayer (aminosilane), and biomolecules can be immobilized onto the sensor surface by various chemical approaches. Specific bindings at the sensing interface result in a mass change and thus a change in the refractive index within the evanescent field. A change in the resonance angle is proportional to the refractive index and can be precisely followed by monitoring the shift in the position of the intensity peak, which permits monitoring macromolecular interactions in real-time without the need for labeling.

The IAsys optical biosensor has been used to study the binding kinetics of various ligand–receptor systems, including chymotrypsin and chymotrypsin inhibitor-2, anti-TNF and TNF (tumor necrosis factor), anti-insulin and insulin, and anti-HSA and HSA (human serum albumin) interactions [5–8]. Recently, the biosensor has been used to detect DNA hybridization where the estimated limit of detection is 19.9 fmol/mm^2 [9]. The interaction at the sensor surface is sequence-specific under conditions of low stringency. Regeneration of the surface-immobilized probe is

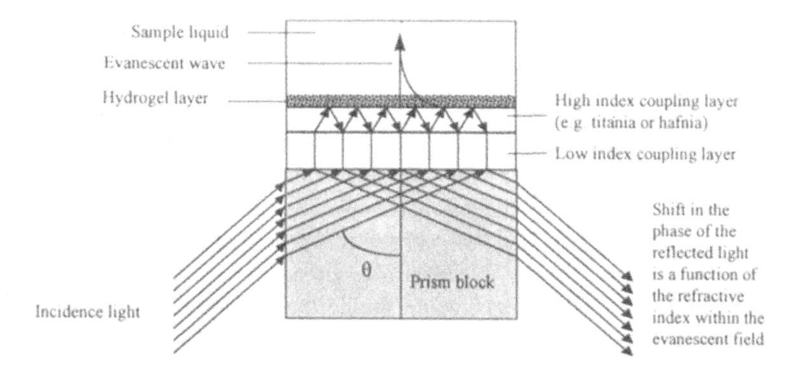

Figure 1. Schematic representation of the evanescent-wave optical biosensor (IAsys Resonant Mirror, Affinity Sensors).

possible, allowing reuse without a significant loss of hybridization activity. A comparison of probes indicates that the relative position of the complementary sequence and the length of the immobilized probe affect the hybridization response obtained. The sensor system was also used to study binding of TATA box-binding protein (TBP) to an immobilized DNA promoter sequence [10]. Promoter-specific recognition by TBP was observed in the presence of poly(dG-dC).

3. IMMOBILIZATION OF BIOMOLECULES ONTO THE SENSOR SURFACE

The surface of the resonant mirror can be modified with a layer of carboxylmethyl dextran or an aminosilane monolayer. To prepare a three-dimensional dextran matrix [11,12], a hydrophilic monolayer of 1,ω-hydroxyalkyl silane is cross-linked onto the resonant mirror surface and is partly a barrier to prevent proteins and other ligands from coming into contact with the sensor surface. The hydroxyl groups can be activated with epichlorohydrin under basic conditions to form epoxides. Covalent linkage formed between the epoxides and the hydroxyl groups of the dextran generates a layer of polymeric hydrogel of 100–200 nm thick, using dextran that has an average molecular weight of 500 kDa. Carboxyl groups are introduced into the dextran molecules by treating the dextran layer with bromoacetic acid to give a layer of carboxymethyl dextran (CMD). The CMD layer is negatively charged and has about one carboxyl group per two glucose residues. The thickness of the CMD layer is within the probing distance of the evanescent field.

A number of approaches may be employed to immobilize biomolecules onto the CMD matrix [12]. The negatively charged dextran-covered surface can be used for direct adsorption of proteins and peptides. The most widely used method for covalently linking proteins is based on EDC/NHS chemistry. The carboxylate groups on CMD are first activated to form *N*-hydroxysuccinimide esters using *N*-ethyl-*N*′-(3-dimethylaminopropyl) carbodiimide (EDC) and *N*-hydroxysuccinimide (NHS). The NHS-activated carboxyl groups form amide bonds with the primary amino groups in proteins (N-terminal and lysine residues). Alternatively, the NHS-activated dextran matrix may be modified to generate dithiol, hydrazine, or avidin/strepavidin groups on the surfaces. Proteins are immobilized via thiol–disulfide exchange, amino–aldehyde reaction, and biotin–avidin binding.

For aminosilane-based immobilization, the cuvette surface is treated with γ-aminohexyl triethoxysilane (AHTES) to generate a two-dimensional monolayer that contains amino groups attached to the surface via an alkyl chain linker [13,14]. Proteins are covalently immobilized on the amino-functionalized surface by using a number of bifunctional cross-linking reagents, such as glutaraldehyde, maleic anhydride, nitrophenyl chloroformate (NPC), and bis(sulfosuccinimidyl) suberate (BS^3). The advantage of using the monolayer chemistry is that the association and dissociation of biomolecules occurs closer to the sensor surface where the evanes-

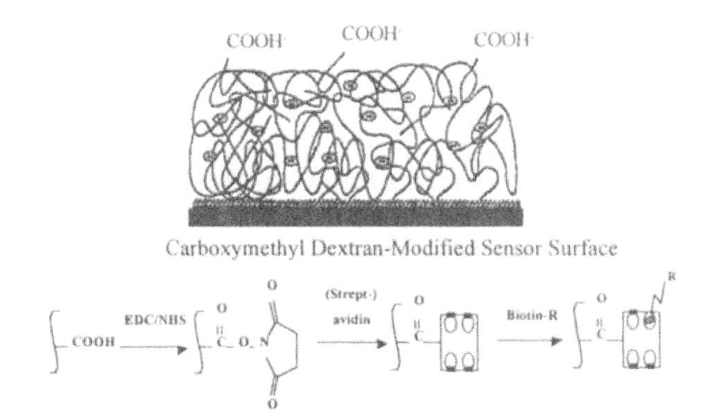

Figure 2. Immobilization of DNA on carboxymethyl-modified dextran (CMD) hydrogel.

cent field is more intense. The disadvantage is that the amount of protein immobilized is small and, therefore, the binding signal is weak, and nonspecific binding of the protein to the sensor surface may occur.

In our experiments, the DNA molecules were immobilized on the sensor surface, only the 5′-end was attached to the CMD layer. Attempts to form the amide bonds between the NHS-activated carboxyl groups on CMD and 5′-NH_2-labeled oligonucleotides (single- or double-stranded) based on the EDC/NHS chemistry resulted in limited success, possibly caused by the electrostatic repulsion between negatively charged groups on the DNA backbone and the CMD layer. The biotin–streptavdin approach provides the most specific and reproducible results for DNA immobilization (Figure 2). In this approach, streptavidin is readily immobilized to the CMD surface using the EDC/NHS chemistry and subsequently is used to capture 5′-biotinlyated oligonucleotides. The affinity of the biotin–streptavidin complex is very high (K_D = 10^{-15} M), therefore, the electrostatic repulsion between the DNA backbone and the CMD layer is easily overcome, and the complex is stable under a wide variety of buffer conditions. The amount of immobilized DNA is controlled by varying the EDC/NHS activation time, pH, ionic strength, and the contact time of the biotinylated DNA with the streptavidin-modified sensor surface. In the experiments, the amount of immobilized DNA was maintained at a relatively low level to ease the kinetic analysis of the binding data but was sufficient to generate detectable sensor signals with a reasonable signal-to-noise ratio.

4. KINETIC ANALYSIS OF BINDING DATA

Biomolecular interactions are complex processes involving multiple steps, such as association of binding partners, conformational changes of biomolecules upon binding, and dissociation of the complexes. Kinetic analysis using biosensors is further complicated because the binding data are obtained under conditions which differ from free solution measurements. In general, for a ligand–receptor interaction (where the ligand is the partner in free solution and the receptor is the immobilized partner), the receptor is irreversibly attached or captured on the sensor surface in a fixed but relatively low absolute amount, whereas the ligand exceeds the absolute amount of the immobilized receptor and can be considered constant in concentration. Additionally, the IAsys optical sensor employs constant stirring to replenish the ligand at the boundary layer and to aid mass transport.

A basic assumption in analyzing the sensor response is that the specific biomolecular interaction at the interface follows a simple Langmuirian model that involves single steps of association and dissociation [15–17]. Although the assumption may oversimplify the situation and may not always be true, the biosensor approach in concert with appropriate experimental design and data analysis yields valuable insights into the function and mechanism of the interacting system [1–4]. For some biomolecular interactions, monophasic association occurs when the sensor response to the addition of ligand increases in an exponentially and has only one distinguishable phase. However, experimentally derived association curves from a wide range of biological systems that have an immobilized receptor and a soluble ligand frequently have at least two distinguishable phases [8]. Although the fast phase of the binding profiles should be directly comparable to those in free solution, it is believed that the slower binding phase is caused by restricted accessibility due to the location of the immobilized receptor within the CMD matrix and/or steric constraints imposed by binding adjacent macromolecules.

Similarly, once a ligand–receptor complex is formed and the ligand in free solution is replaced with a buffer, dissociation occurs. For a simple system, the time for a dissociation event to occur depends on the time required to break the various classes of noncovalent bonds that stabilize the complex and is a feature intrinsic to the particular interacting biomolecules. Therefore, the time taken is independent of the concentration of the complex, and the dissociation curve should be monophasic. However, as is the case for association curves, the dissociation may be biphasic as a result of different phases of dissociation caused by different degrees of steric hindrance and any reassociation (ligand rebinding) within the matrix. For both biphasic association and dissociation curves, the slower phase has various contributing factors and contains no readily interpretable kinetic information. Therefore, the faster association and dissociation rates are considered more closely related to binding in solution.

In the ligand–DNA binding studies, for example, there are at least two possible modes of interaction for addition of a ligand (drugs or proteins) at a defined concentration to a sensor surface that contains immobilized DNA: monophasic association $A + B \rightarrow AB$, where A and B represent the immobilized DNA and the free ligand in solution, respectively, and biphasic association $A_1 + A_2 + B \rightarrow A_1B + A_2B$, where A_1 and A_2 represent the immobilized DNA that has different accessibilities and B is the free ligand, respectively. (In certain cases A_1 and A_2 may represent the different DNA-binding domains of the same ligand and B the immobilized DNA.) Upon binding of a ligand to the immobilized DNA on the sensor surface, the sensor response R can be expressed by Eqs. (1) and (2):

$$R_t = (R_{eq} - R_0)[1 - \exp(-k_{on}t)] + R_0 \quad \text{(monophasic)} \tag{1}$$

$$R_t = \alpha[1 - \exp(-k_{on(1)}t)] + \beta[1 - \exp(-k_{on(2)}t)] \quad \text{(biphasic)} \tag{2}$$

where R_t, R_0, and R_{eq} represent the sensor response at a particular time, at the initial point, and at the maximum level, respectively. For the biphasic curves, the contribution from the faster and the slower phases to the overall binding kinetics are represented by α and β, respectively. The pseudo-first-order rate constants k_{on}, $k_{on(1)}$, and $k_{on(2)}$ can be recovered by fitting the experimental binding curves to the respective equations by using nonlinear regression analysis. The intrinsic association and dissociation rates (k_{ass} and k_{diss}) may be obtained from the plot of k_{on} versus the ligand concentration $[B]$ according to Eqs. (3)–(5):

$$k_{on} = k_{ass}[B] + k_{diss} \quad \text{(monophasic)} \tag{3}$$

$$k_{on(1)} = k_{ass(1)}[B] + k_{diss(1)} \quad \text{(biphasic, faster phase)} \tag{4}$$

$$k_{on(2)} = k_{ass(2)}[B] + k_{diss(2)} \quad \text{(biphasic, slower phase)} \tag{5}$$

although the values of k_{diss} obtained from these plots are not well defined and often cannot be used to describe the kinetics of the dissociation processes.

The dissociation rate constants can be derived directly from the experimental dissociation data. Similar to association, dissociation may also occur as two distinct processes and can be described by

$$R_t = R_0 \exp(-k_{diss}t) \quad \text{(monophasic)} \tag{7}$$

$$R_t = \alpha \exp(-k_{diss(1)}t) + \beta \exp(-k_{diss(2)}t) \quad \text{(biphasic)} \tag{8}$$

and the values of k_{diss}, $k_{diss(1)}$, and $k_{diss(2)}$ can be calculated accordingly. The equilibrium binding constant can be calculated as $K_D = k_{diss}/k_{ass}$, although it may also be obtained directly by using Scatchard analysis [18] by determining the maximum sensor response at a variety of receptor concentrations.

5. DRUG–DNA BINDING

The rational design of new anticancer drugs targeted at DNA requires thorough understanding of the interaction mechanisms of existing antitumor drugs. The most studied anthracycline antibiotics, specifically doxorubicin (adriamycin) and its analogs, represent an important class of anticancer drugs currently in clinical use

[19]. It is generally accepted that doxorubicin binds with DNA by intercalative base stacking and electrostatic interactions at the minor groove [20]. The kinetic parameters of anthracycline–DNA interactions have been obtained by a number of studies [21–25] using stopped-flow and laser time-resolved fluorescent methods. Previous studies in our laboratory have shown that polymeric dextran-conjugated doxorubicin (Figure 3) combined with other antitumor drugs exhibits enhanced cytotoxicity against multidrug-resistant cells [26]. To further characterize the DNA-binding properties of the polymerically-modified doxorubicins, we have synthesized a series of dextran-conjugated doxorubicins that have various molecular weights and determined the association and dissociation kinetics of the binding between these conjugates and an immobilized dsDNA using the IAsys optical biosensor.

5.1 Dextran–Doxorubicin Conjugates

Dextrans of various molecular weights (70 kDa, 200 kDa, and 500 kDa) are first oxidized by sodium periodate to form dialdehyde dextrans in the presence of sodium bisulfite. Doxorubicin is coupled with the dialdehyde dextran polymer under basic conditions [26] and the unreacted aldehyde groups are reduced with excess sodium borohydride. The doxorubicin–dextran conjugates are purified by Sephadex G25 gel filtration and dialysis against distilled water. The coupling efficiency is estimated by UV absorbance at 478 nm (ε_{478} = 13,600 $M^{-1}cm^{-1}$ for doxorubicin), and the concentration of the dextran–doxorubicin conjugate is expressed in terms of doxorubicin content.

The emission of doxorubicin is quenched when the molecule is bound to DNA [27]. Therefore, the affinity of the doxorubicin–dextran conjugates for DNA was examined by fluorescent spectroscopy. The fluorescent spectra of the T200-Dox

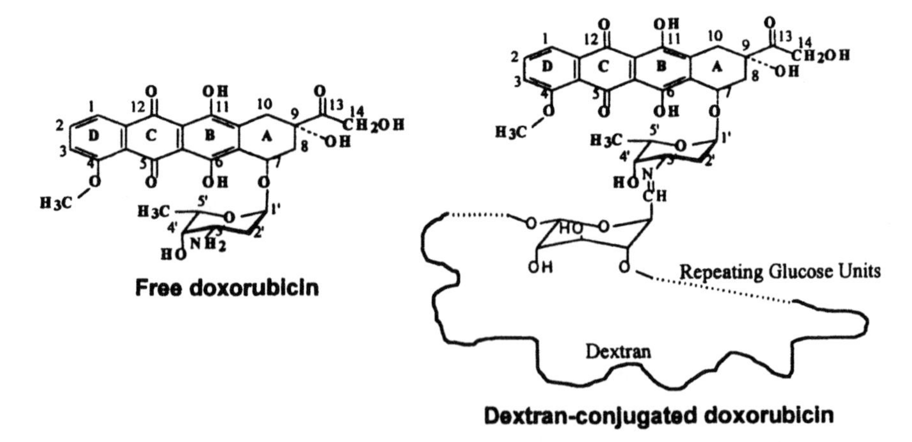

Figure 3. Structures of doxorubicin and dextran-conjugated doxorubicin.

conjugate (doxorubicin conjugated with 200-kDa dextran) in the absence and the presence of calf-thymus double-stranded DNA (ctDNA) are shown in Figure 4. In the presence of ctDNA, the emission intensity of the conjugate decreases as the DNA concentration increases, indicating that the dextran-conjugated doxorubicin still can bind with DNA. The ratios of the fluorescent intensity of the conjugates in the absence of DNA (I_0) and in the presence of DNA (I) were used to calculate the Stern–Volmer quenching constant, k_{sv}, according to $I_0/I = k_{sv}[DNA] + 1$. The value of k_{sv}^{-1} represents the DNA concentration at which 50% of the initial doxorubicin fluorescence is quenched. The k_{sv}^{-1} values estimated for 10 μM solutions of the free, T70-, T200-, and T500-conjugated doxorubicins are 1.2×10^{-5}, 1.5×10^{-4}, 1.0×10^{-4}, and 2.0×10^{-4} M, respectively. The concentration of ctDNA required to reduce 50% of the initial fluorescent intensity of doxorubicin is only about one order of magnitude greater than that for free doxorubicin, despite the considerable steric and dynamic constraints imposed by the dextran in the conjugates. The equilibrium affinity constants of the conjugate-DNA complexes obtained by Scatchard analysis of the fluorescent titration data [18,28] are shown in Table 1. The fluorescent titration experiments demonstrated that the dextran–doxorubicin conjugates bind to DNA with reduced affinities compared to free doxorubicin.

5.2 Kinetic Studies and Data Analysis

The association and dissociation rate constants and the equilibrium binding constants of the conjugated doxorubicin–DNA interactions are determined by using the IAsys evanescent-wave optical biosensor according to the methods described

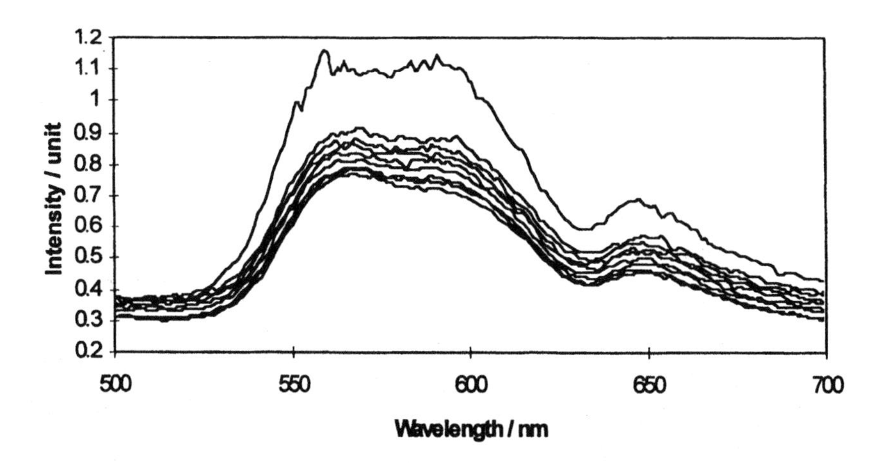

Figure 4. Fluorescent emission spectra of 10 μM of 200-kDa dextran–doxorubicin conjugate in various concentrations of calf thymus DNA. Fluorescent titration experiments were performed on an SLM-Aminco 4800C spectrofluorometer at ambient temperature. λ_{ex} = 480 nm and λ_{em} = 590 nm; DC gain: 10, HV: 800.

Table 1. Kinetic Binding Constants of Various Dextran-Doxorubicin Conjugates and Immobilized dsDNA Determined by Biosensor and Fluorescence Measurements

Dex-Dox Conjugates	$k_{ass}/M^{-1}s^{-1}$ [a]	k_{diss}/s^{-1} [a]	K_D/M [a]	K_D/M [b]
70kDa-Dox	$3.70 \pm 1.09 \times 10^3$	$4.27 \pm 1.36 \times 10^{-2}$	$1.2 \pm 0.5 \times 10^{-5}$	$1.1 \pm 0.6 \times 10^{-5}$
200kD-Dox	$2.45 \pm 0.90 \times 10^3$	$3.68 \pm 0.66 \times 10^{-2}$	$1.5 \pm 0.6 \times 10^{-5}$	$1.3 \pm 0.4 \times 10^{-5}$
500kD-Dox	$3.02 \pm 1.24 \times 10^3$	$3.76 \pm 0.77 \times 10^{-2}$	$1.3 \pm 0.6 \times 10^{-5}$	$0.9 \pm 0.4 \times 10^{-5}$
free	Dox	7.0 ± 10^6 [c]	30 [c]	1.4×10^{-6} [d]

Notes: [a] Determined by the evanescent wave biosensor.
[b] Determined by fluorescence titration experiments.
[c] Ref. 25.
[d] Ref. 28.

previously. Briefly, the CMD matrix is first activated with NHS and the activated carboxyl groups form amide bonds with amine residues in streptavidin in the presence of EDC. Any unoccupied NHS-activated sites are blocked by ethanolamine. A dsDNA that has one of the two strands biotinylated at the 5′-end is prepared by polymerase chain reaction (PCR), using two primers, one of which is 5′-labeled with biotin, and total DNA extracts from *E. coli* strain H10407 as the template [29]. The resulting 275-bp 5′-biotinylated PCR product is captured on the streptavidin-modified sensor surface and the amount of immobilized DNA (334 arc second) is determined by measuring the change in the resonance angle before and after the reaction. Different concentrations of the conjugates in a phosphate buffer (0.0125M Na_2HPO_4, 0.0125M NaH_2PO_4, and 0.1M NaCl at pH 7.0) are injected into the sensor cuvette and allowed to equilibrate for 10–15 min, followed by washing with a phosphate-buffered silane solution containing Tween 20™ at 0.05%(v/v) (PBS/T) until the sensor signal returns to the original baseline. The association and the dissociation events are monitored continuously for different conjugate solutions in real-time.

A typical sensorgram following the binding of the T500-Dox conjugate to the immobilized 275-bp dsDNA is shown in Figure 5. The experimental association data were fitted using equations having one (monophasic) and two (biphasic) exponential terms [Eqs. (1) and (2)]. At all concentrations of the conjugate, the binding curves may be fitted by both monophasic and biphasic models, as judged by the values of the reduced chi-square and the random distribution of the errors. Consequently, the simpler monophasic model was used to obtain the pseudo-first-order rate constants, k_{on}, by fitting individual binding curves to Eq. (1), using a selected "window" of binding data for all of the concentrations (Figure 6). Because the k_{on} value depends on the size of the data set used, the fitting "window" was selected so that the value of the reduced chi square and the random distribution of

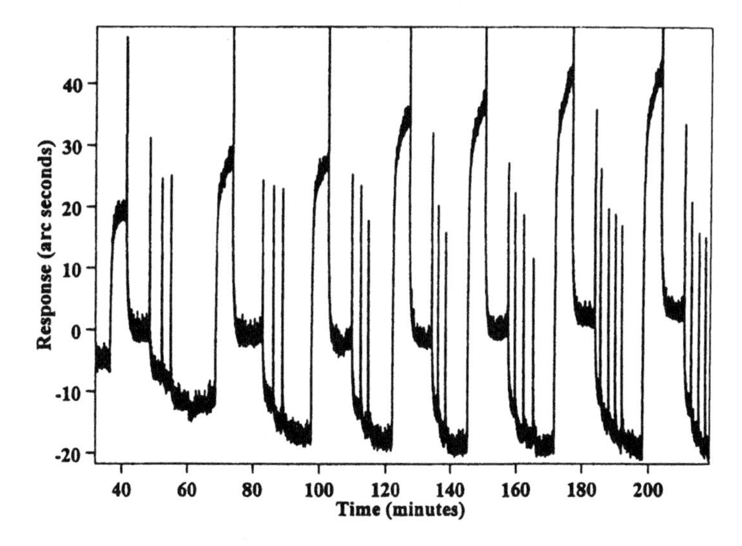

Figure 5. Real-time sensorgram of the interaction between an immobilized dsDNA and various concentrations of dextran(T500)-doxorubicin. The sharp lines correspond to changes of solution/buffer.

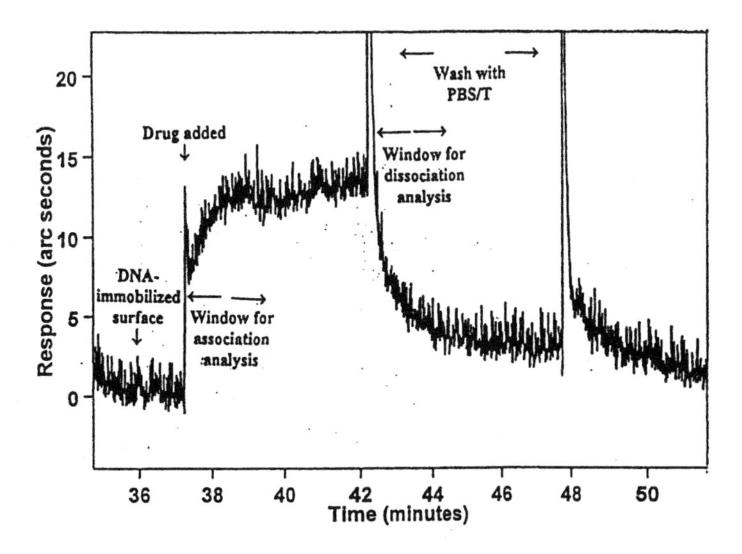

Figure 6. Representative sensorgram showing the binding profile of the dextran(T500)-doxorubicin and an immobilized dsDNA. Data "windows" for kinetic analysis were selected from the association and the dissociation curves.

the errors indicated a best fit (Figure 7). The association constants, k_{ass}, were found from plots of k_{on} versus ligand concentration [Eq. (5)]. Similarly, the experimental dissociation data were also adequately fitted by both monophasic and biphasic models, and the apparent off-rates, k_{off}, were obtained from the monophasic dissociation equation [Eq. (7), Figure 8]. Under monophasic conditions, the dissociation rate constants, k_{diss}, are independent of the concentration of the complex. Although theoretically k_{diss} may be obtained from a single dissociation curve, the values of k_{diss} are better represented by averaging the k_{off} values determined for different ligand concentrations. The values of k_{ass} and k_{diss} and the equilibrium binding constants, $K_D = k_{diss}/k_{ass}$, for the interactions between the various Dex-Dox conjugates and DNA are listed in Table 1. The DNA-binding parameters for free doxorubicin and the conjugates measured in free solution, as determined by Scatchard analysis using the fluorescent titration method, are also listed in Table 1 for comparison.

It is apparent that the doxorubicin-dextran conjugates exhibit different DNA-binding kinetics from those of free doxorubicin. Both the association and the dissociation rate constants of the conjugates are more than three magnitudes smaller than those for free doxorubicin, probably caused by the reduced diffusional mobil-

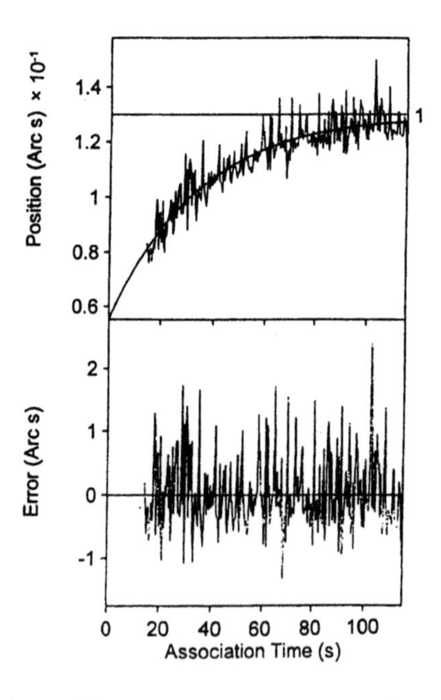

Figure 7. Kinetic analysis of the association curve using the monophasic model (Eq. 1). The residual errors were calculated as the difference between the experimental data and the fitted data.

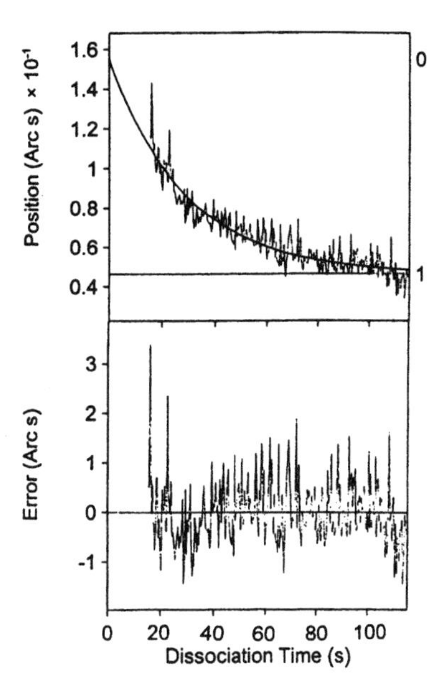

Figure 8. Kinetic analysis of the dissociation curve using the monophasic model [Eq. (7)].

ity of the conjugates. On the other hand, the equilibrium dissociation constants are only about ten times larger for all the conjugates compared to that for free doxorubicin. This indicates that although the conjugate–DNA complexes are less stable than the doxorubicin–DNA complex because of the significant steric constraints presented by the dextran matrices, each conjugate molecule may offer more than one doxorubicin group for DNA binding. Within experimental error, the rate constants and the affinity constants are similar for all of the conjugates, despite the difference in the molecular weights of the various dextrans. This suggests that binding of conjugates to immobilized DNA is controlled by diffusion of the conjugates at the solution-sensor interface and that the diffusion coefficients are similar for the conjugates in the same range of the molecular weights. The affinity constants obtained from biosensor measurements are similar to those from the fluorescent titration experiments and further verify the validity of the biosensor approach.

The kinetic parameters obtained from the nonlinear analysis vary as the amount of binding data selected for curve fitting changes. The number of binding data used for kinetic analysis should be such that a good fit is obtained and no useful information is lost. Typically, the window of the binding data selected for curve fitting starts from about ten seconds after introducing the sample to the sensor surface, allowing the solution-sensor interface to reach a homogeneous state under constant stirring. Data windows of 50–250 second intervals were fitted to the kinetic models and the goodness of fit was determined by the values of the reduced chi-square and the randomness of the errors. In general, a smaller data set gives a better fit, particularly for the monophasic model. For the conjugate-DNA binding system, a window size of 100 seconds was used to conduct both association and dissociation analyses for all of the conjugates.

It is interesting to note that all the association and dissociation curves obtained in this study are fitted satisfactorily by monophasic models for all of the data windows (50–250 seconds), despite the relatively high concentrations (0.5–8.0 μM) of the conjugates used. Many of the biomolecular interactions analyzed by using biosensors, which usually consist of antibody-antigen or receptor-protein bindings, have monophasic association curves only at low free ligand concentration only (<100 nM) and biphasic curves at moderate to high ligand concentration (>100 μM) [8]. There are two possibilities for the observed monophasic ligand-DNA bindings in this study: (1) Because the biotinylated DNA could not diffuse into the dextran matrix easily due to electrostatic repulsion and steric hindrance, the DNA was captured only by the outer layers of the streptavidin within the incubation period. Binding at these outer layers might exhibit monophasic characteristics. (2) The dextran-doxorubicin conjugates were too bulky to penetrate the detran matrix within the time frame of the observation, so that the contribution of the slower binding phase is minimized.

6. PROTEIN–DNA BINDING

In this section, we report some of the preliminary work being carried out in our laboratory in using the IAsys evanescent-wave biosensor to study the kinetics of polymerase-DNA binding. DNA replication is a fundamental growth process for all living organisms [30–33], and DNA polymerase catalyzes the replication of a template DNA chain by extending the primer chain in the $5' \rightarrow 3'$ direction in the presence of deoxyribonucleoside triphosphate (dNTP) (Figure 9). The fidelity of replication is protected by the proofreading activity of the enzyme. Although the kinetic mechanisms of DNA replication have been extensively investigated by a variety of kinetic [34,35], genetic [36–38], chemical [39], and spectroscopic approaches [40–43], most of the methods employed for kinetic studies are discontinuous and usually require nucleotide analogs and radioactive or fluorescent labels. In this study, however, we have prepared sensor surfaces containing DNA template–primer duplexes as substrates and used the IAsys biosensor to measure the kinetic profiles of polymerase–DNA binding. By modifying the DNA substrates with various sequences and determining the binding characteristics of the protein–DNA interaction in the presence of different polymerases, it is hoped that the biosensor technique may provide additional information about the molecular basis of DNA replication and proofreading.

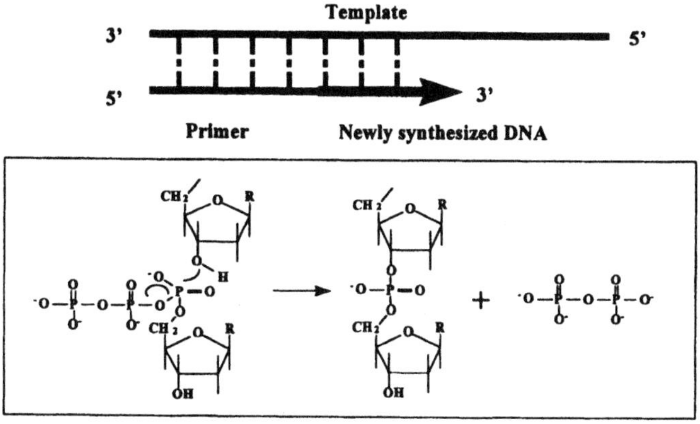

Figure 9. Template-directed DNA replication.

6.1 Experimental Design

DNA polymerase catalyzes the formation of a phosphodiester bond between the 3′-hydroxyl group at the growing end of the DNA primer and the 5′-phosphate group of the incoming dNTP. Growth is in the 5′(3′ direction, and the order in which dNTPs are added is dictated by base pairing to a template DNA chain (Figure 9). The simplest and the best understood DNA polymerase is the DNA polymerase I from *E. coli* [30–33]. DNA polymerase I contains three different active sites on its 103-kDa single polypeptide chain. In addition to its template-directed 5′ → 3′ polymerase activity, DNA polymerase I has a proofreading 3′ → 5′ exonuclease domain that removes mismatched residues at the primer terminus before proceeding with polymerization. It also has an error-correcting 5′ → 3′ exonuclease activity that removes RNA primer and participates in excising of UV-induced pyrimidine dimers. This trifunctional enzyme is split by protease into a 36-kDa small fragment that has all of the original 5′ → 3′ exonuclease activity and a large 67-kDa fragment (the Klenow fragment) that has all of the polymerase and 3′ → 5′ exonuclease activities. Crystal structures of the Klenow enzyme complexed with dNTP [44] and dTMP [45] substrates or with DNA bound at the exonuclease site of the enzyme [46,47] have been determined, and the basic kinetic scheme of Klenow–DNA interactions has been proposed [40–43, 48–50].

The experimental design of this study is shown schematically in Figure 10. Briefly, we used proteins that have various polymerase and exonuclease activities and template–primer duplex DNA with perfectly matched or mismatched sequences for the DNA binding studies. The biotinylated DNA template–primer duplex was immobilized on a streptavidin-activated CMD sensor surface, and the binding profiles of various polymerases and the immobilized DNA are recorded using the optical biosensor in the absence of dNTPs necessary for DNA polymeri-

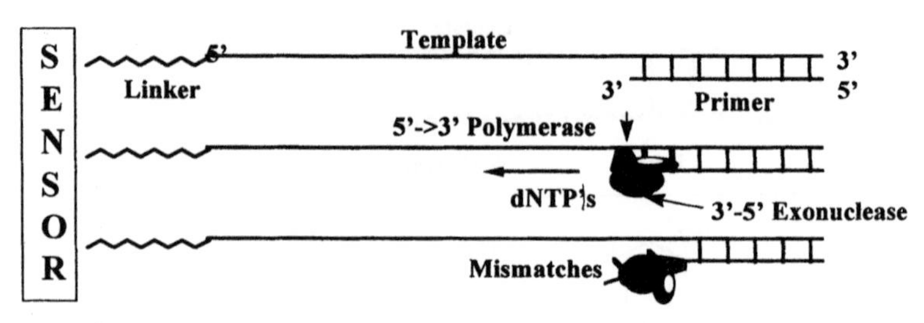

GC-terminus
Template (50-mer)
3′-<u>CGTCGGCAGGTTCCC</u>GTCAGAACCATAGTCAGCGATTGCACGCAACTACG **B**-5′
Primer (15-mer)
5′-<u>GCAGCCGTCCAAGGG</u> (matched)
5′-<u>GCAGCCGTCCAATTT</u> (mismatched)

Figure 10. Top: Schematic representation of the experimental design. Bottom: Synthetic oligonucleotides for constructing the DNA template-primer at the sensor surface. The underlined base pairs indicate where the duplex was formed. The bold face bases indicate where a multiple mismatch was introduced.

zation. In this preliminary study, we prepared two template–primer DNA duplexes for the binding studies. One contained a perfectly matched 15-mer oligonucleotide hybridized with a 5′-biotin-labeled 50-mer oligonucleotide. The other contained three consecutive G-T mismatches at the 3′-terminus of the primer (Figure 10). A fully complementary double-stranded DNA was also prepared as a control. The template–primer DNA that contains two or more consecutive mismatches binds exclusively at the exonuclease site of the Klenow fragment [42]. Therefore, it is expected that the enzyme will bind to the perfectly matched DNA that has the polymerase domain and to the mismatched DNA that has the exonuclease domain. A difference in binding kinetics is expected for enzymes from different sources that vary in activity. Three DNA polymerases were used in the present study, including the well-known Klenow fragment, the T7 polymerase, and the Taq polymerase. T7 DNA polymerase lacks $5' \rightarrow 3'$ exonuclease activity but has $5' \rightarrow 3'$ polymerase activity and $3' \rightarrow 5'$ exonuclease activity that is 250-fold stronger than that of the Klenow fragment. The enzyme digests both single-stranded and double-stranded DNA. In the experiments, the removal of the terminal bases of the primer by the $3' \rightarrow 5'$ exonuclease of the T7 and the Klenow enzymes was prevented by using an excess of the respective dNTP. Taq polymerase is well known for its application in PCR. It has an optimum temperature of 75 °C and withstands temperatures up to 95 °C. It has $5' \rightarrow 3'$ but not $3' \rightarrow 5'$ exonuclease activity. Both its enzymatic activities are greatly reduced at ambient temperature.

6.2 Kinetic Studies

The template-primer DNA is immobilized on the sensor surface according to the biotin-streptavidin chemistry described previously. For ease of comparison, a similar amount of the perfectly matched, mismatched, and fully complementary DNA duplexes is immobilized on the sensor surface for subsequent experiments. The amount of DNA immobilized on the sensor surface (approximately 200 arc second) is estimated by measuring the changes in the sensor response (resonance angle) before and after the reaction. Individual DNA polymerase is injected over the sensor surface in a buffer (50 mM KCl, 1.5 mM $MgCl_2$, and 10 mM tris-HCl, pH 9) with an excess of the base at the 3′-primer terminal (dGTP for the perfectly matched DNA and dTTP for the mismatched DNA) to prevent the excision of the primer terminal bases by the enzymes that have $3' \rightarrow 5'$ exonuclease activity. An equal amount of the polymerase (10 units) is added for all of the experiments. Each polymerase is allowed to equilibrate for 10 min followed by washing with tris and then PBS/T buffers. The association and the dissociation are monitored continuously for each polymerase that has a different immobilized DNA.

The sensorgrams upon binding of Taq, Klenow, and T7 polymerases to the immobilized, perfectly matched template–primer DNA duplex are shown in Figure 11. A fraction of the rapid increases and decreases in the sensor response immediately after injection of the sample and the washing buffer was due to the change in the refractive index of the solutions and nonspecific adsorption/electrostatic interactions. Consequently, kinetic analysis of the association or dissociation data was usually carried out following a 10 second delay after each injection. Because of the differences in the molecular weights of these polymerases and in the refractive indices of respective solutions upon adding the polymerase to the buffer, the relative changes in the resonance angles cannot be used to calculate the absolute amount of protein bound to the DNA. However, the kinetic profiles of the polymerase-DNA interactions shown in Figure 11 clearly demonstrate the different DNA-binding properties among the different enzymes. For Taq polymerase, where little DNA-binding activity is expected at room temperature, the sensorgram shows a gradual association curve following the sample injection, and the enzyme can be readily washed away by repeated washing with the tris buffer. On the other hand, the Klenow enzyme binds to DNA fairly rapidly and dissociates gradually when the sensor surface is rinsed with tris buffer. A certain fraction of the surface-bound Klenow enzyme does not dissociate in the tris buffer but is washed away with the PBS/T buffer. Prolonged association and dissociation curves were recorded for the T7 polymerase, consistent with enzymatic activities much stronger than those of the Klenow enzyme. Similarly, a fraction of the surface-bound T7 enzyme does not dissociate upon repeated rinsing with the tris buffer but is washed away with the PBS/T buffer.

Direct comparison of binding between a specific polymerase with different DNA substrates is possible if the amounts of the immobilized DNA substrates are similar.

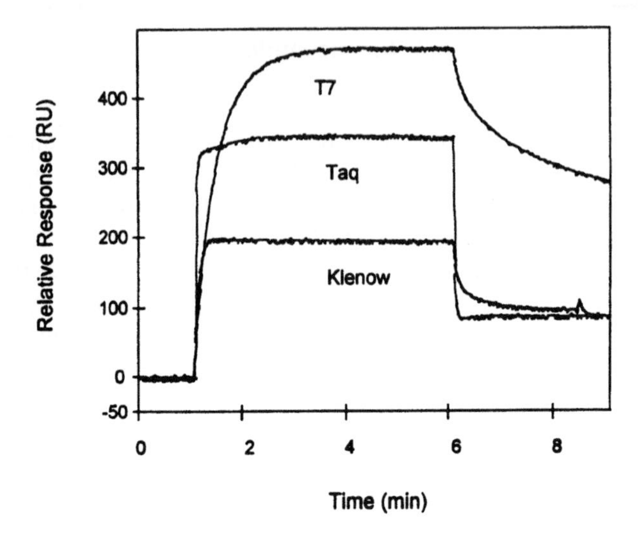

Figure 11. Sensorgrams of the interactions between an immobilized, perfectly matched primer-template duplex DNA and various DNA polymerases.

Figure 12 shows the real-time binding profiles of the Klenow polymerase to the perfectly matched, mismatched, and fully complementary DNA substrates. In each binding experiment, approximately 200 arc seconds of DNA was immobilized on the sensor surface. For the binding of the Klenow enzyme to the mismatched DNA, the sensorgram showed faster association and dissociation rates than to the matched DNA. The amount of the Klenow protein bound to the mismatched substrate after

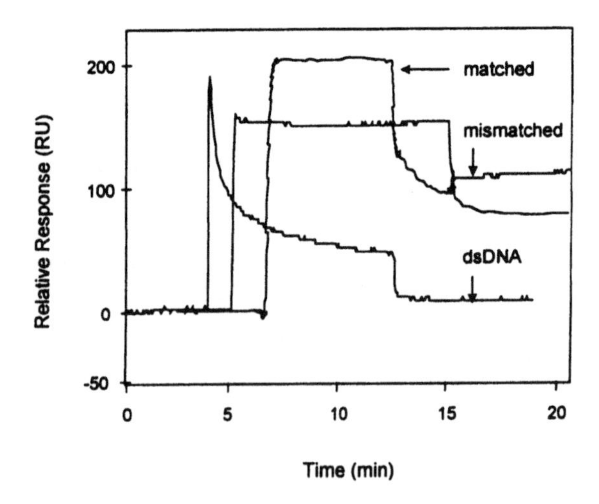

Figure 12. Sensorgrams of the interactions between the Klenow enzyme and various immobilized DNA substrates.

Table 2. Dissociation Constants of Polymerase-DNA Complexes Determined by the Evanescent Wave Biosensor

	k_{diss} / s^{-1}	
	Matched DNA	*Mismatched DNA*
Klenow	$5.2 \pm 0.2 \times 10^{-2}$	$10.3 \pm 1.4 \times 10^{-2}$
T7	$2.1 \pm 0.1 \times 10^{-2}$	$1.5 \pm 0.3 \times 10^{-2}$

washing with the tris buffer was about 30% more than that bound to the matched DNA, as estimated from the increase in the sensor signal. Similarly, the surface-bound protein is completely washed away only by using the PBS/T buffer. This seems to indicate that the exonuclease domain of the Klenow fragment has a higher binding affinity to the mismatched DNA than the binding of the polymerase domain to the perfectly matched primer–template. It is interesting to note that after the Klenow enzyme is injected to the sensor surface with the fully complementary dsDNA, a rapid increase in the sensor response was observed, followed by a gradual decrease in the sensor signal over a period of more than five minutes. The protein is readily washed away with the tris buffer. The observation is consistent with the random-walk model of protein–DNA recognition. Initially, the Klenow enzyme is nonspecifically bound to the DNA, because it is searching for a specific primer–template site. The protein dissociates from the DNA when no specific site is recognized.

Similar sensorgrams were obtained for the binding of the T7 polymerase to different DNA substrates (data not shown). The T7 polymerase has a much higher affinity to the mismatched substrate, consistent with its much stronger exonucleoase activity compared to the Klenow enzyme. It also binds rapidly to the fully complementary dsDNA and dissociates gradually in the same buffer solution. In contrast, the affinity of the Taq polymerase to different DNA substrates is very weak. The enzyme is readily washed away by the tris buffer regardless of the DNA substrate. Because the dissociation rate constant, k_{diss}, is independent of the concentration of the complex, k_{diss} values for different polymerase–DNA complexes were estimated by fitting the individual dissociation curves to the monophasic kinetic equation [Eq. (7)]. A data window of 50 s was selected for each dissociation curve, and the fitting was adequate as judged by the reduced chi-square and the randomness of the error. The dissociation rate constants, k_{diss}, for the interactions between the Klenow and the T7 polymerases with the perfectly matched and the mismatched primer-template DNA duplexes are listed in Table 2.

7. SUMMARY

We have demonstrated the application of the IAsys evanescent-wave biosensor for determining thermodynamic and kinetic parameters of ligand–DNA interactions.

The immobilization of DNA on the sensor surface is readily accomplished by using a biotin–streptavidin linkage on a carboxymethyl-modified dextran hydrogel. The binding profiles of drug–DNA and protein–DNA interactions are monitored in real time without the need for a label. The affinity and kinetic parameters are obtained by using various kinetic models and are consistent with solution measurements. Because of the nature of the dextran hydrogel, monophasic and biphasic kinetic models and different data windows should always be compared to obtain the best fit of the experimental binding data. Nonspecific adsorption of small molecules (such as free doxorubicin) by the matrix was observed, which prevents the use of the biosensor for studying small DNA-binding ligands. The stability of the sensor system needs to be further improved to avoid drift of the baseline observed in some experiments and to achieve more sensitive detection of minute amounts of binding. Finally, development of alternative chemical methods for efficient DNA immobilization will enable the study of ligand–DNA interactions on various types of sensor surfaces.

ACKNOWLEDGMENTS

This work is supported by grants from the Research Grants Council, Hong Kong; the Biotechnology Research Institute, Hong Kong University of Science and Technology; and the Croucher Foundation.

REFERENCES

[1] Lonberg, N., Taylor, L.D. et al. (1994). Antigen-specific human antibodies from mice comprising four distinct genetic modifications. *Nature*, **368**, 856–859.

[2] Calakos, N., Bennett, M.K. et al. (1994). Protein-protein interaction contributing to the specificity of intracellular vesicular trafficking. *Science*, **263**, 1146–1149.

[3] Wu, H., Yang, W.-P., Babbas, C.F. (1995). Building zinc fingers by selection: Toward a therapeutic application. *Proc. Natl. Acad. Sci. USA*, **92**, 344–348.

[4] Colas, P.C.B., Jessen, T. et al. (1996). Genetic selection of peptide aptamers that recognize and inhibit cyclin-dependent kinase 2. *Nature*, **380**, 548–550.

[5] Cush, R., Cronin, J.M., Stewart, W.J., Maule, C.H., Molloy, J., Goddard, N.J. (1993). The resonant mirror: A novel optical biosensor for direct sensing of biomolecular interactions, Part I: Principle of operation and associated instrumentation. *Biosensors Bioelectronics*, **1993**, *8*, 347–351.

[6] Buckle, P.E., Davies, R.J., Kinning, T., Yeung, D., Edwards, P.R., Pollard-Knight, D., Lowe, C.R. (1993). The resonant mirror: a novel optical biosensor for direct sensing of biomolecular interactions, Part II: Applications. *Biosensors Bioelectronics*, **8**, 355–363.

[7] Davies, R.J., Edwards, P.R., Watts, H.J., Lowe, C.R., Buckle, P. E., Yeung, D., Kinning, T., Pollard-Knight, D. (1994). In *Techniques in Protein Chemistry V*, pp. 285–292, Academic Press, San Diego.

[8] Edwards, P.R., Gill, A., Pollard-Knight, D., Hoare, M., Buckle, P. E., Lowe, P.A., Leatherbarrow, R.J. (1995). Kinetics of protein-protein interactions at the surface of an optical biosensor. *Anal. Biochem.*, **231**, 210–217.

[9] Watts, H.J., Yeung, D., Parkes, H. (1995). Real-time detection and quantification of DNA hybridization by an optical biosensor. *Anal. Chem.*, **67**, 4283–4287.

[10] Application Note 3.4, Affinity Sensors.

[11] Lofas, S., Johnsson, B. (1990). A novel hydrogel matrix on gold surfaces in surface plasmon resonance sensors for fast and efficient covalent immobilization of ligands. *J. Chem. Soc., Chem. Commun.*, 1526–1528.

[12] Johnsson, B., Lofas, S., Lindquist, G. (1991). Immobilization of proteins to a carboxymethyldextran-modified gold surface for biospecific interaction analysis in surface plasmon resonance sensors. *Anal. Biochem.*, **198**, 268–276.

[13] Jonsson, U., Ivarsson, B., Lundstrom, I., Berghem, L. (1982). Adsorption behaviour of fibronectin on well-characterized silica surfaces. *J. Colloid Interface Sci.*, **90**, 148–158.

[14] Ghosh, S.S., Musso, G.F. (1987). Covalent attachment of oligonucleotides to solid supports. *Nucleic Acids Res.*, **15**, 5353–5358.

[15] Winzor, D.J., Sawyer, W.H. (1995). *Quantitative Characterization of Ligand Binding*. Wiley-Liss, New York.

[16] O'Shannessy, D.J., Brigham-Burke, M. (1993). Determination of rate and equilibrium binding constants for macromolecular interactions using surface plasmon resonance: use of nonlinear least squares analysis methods. *Anal. Biochem.*, **212**, 457–468.

[17] O'Shannessy, D.J. (1994). *Curr. Opinions Biotechnol.*, **5**, 65–71.

[18] Dahlquist, F.W. (1978). *Methods Enzymol.*, **48**, 270.

[19] Lowen, J.W. (Ed.) (1988). *Anthracycline and Anthracenechone Based Anticancer Agents*, Elsevier, Amsterdam.

[20] Quigley, G.J., Wang, A.H-J. et al. (1980). Molecular structure of an anticancer drug-DNA complex: Daunomycin plus d(CpGpTpApCpG). *Proc. Natl. Acad. Sci. USA* **77**, 7204.

[21] Chaires, J.B., Dattagupta, N., Crothers, D.M. (1985). Kinetics of the daunomycin-DNA interaction. *Biochemistry*, **24**, 260–267.

[22] Chaires, J.B. (1990). Biophysical chemistry of the daunomycin-DNA interaction. *Biophys. Chem.*, **35**, 191–202.

[23] Krishnamoorthy, C.R., Yan, S-F., Smith, J.C., Lown, J.W., Wilson, W.D. (1986). Stopped-flow kinetic analysis of the interaction of anthraquinone anticancer drugs with calf thymus DNA, poly[d(G-C)]·poly[d(G-C)], and poly[d(A-T)]·poly[d(A-T)]. *Biochemistry*, **25**, 5933–5940.

[24] Malatesta, V., Andreoni, A. (1988). Dynamics of anthracyclines/DNA interaction: A laser time-resolved fluorescence study. *Photochemistry and Photobiology*, **48**, 409–415.

[25] Rizzo, V., Sacchi, N., Menozzi, M. (1989). Kinetic studies of anthracycline-dna interaction by fluorescence stopped flow confirm a complex association mechanism. *Biochemistry*, **28**, 274–282.

[26] Fong, W.F., Lam, W., Yang, M., Wong, J.T-F. (1996). Partial synergism between dextran-conjugated doxorubicin and cancer drugs on the killing of multidrug resistant KB-V1 cells. *Anticancer Res.*, **16**, 3773–3778.

[27] Barcelo, F., Barcelo, I., Bavilances, F., Ferragut, J.A., Yanovich, S., Gonzalez-Ros, J.M. (1986). Interaction of anthracyclines with nucleotides and related compounds studied by spectroscopy. *Biochim. Biophys. Acta*, **884**, 172–181.

[28] Chaires, J.B., Dattagupta, N., Crothers, D.M. (1982). Self-association of daunomycin. *Biochemistry*, **21**, 3927–3932.

[29] Kong, Y.C., Dung, W.F. et al. (1995). Co-detection of waterborne bacterial species by multiplex PCR. *Mar. Pollut. Bull.*, **31**, 317–324.

[30] Kornberg, A., Baker, T.A. (1992). *DNA Replication*, W. H. Freeman, New York.

[31] Modrich, P.M. (1987). DNA mismatch correction. *Ann. Rev. Biochem.*, **56**, 435–466.

[32] Kunkel, T.A. (1993). Misalignment-mediated DNA synthesis errors. *Cell*, **267**, 18251–18254.

[33] Goodman, M.F., Creighton, S. et al. (1993). Biochemical basis of DNA replication fidelity. *Crit. Rev. Biochem. Mol. Biol.*, **28**, 83–126.

[34] Kuchta, R.D., Mizrahi, V. et al. (1987). Kinetic mechanism of DNA polymerase I (Klenow). *Biochemistry*, **26**, 8410–8417.
[35] Gupta, A.F., Benkovic, S.J. (1984). *Biochemistry*, **23**, 5874–5881.
[36] Freemont, P.S., Ollis, D.L., Steitz, T.A., Joyce, C.M. (1986). *Proteins*, **1**, 66–73.
[37] Derbyshire, V., Freemont, P.S. et al. (1988). Genetic and crystallographic studies of the 3′,5′-exonucleolytic site of DNA polymerase I. *Science*, **240**, 199–201.
[38] Polesky, A.H., Steitz, T.A., Grindley, N.D.F., Joyce, C.M. (1990). Identification of residues critical for the polymerase activity of the klenow fragment of DNA polymerase I from *Escherichia coli*. *J. Biol. Chem.*, **265**, 14579–14591.
[39] Mohan, P.M., Basu, A., Abraham, K.I., Modak, M.J. (1988). DNA binding domain of *Escherichia coli* DNA polymerase I: Identification of arginine-841 as an essential residue. *Biochemistry*, **27**, 226–233.
[40] Allen, D.J., Darke, P.L., Benkovic, S.J. (1989). Fluorescent oligonucleotides and deoxynucleotide triphosphates: Preparation and their interaction with the large (Klenow) fragment of *Escherichia coli* DNA polymerase I. *Biochemistry*, **28**, 4601–4607.
[41] Guest, C.R., Hochstrasser, R.A. et al. (1991). Interaction of DNA with the Klenow fragment of DNA polymerase I studied by time-resolved fluorescence spectroscopy. *Biochemistry*, **30**, 8759–8770.
[42] Carver, T.E., Hochstrasser, R.A., Millar, D.P. (1994). Proofreading DNA: Recognition of aberrant DNA termini by the Klenow fragment of DNA polymerase I. *Proc. Natl. Acad. Sci. USA*, **91**, 10670–10674.
[43] Frey, M.W., Sowers, L.C., Millar, D.P., Benkovic, S.J. (1995). The nucleotide analog 2-aminopurine as a spectroscopic probe of nucleotide incorporation by the Klenow fragment of *Escherichia coli* polymerase I and bacteriophage T4 DNA polymerase. *Biochemistry*, **34**, 9185–9192.
[44] Beese, L.S., Friedman, J.M., Steitz, T.A. (1993). Crystal structures of the Klenow fragment of DNA polymerase I complexes with deoxynucleoside triphosphate and pyrophosphate. *Biochemistry*, **32**, 14095–14101.
[45] Ollis, D.L., Brick, P., Hamlin, R., Xuong, N.G., Steitz, T.A. (1985). Structure of large fragment of *Escherichia coli* DNA polymerase I complexed with dTMP. *Nature*, **313**, 762–766.
[46] Freemont, P.G., Friedman, J.M. et al. (1988). Co-crystal structure of an editing complex of flenow Fragment with DNA. *Proc. Natl. Acad. Sci. USA*, **85**, 8924–8928.
[47] Beese, L.S., Derbyshire, V., Steitz, T.A. (1993). Structure of DNA polymerase I Klenow fragment bound to duplex DNA. *Science*, **260**, 352–355.
[48] Kuchta, R.D., Benkovic, P.A., Benkovic, S.J. (1988). Kinetic mechanism whereby DNA polymerase I (Klenow) replicates DNA with high fidelity. *Biochemistry*, **27**, 6716–6725.
[49] Donlin, M.J., Patel, S.S., Johnson, K.A. (1991). Kinetic partitioning between the exonuclease and polymerase sites in DNA error correction. *Biochemistry*, **30**, 538–546.
[50] Hochstrasser, R.A., Carver, T.E., Sowers, L.C., Millar, D.P. (1994). Melting of a DNA helix terminus within the active site of a DNA polymerase. *Biochemistry*, **33**, 11971–11979.

RECENT DEVELOPMENTS OF BIOSENSORS

Xian-En Zhang

OUTLINE

ABSTRACT

A review of selected biosensors developed in recent years is presented. These include DNA biosensors, biochemical oxygen demand (BOD) microbial sensors and multien-

Advances in Biosensors
Volume 4, pages 179–193.

ISBN: 0-7623-0073-6

zyme sensors for simultaneously determinating glucose and analogous disaccharides, such as maltose, sucrose, and lactose, on-line control of bioprocesses, and a portable glucose sensor (*GluTest*) for diabetes. Study of microbial respiration is a novel application of biosensor technology, which is evaluated in this chapter.

1. INTRODUCTION

Biosensor research has been carried out at the Wuhan Institute of Virology at the Chinese Academy of Sciences since the early 1980s. This chapter summarizes selected biosensors developed at the Institute, including a DNA sensor, a BOD microbial sensor, multienzyme sensors for the simultaneous determination of glucose and disaccharides, on-line fermentation control, and a portable glucose sensor for diabetes. The principles and methods of biosensor technology have also been used to study microbial physiological subjects, which have been reported only rarely to date.

2. DNA SENSORS

Nucleic acid hybridization is a basic method in molecular biology that provides new possibilities in various biomedical and biotechnological fields. Radiolabeled DNA probes have long been used in conventional hybridization. Unfortunately, radioactive labels are shortlived and require special handling. Recently, various techniques have been developed to replace traditional methods that use radioactive substances, including DNA biosensors which became attractive for their speed, low cost, and direct detection. Many technologies and principles, such as piezoelectrocity [1], surface plasmon resonance [2,3], optic fiber and chemiluminescence [4,5], and electrochemistry [6–8], were employed to develop DNA sensors.

Our DNA sensor employs nonradioactive probes, sandwich solution hybridization, magnetic bead capture, flow-injection analysis, and a chemiluminescence reaction. Two sets of deoxyribonucleic acid probes were exploited. One set was biotinylated probes prepared by using the method described by Reisfeld et al. [9]. We also used calf alkaline phosphatase (CAP) probes prepared according to Renz et al. with modification [10]. The biotinylated probes were produced to capture the probe-analyte complex on the streptavidin-coated magnetic beads in sandwich-type nucleic acid analysis. The CAP-labeled probes were used to ultimately detect the analyte. Both sets of labeled ssDNA (single-strand DNA) were added in a five-fold molar excess over the maximum target fragment concentration. Hybridizations were carried out in plastic Eppendorf tubes under the following conditions: 10 μL of 20 × SSC (NaCl and sodium-citrate solution, pH 7.0), 5 μL of 100 × Dengardt's solution, shaken for 100 min at 37 °C in a final volume of 20 μL. A "no target" control was run with each series. Paramagnetic streptavidin-coated particles were prewashed with 0.5 × SSC. A 15 μL aliquot of beads (0.15 mg) was added to each sample (hybridization products) and incubated for 30 min at room temperature. After washing, the mixture was thoroughly mixed, and the suspension was pumped into the flow injection cell of the working system of the DNA biosensor. Beads with captured DNA were trapped at the bottom of the cell by using a magnetic separation stand, which caused them to adhere to the cell. The time taken to stop the flow depends on the flow rate and distance and was optimized to ensure that the bead-capture mixture could be thoroughly incubated in the chemiluminescent buffer containing CSPD™ (disodium 3-(4-methoxyspiro(1,2-dioxetane-3,2′-(5′-chloro)tricyclodecan)-4-yl)phenyl phosphate), a chemiluminescent substrate of CAP. Light output was obtained on a photodynamometer. Output was given as a full integral of the light produced during a specific time. After measurement, the magnet was removed and beads were ejected by washing with buffer, enabling the subsequent sampling and detection.

The DNA sensor was used to detect the phoA gene (*E.coli* alkaline phosphatase gene) and HBV (hepatitis B viral) DNA. Ten pg of phoA and 20 pg HBV DNA were detected with the method. There was a positive output signal only in the presence of the target DNA. The technology presented here is superior to traditional methods in its speed, simplicity, stability, and relative safety. All of these make it especially valuable in routine medical diagnosis and in basic research.

3. BOD SENSORS

Since the first BOD sensor was reported by Karube et al. [11], subsequent investigations have been made by many authors (for example, [12–16]). Compared to the conventional method (BOD_5, five-day measuring cycle) and some other modified methods, the BOD sensor features mainly a short measuring time, normally less than 20 minutes. However, although commercial BOD sensors were launched some time ago by some Japanese and German companies and the Industrial Standard BOD sensor was built in Japan, sales of BOD sensors have been limited. One of the reasons is that the composition of wastewater varies from site to site whereas the working microorganisms are limited to a few strains. Each of the strains, determined by a genome, can take up a certain range of organic substances, so there is plenty of room for developing a series of BOD sensors.

The BOD sensor we studied worked well for measuring petrochemical industry effluent. The bacterial strain *Pseudomonas sp.* A4, isolated from the active sludge of a petrochemical wastewater control plant, was the working strain. The respiratory activity of the strain in response to the petrochemicals was induced by culturing the strain in an artificial medium composed of domestic sewage and industrial wastewater (1:1). The sandwich method was selected to prepare the sensor membrane. Typically, one milligram of cells was adsorbed onto a sterile filter membrane (0.2 μm pore size) by slight suction. Then, the membrane was fixed onto the tip of an oxygen electrode. The bacterial layer was between the filter membrane and the gas-permeable membrane of the oxygen electrode. After stabilization, the sensor was ready for detection.

The working system consisted mainly of a microbial electrode, a thermal bath with a measuring cell, a magnetic stirrer, sampling and washing tubing, an amplifier, an A/D converter, a microprocessor and digital circuit, a digital display and a printer [18]. The working temperature was 30 °C, and GGA solution [a mixture of glucose (150 mg/L) and glutamic acid (150 mg/L)] was used as the BOD standard. When using the steady state measurement method, the response time was 5 to 15 minutes, depending on the concentration of the sample solution. The washing time was about equal to the response time or longer.

More than 50 pure chemicals were selected to estimate their BOD values using the BOD sensor. Table 1 compares selected data produced by the BOD sensor (BOD_e) and data recorded in the literature using the conventional method (BOD_5). From the data, we concluded that the average BOD_e for glucose and glutamic acid is 0.68 (g/g), and for GGA is 0.67 (g/g). This showed that GGA is a suitable standard for the BOD test. BOD_e and BOD_5 were comparable for about 50% of the compounds tested. There was significant difference between BOD_e and BOD_5 for some compounds because of the complexity and variety of microbial metabolism.

The sensor was used to estimate the BOD of wastewaters from various sources. The correlation between BOD_e and BOD_5 is shown in Table 2.

Other features of the microbial sensor were closely examined. The linear range is up to 30 mg/L, and the coefficient of variation (CV) is 5% at a concentration of 10 mg/L BOD in GGA solution. The sensor activity is not inhibited by Cr^{2+}, Mn^{2+}, Fe^{2+} or Co^{2+} at a concentration of 50 mg/L but it decreases when exposed to Zn^{2+}, Hg^{2+}, or Cu^{2+} at the same concentration. The critical points of concentrations that inhibit the sensor response are 20 mg/L for Hg^{2+} and 5 mg/L for Cu^{2+}. The sensor was used for BOD estimation at the petrochemical waste control plant for at least three months without changing the microbial membrane, thus demonstrating its long lifetime.

4. SIMULTANEOUS DETERMINATION OF GLUCOSE AND ANALOGOUS DISACCHARIDES

Determination of glucose and disaccharides is important in the food and brewing industries. The glucose electrode is well developed and has been widely used, but sensors for disaccharides, such as maltose, sucrose, and lactose, are still at the laboratory stage. Some sequential enzyme electrodes are proposed for determining disaccharides that contain at least one glucose residue. The principle is as follows:

$$\text{Disaccharide} + H_2O \xrightarrow{\text{Appropriate enzyme}} \text{Glucose}$$

Table 1. Comparison of BOD Values of Pure Compounds Determined by the Microbial Sensor and Those Determined by the Conventional Method (Literature Data)

Compound	BOD_e, *g/g*	BOD_5,[a] *g/g*
o-Cresol	1.50	1.54
p-Dehydroxybenzene	0.45	0.48
o-Dehydroxybenzene	1.09	0.49–1.20
m-Dehydroxybenzene	0.51	1.15
Dodecane	0.04	0.28
Ethanol	1.28	1.07–1.18
Galactose	0.32	0.21–0.67
Glucose	0.30	0.3–0.60
Glutamate	1.06	0.52–0.64
Glycerol	0.79	0.37–0.81
Glysine	0.23	0.55–1.03
Hexadecane	0.08	–
Lactose	0.00	0.4–0.63
Maltose	0.45	0.2–0.62
Mannitol	0.62	0.68
Methanol	1.09	0.85–0.91
Octane	0.26	–
Phenol	1.31	0.93–1.10
Proline	0.83	0.23
Sodium benzoate	0.26	1.25
Sodium SDS	0.06	1.20
Sodium oleate	0.50	1.2–1.76
Sodium acetate	1.01	0.63–1.23
Sodium citrate	0.00	0.23–0.61
Sodium calicylate	0.19	0.97
Soluble starch	0.06	0.51
Sucrose	0.11	0.28–0.44

Note: [a]Data collected from [31–33].

$$\text{Glucose} + O_2 \xrightarrow{\text{GOD(glucose oxidase)}} \text{Gluconate} + H_2O_2$$

Glucose is detected by a GOD electrode based on the measurement of O_2 consumed or H_2O_2 produced. Disaccharides can be detected by sequence enzyme electrodes that are made of a GOD electrode combined with the appropriate enzymes (Table 3).

Table 2. Comparison of BOD Values Determined by Microbial Sensor and Conventional Method

Sample	*BOD, mg/L*	*Sample Size*	*Correlation Coefficient of BOD_e and BOD_5*
Wood preservative waste	< 100	n = 9	0.767
Petrochemical waste	< 400	n = 14	0.979
City domestic waste	> 200	n = 16	0.821
Food industry waste	> 1000	n = 8	0.824

When measuring any of these disaccharides, however, sequence enzyme electrodes, suffer significantly from interference of the coexisting glucose. We found that some di- or polysaccharides produce glucose by spontaneous hydrolysis after dissolving for a period of time, causing difficulty in calibrating the sequence electrode and in measurement. Distinguishing glucose from disaccharides becomes essential. Another point is that the sequence enzyme electrodes cannot meet the requirement of total sugar measurement, which is important in many cases. For example, in the fermentation industry total reducing sugar (mainly glucose and maltose or sucrose) is measured to estimate the kinetics and balance of bioreaction processes.

To solve these problems, the dual-electrode enzyme sensor method is suggested, which is the concurrent use of a bienzyme maltose electrode and a GOD electrode. Because the GOD electrode measures only glucose and the sequence electrode measures the sum of sugars, the concentration of either sugar can be calculated from the following formula [19]:

$$C_g = \frac{R_g}{A'_g}$$

$$C_d = \frac{\left(R_t - \frac{A_g \times R_g}{A'_g}\right)}{A_d}$$

Table 3. Enzymes Used in Sequence Enzyme Electrodes

Analyte	*Enzyme*
Glucose	GOD
Maltose	GOD + amyloglucosidase or maltase
Sucrose	GOD + invertase + mutarotase
Lactose	GOD + galactase

Table 4. Summary of Various of Dual-Electrode Enzyme Sensors

Analyte	*Transducer*	*Working System*	*Detection Limit mmol/L*		*Reference*
			Glucose	*Disaccharide*	
Glucose and maltose	DO electrode	FIA	0.3	0.3	[19]
	Mediator electrode	Disposable	1.5	2.0	[34]
Glucose and sucrose	Carbon electrode	FIA	0.5	1.0	[35]
Glucose and lactose	Carbon electrode	FIA	1.0		[36]

where C_d and C_g are the respective concentrations of disaccharide and glucose in the mixture; R_t and R_g are the respective responses of the sequence electrode and the GOD electrode; A_g and A'_g are the respective calibration coefficients of the sequence electrode and the GOD electrode to glucose; and A_d is the calibration coefficient of the sequence electrode to disaccharide. The principle of this method was first proposed in 1980 [20] and has been used in our laboratory to develop a series of biosensor systems.

The working systems are basically electrochemical and designed either as FIA (flow-injection analysis) or disposable types. Table 4 summarizes the features of the sensors. Simultaneous determinations of glucose and disaccharides were carried out with mixtures of glucose, disaccharides, and real samples. As predicted, disaccharides have no effect on glucose determination. The maximum concentration of disaccharides determined depends strongly on the concentration of the coexisting glucose. The higher the glucose concentration, the lower the maximum disaccharide concentration that can be measured accurately. Variation coefficients of the sensors range from 2% to 7%. Factors affecting sensor response were investigated closely. These sensor systems were used to simultaneously determine glucose and sucrose daily in sweet soft drinks, glucose and maltose in starch hydrolysis [21], and glucose and lactose in milk. We were informed by the Dongxihu Brew Company that by adjusting the ratio of glucose and maltose with our sensor system, the utilization of starch hydrolytes increased from 62–68% to 66–72% in fermentation and that the quality of dry beer is improved greatly.

5. GLUCOSE ON-LINE CONTROL BASED ON THE ENZYME ELECTRODE SYSTEM

Feed-batch culture is an important technique for obtaining high yield, and it is widely used in the fermentation industry, where the substrate is fed to the bioreactor at appropriate stages either continuously or at intervals during the fermentation process. There are two categories of feeding [22], nonfeedback control and feedback control. The former depends greatly on the experience of the operator. Ideally, mathematical models should be developed for optimum operation. Unfortunately, it is not always the case because many factors affect fermentation. Therefore, the course of fermentation often deviates from the predicted. Feedback control is based on sensor technology. If the growth rate or substrate consumption is related to the pH change, dissolved oxygen tension, or CO_2 concentration in exhaust, the appropriate sensor for one of these parameters can be used to control the feed rate so as to control the cell growth or substrate uptake rate indirectly. In recent years, the glucose sensor has been used to measure glucose directly in fermentation. Glucose is the most widely used carbon source for microorganisms in fermentation, and therefore, direct on-line control of glucose is of great interest.

The author reported on a glucose electrode [23] and an on-line measuring system [24], from which the on-line control system (FAS/S) was later built [25]. Figure 1 is a diagram of the working system for an on-line feed-batch control. One can set the control point of glucose concentration through the system keyboard. Fermentations were carried out in 2 or 5 liter reactors. The fermentation broth was pumped continuously through an organic membrane sampling filter (ABC), a ceramic membrane sampling filter [26] or a membrane flow cell [27] at a flow rate of 0.5–1.6 mL/min. Fifty μL of the filtrate is injected into the carrier stream (phosphate buffer or water) at three minute intervals by an autoinjection valve. The main stream of the filtrate flowed back to the bioreactor through a sterile filter (pore size 0.45 μm). After sample injection, the glucose electrode responded immediately, the signals were processed by the computer built into the system. If the measured glucose concentration is lower than the set point, the feeding pump is activated by the output

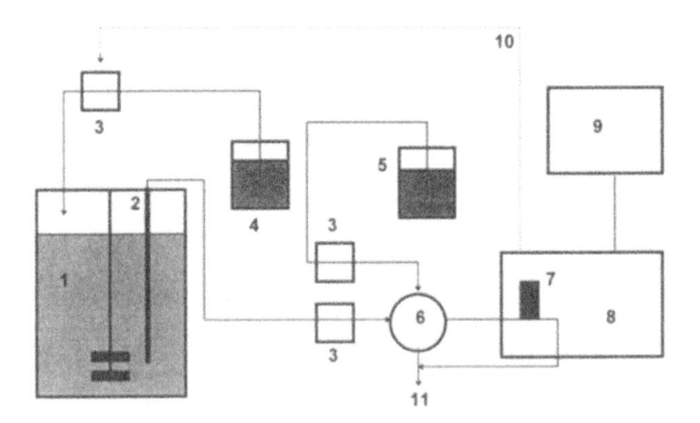

Figure 1. Diagram of the enzyme sensor system for on-line feedback control of fermentation. **1.** Fermentation; **2.** Sampling filter; **3.** Peristaltic pumps; **4.** Stock solution for feeding; **5.** Maintenance buffer; **6.** Injection valve; **7.** Enzyme electrode; **8.** Analyzer/controller; **9.** Chart recorder; **10.** Feedback control signal; **11.** Waste.

Table 5. Relationship of Glucose Level and Growth Parameters of *E.coli* 11130

Glucose Concentration, mmol/L	*Specific Growth Rate, 1/h*	*Lag Phase, h*	*Exponential Phase, h*	*Biomass Production Rate, g/L/h*[a]
2	0.470	2	7	0.42
5	0.655	3	5	0.52
10	0.965	4	3	0.46
15	1.210	5	3	0.36

Note: [a]Calculated at the 12th hour of fermentation.

signal of the system. The feeding of glucose is by an on-off control model. The precision of control depends mainly on the measurement delay, including the distance of sample flow between the bioreactor and the enzyme electrode and the response and recovery time of the enzyme electrode. In investigating the working system, the delay time was about 3 minutes and the precision of glucose control was about 1.5 mmol/L at the control levels of 2, 5, 10, and 15 mmol/L.

The working system was tested with an *E.coli* fermentation. The set points of glucose control were 5 and 10 mmol/L, respectively. Glucose concentrations determined on-line and off-line agreed well with each other. The concentration of feeding glucose was ten times that of the starting culture, which was very concentrated so that there was no significant change in the working volume of the fermentation during initial feeding. This allows us to investigate feed-batch culture kinetics which is important but difficult by the standard method. Table 5 shows some growth kinetic data for *E.coli* fermentation under glucose control. By the couple reciprocal method, using the data in Table 5, the calculated substrate (glucose) saturation constant (K_s) of the microorganism under this condition is 10.39 mmol/L. The working system was also used in glutamate fermentation, the process that employs feed-batch culture to increase productivity. More features of the working system were explained elsewhere [25].

6. STUDY OF MICROBIAL RESPIRATION BASED ON THE BIOSENSOR METHOD

In addition to detecting specific substances, the biosensor method can also be used to study microbial physiology, e.g., K_s (substrate saturation constant) estimation [28]. To optimize wastewater control, many investigators have long been interested in the microbial cometabolism of industrial and domestic waste. Phenol and glucose were selected for study as typical substances of both types of waste.

Three phenol-degrading microorganisms, *Pseudomonas* sp. A4, *Nocardia* sp. N4-1, and an unidentified bacteria S4, were selected for the experiment. Each strain was cultured in a medium containing glucose and phenol, allowing full induction of the necessary enzymes. The harvested cells were washed and used to make individual microbial sensors. Preparation of the sensors and working system was the same as in the method for the BOD sensor. Three kinds of solutions, phenol (1 mg/L), glucose (1 mg/L) and a mixture of the two substances (1 mg/L each) were prepared with a phosphate buffer. The microbial sensors were equilibrated in a phosphate buffer with constant stirring. After the output of the microbial sensors became steady, the sample was injected and the initial response slope ($R = dI/dt$) was recorded for 1 min. The sensor and measuring cell were washed after each measurement. R_p, the response value to phenol, R_g, the response value to glucose, and R_{p+g}, the response value to the mixture of phenol and glucose were obtained separately. Then, $(R_p + R_g)$ and R_{p+g} were compared. If there was no obvious difference between $(R_p + R_g)$ and R_{p+g}, both phenol and glucose would be taken up by the experimental strain without influencing each other. Otherwise, the utilization of glucose by cells might be influenced by phenol, or vice versa.

The results are shown in Table 6. Three types of interactions between coexisting phenol and glucose, when used as microbial substrates, were found in the investigation: noninteraction, coordination, and depression. These inferences were confirmed in an investigation by the conventional flask culture method. We found that strain A4 utilized both substrates simultaneously in the medium containing phenol and glucose, whereas strain S4 utilized phenol first and then glucose. This means that phenol depresses the degradation of glucose in the S4 culture. Degradation of phenol and glucose, when coexisting in strain N4-1, slightly increases compared to their degradation when they are used as sole carbon sources in culture medium. The results obtained from both the sensor method and conventional method tally with each other, revealing the range of metabolic pathways in the microbial world. Degradation of glucose and phenol in bacterial S4 cultures is shown as an example

Table 6. Comparison of Response Values

Working Strain	*Substrate*	R_x[a]	R_{p+g}	$(R_p + R_g) - R_{p+g}$	*Interaction*
A4	Phenol	33.3			
	Glucose	54.0			
	Phenol & Glucose	87.7	87.3	0.4	None
N4-1	Phenol	22.7			
	Glucose	67.3			
	Phenol & Glucose	101.3	90.0	11.3	Coordination
S4	Phenol	41.5			
	Glucose	31.2			
	Phenol & Glucose	57.2	74.2	–16.7	Depression

Note: [a]R_x represents R_p or R_g. Refer to the text for the definitions of R_p, R_g, and R_{p+g}.

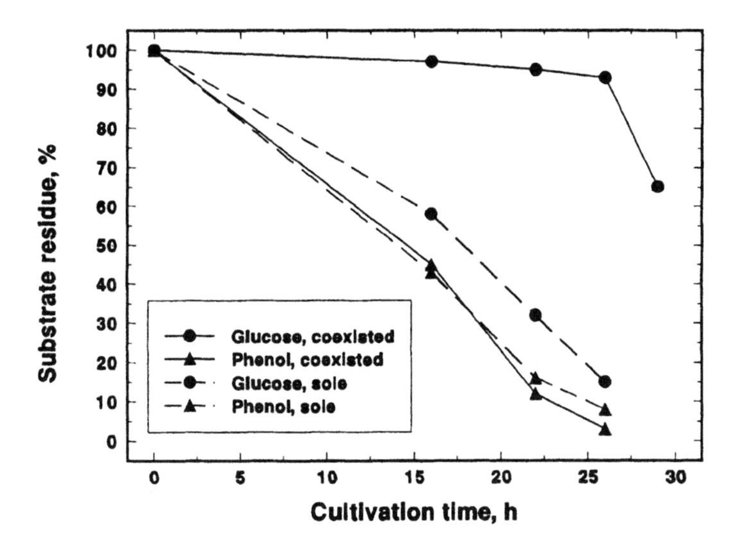

Figure 2. Degradation of glucose and phenol in bacterial S4 cultures. The data presented provide evidence that the results obtained from the biosensor method tally with the conventional shaking flask method. Solid lines represent the degradation of both substrates as carbon sources in the same culture medium. Short dashed lines represent the degradation of both substrates as sole carbon sources in different culture media.

in Figure 2. Furthermore, the question why people obtain different results from their own microorganisms can be answered. Compared with the shaking flask experiment, the biosensor method is fast, easy, sensitive, and automatic. This method can be expected to be exploited for many similar purposes.

7. *GLUTEST*—A PORTABLE GLUCOSE SENSOR FOR DIABETICS

According to a WHO report [29], the incidence of diabetes has been increasing. The total number of cases may reach 100 million around the world. Thus, the portable glucose sensor is important for family care of diabetics. There are already some types of portable glucose sensors commercially available. Because these sensors are disposable, the measurement may be incorrect if the working temperature varies much from the temperature used to calibrate the sensors. We have solved this problem by a temperature compensation method. *GluTest*, our solution (Figure 3), is manufactured by Comet (Keyuan) Bioelectronics Technology Co. Ltd. in Wuhan, China. The sensor strips are made by combining the screen printing technique and the mediated enzyme electrode method. All features of the analyzer

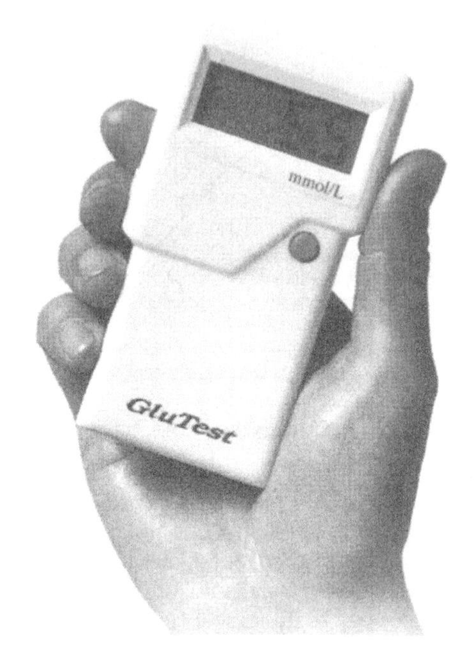

Figure 3. *GluTest,* a potable glucose analyzer for diabetics.

Table 7. Temperature Compensation from One Batch Measurement of Standard Glucose Sample Using *GluTest*

Temperature, °C	*Measurement of Glucose, mmol/L*	*Relative Error,[a] %*
4	5.59	101
12	5.45	99
25	5.55	100
40	5.34	96

Note: [a]Compare to the result obtained at 25 °C.

meet the requirements of the clinical analysis standard. Table 7 shows the temperature compensation result in one batch measurement [30].

8. CONCLUSIONS

The biosensors presented in this chapter show their great potential in applications. Some of them are close to mass production, e.g., the disposable electrode for simultaneously determining glucose and analogous disaccharides based on the screen printing technique. *GluTest* is already manufactured. The method proposed for studying microbial respiration can be used to investigate a series of physiological phenomena.

ABBREVIATIONS

BOD	Biochemical oxygen demand
BOD_5	BOD value determined by the conventional five-day method
BOD_e	BOD value determined by the biosensor method
CAP	Calf alkaline phosphatase
CSPD™	Disodium 3-(4-methoxyspiro(1,2-dioxetane-3,2′-(5′-chloro)tricyclodecan)-4-yl)phenyl phosphate
FIA	Flow-injection analysis
GGA	Mixture of glucose (150 mg/L) and glutamic acid (150 mg/L)
GluTest	Commercial name for a portable glucose analyzer for diabetics
GOD	Glucose oxidase
HBV	Hepatitis B viral
K_s	Substrate saturation constant
phoA gene	*E.coli* alkaline phosphatase gene
SSC	NaCl and sodium-citrate solution, pH 7.0
ssDNA	Single-strand DNA

ACKNOWLEDGMENTS

The research was supported mainly by the Chinese Academy of Sciences, the National Natural Science Foundation, the Wuhan Committee for Science and Technology, and the National Committee for Planning. The author would also like to thank his group members for their contributions to the research.

REFERENCES

[1] Fawcett, N.C., Evans, J.A., Chien, L.C., Flowers, N. (1988). A piezoelectric biosensor for gene-probe. *Anal. Lett.*, **21**, 1099–1114.

[2] Devries, E.F.A., Schasfort, R.B.M, Vanderplas, J., Greve J. (1994). Nucleic acid detection with surface plasma resonance using cationic latex. *Biosensors & Bioelectronices*, **9**, 509–514.

[3] Evans, A.G., Charles, S.A. (1990). The application of rapid, homogeneous biosensor based on surface plasma resonance to clinical chemistry, DNA probes and immunoassay. *1st World Congress on Biosensors (Abstracts)*, Elsevier, p. 223.

[4] Piunno, P., Krull, U.J., Hudson, R., Damha, M.J., Cohen, H. (1995). Fiber optical DNA sensor for fluorometric nucleic acid determination. *Anal. Chem.*, **67**, 2635–2643.

[5] Wu, M., Xia, S.Z., Ren, S. (1996). Novel oligodeoxynucleotide conjugatives for the fiber optical sensor. *4th World Congress on Biosensors (Abstracts)*, Elsevier, p. 26.

[6] Hashimoto, K., Ishimori, Y. (1994). Novel DNA sensor for electrochemical gene detection. *Anal. Chim. Acta*, **286**, 219–224.

[7] Millan, K.M., Saraullo, A., Mikelsen, S.R. (1994). Voltammetric DNA biosensor for cystic fibrosis based on a modified carbon-paste electrode. *Anal. Chem.* **66**, 2943–2948.

[8] Palanti, S., Marrazza, G., Mascini, M. (1996). Preparation of a voltammetric DNA sensor based on electroactive hybridization intercalator. *4th World Congress on Biosensors (Abstracts)*, Elsevier, p. 69.

[9] Reisfeld, A., Rothenberg, F.M., Bayer, E.A., Wilchek, M. (1987). Nonradioactive hybridization probes prepared by the reaction of biotin hydrazide with DNA. *Biochem. Biophys. Res. Commun.*, **142**, 519–526.

[10] Renz, M., Jurz, C. (1984). A colorimetric method of DNA hybridization. *Nucleic Acids Res.*, **12**, 3435–3444.

[11] Karube, I., Matsunaga, T., Mitsuda, S., Suzuki, S. (1977). Microbial electrode BOD sensors. *Biotechnol. Bioeng.*, **19**, 1535–1547.

[12] Hikuma, M., Suzuki, H., Yasuda, T., Karube, I., Suzuki, S. (1979). Amperometric estimation of BOD by using living immobilized yeasts. *Eur. J. Appl. Microbiol. Biotechnol.*, **8**, 289–297.

[13] Kulys, J., Kadziauskiene, K. (1980). Yeast BOD sensor. *Biotechnol. Bioeng.*, **22**, 221–226.

[14] Strand, S.E., Carlso, D.A. (1984). Rapid BOD measurement for municipal wastewater samples using a biofilm electrode. *JWPCF*, **56**, 464–467.

[15] Riedel, K., Renneberg, R., Kuehn, M., Scheller, F. (1989). A fast estimation of biochemical oxygen demand using microbial sensors. *Appl. Microb. Biotechnol.*, **28**, 316–318.

[16] Qian, Z.R., Tan, T.C. (1998). Response characteristics of a dead-cell BOD sensor, *Water Research*, **32**, 801–807.

[17] Zhang, X.-E., Wang, Z.-T., Jian, H.-R. (1986). Study on microbial BOD sensor. *Acta Scientiae Circumstantiae* (Chinese), **6**, 184–192.

[18] Wang, Z.-T., Zhang, X.-E., Shong, D.-L., Wang, K.-F., Zhang, S.-Q., Jiang, X.-Y., Zhang, X.-L. (1987). BOD analyzer based on microbial sensor. Contract report (unpublished).

[19] Qu, H.-B., Zhang, X.-E., Zhang, S.-Z. (1995). Simultaneous determination of maltose and glucose using a dual-electrode flow injection system. *Food Chem.* **52**, 187–192.

[20] Pfeiffer, D., Scheller, F., Janchen, M., Bertermann, K., Weise, H. (1980). Bienzyme electrodes for ATP, NAD^+, starch and disaccharides based on a glucose sensor. *Anal. Lett.*, **13**, 1179–1120.

[21] Qu, H.-B., Zhang, X.-E., Chui, W.-W., Zhang, S.-Z., Li, G.-X., Oyang, F. (1995). Kinetic study on starch hydrolysis using a glucose/maltose sensing system. *Biotechnol. Tech.*, **9**, 445–450.

[22] Zhang, X.-E. (1992). Feed-batch culture: Principle, methods and applications. *Trends Biotechnol.* (Chinese), **2**, 31–34.

[23] Zhang, X.-E., Zhang, X., Xia, X.-M., Zhang, Y.-Q., Hu, G.-S. (1991). Determination of glucose by an enzyme electrode flow injection analysis system. *Chin. J. Biochem.*, **6**, 294–300.

[24] Zhang, X.-E., Zhang, Z.-P., Zhang, X.-M., Hu, W.-P., Tang, J.-K. (1995). On-line measurement of glucose based on enzyme electrode system. In *Collected Papers in Biotechnology* (Chinese), Chinese Chemical Engineering Publisher, p. 22–27.

[25] Jia, Z.-J., Zhang, X.-E. (1996). On-line control of glucose concentration based on enzyme electrode system. *Chin. J. Biotechnol.*, **12**, 355–358.

[26] Zhang, Y., Zhang, X.-E., Zhang, Z.-P., Zhang, X.-M., Wei, H.-P., Zhou, Y.-H. (1996). Multifunctional enzyme sensor system. Contract report (unpublished).

[27] Zhang, X.-E., Hu, W.-P., Zhao, G.-Q., Zhang, Z.-P., Zhang, X.-M., Gui, Y.-Q., Zhang, X., Wei, H.-P. (1994). A flow dialysis cell for on-line measurement of glucose in fermentation broth. *Chin. J. Biotechnol.*, **9**, 161–170.

[28] Zhang, X.-E., Shong, D.-L. (1991). Ks estimation using biosensor method. *J. Microbiol.* (Chinese), **11**, 34–37.

[29] World Health Organization (Ed.). (1994). Prevention of diabetes. *WHO report* No. 844, Translated in Chinese by Wu, G.-H. et al., People's Health Publisher, 1996.

[30] Zhang, X.-E., Zhang, Z.-P., Ge, F., Hu, T., Xia, C.-L. Huang, Y.-Y., Zhang, X.-M., Sun, J. (1997). *GluTest*: A full temperature compensated portable glucose analyzer for diabetes. Contract report (unpublished).

[31] Gu, Q.-X. (1984). Biodegradable organic substances. *Inf. Environ. Sci.* (in Chinese), **2**, 67.

[32] Shi, Z.-T. (1984). Oxygen demand during chemical degradation. *Inf. Environ. Sci.* (in Chinese), **5**, 11.

[33] Shun, Y.-X. (1984). BOD values of some chemicals. *Coll. Pap. Environ. Sci.* (in Chinese), **5**, 18.

[34] Ge, F., Zhang, X.-E., Zhang, Z.-P., Zhang, X.-M. (1988). Simultaneous determination of maltose and glucose using a screen-printed electrode system. *Biosensors & Bioelectronics*, **13**, 333–339.

[35] Zhang, X.-E., Rechnitz, G.G. (1994). Simultaneous determination of glucose and sucrose by a dual-working electrode multienzyme sensor flow-injection system. *Electroanalysis*, **6**, 361–370.

[36] Takashi, K., Zhang, X.-E., Rechnitz, G.G. (1995). Simultaneous determination of lactose and glucose in milk using two working enzyme electrodes. *Talanta*, **41**, 361–367.

RAPID MEASUREMENT OF BIODEGRADABLE SUBSTANCES IN WATER USING NOVEL MICROBIAL SENSORS

Reinhard Renneberg, Wai-Kuen Kwong, Chiyui Chan, Gotthard Kunze, and Klaus Riedel

OUTLINE

Advances in Biosensors
Volume 4, pages 195–213.

ISBN: 0-7623-0073-6

1. INTRODUCTION

Microbial sensors based on microorganisms in immediate contact with a transducer, which converts the biochemical signal into a quantifiable electrical response, are used for sensitive determinations of a large spectrum of substances, such as amino acids, saccharides, organic acids, aromatics, alcohols, vitamins, antibiotics, steroids, peptides, and inorganic molecules [1–3]. Each microorganism recognizes a specific set of substances, and this ability helps to determine many complex variables, such as the sum of biodegradable compounds in wastewater [4] and mutagenic compounds [5]. Moreover, the application of biosensor techniques additionally offers an elegant possibility for physiologically characterizing microorganisms, as has been demonstrated by the investigations of glucose uptake [6], peptide metabolism [7], substrate adaptation [8] and by the determination of the alteration of the physiological state during fermentation [9]. Such physiological studies are based on determining the respiration of microorganisms. In these cases a combination of microorganism with an amperometric oxygen electrode is used in the biosensor [10].

The function of such microbial sensors is as follows (Figure 1). Oxygen diffuses from an air-saturated solution through the dialytic membrane, the membrane that contains the microorganisms, and the teflon membrane and then is reduced at the platinum cathode at −600 mV versus Ag/AgCl. Part of the oxygen is consumed by the microorganism. The steady-state current (J_E) represents the oxygen that diffuses through the composite membrane and reflects the endogenous respiration of the microorganism. If an assimilative substrate is added to the measuring solution, the substrate permeates through the dialysis membrane. Some of it is taken up by the microbial cells and is subsequently degraded. The respiration rate A increases, resulting in a decrease in the dissolved oxygen concentration and in the current until a new steady state (J_S) is reached.

In principle, there are two possible types of measurement:

1. end-point measurement (where the difference of currents ΔI reflects the respiration rate of the substrates), and
2. kinetic measurement (first derivative of the current-time curve corresponding to the acceleration of respiration A).

For physiological studies, kinetically controlled sensors, which have very low microbial loadings, and suitable immobilization of microorganisms, and thin membranes, have to be used [9,10]. The sensitivity of this type of sensor is determined mostly by the cell activity, but not by diffusion limitation. This means that the substrate transport into the cell and substrate assimilation should be the rate- limiting process [7].

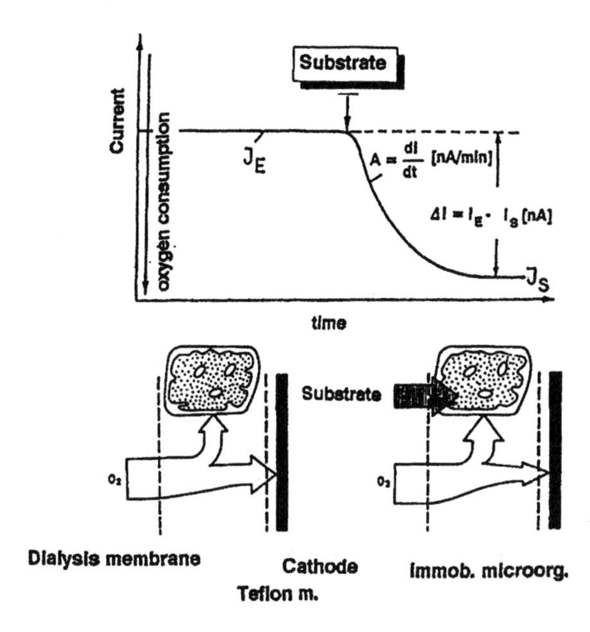

Figure 1. Principle of measurement with microbial sensor. Oxygen is diffusing into the sensor membrane and the immobilized microorganisms at steady state J_E. When the substrate is added, the immobilized microorganisms take up more oxygen by respiration and thus decrease the oxygen concentration in the layer. A decrease in current can be obtained by reaching a steady state J_S. Measurement is achieved by (1) the difference of currents (ΔI) or (2) the first derivative of the current-time curve (A).

2. APPLICATION OF MICROBIAL SENSORS IN POLLUTION CONTROL

Increasing attention is being paid to the use of microbial sensors in environmental monitoring, such as those for estimating biochemical oxygen demand (BOD). BOD, a widely used parameter for controlling wastewater, indicates the amount of biodegradable organic compounds. The conventional Biochemical Oxygen Demand method, BOD_5 [11], requires five days to determine BOD and thus is not suitable for process control. Therefore, a fast and precise sensor that is highly correlated to BOD_5 is needed.

Microbial sensors have been developed to determine BOD or a similar parameter (biodegradable substances) using cells that have broad substrate ranges. Several microbial sensors have been described by Karube [12], Hikuma et al. [13] and Tan [14]. The microorganism used most has been *Trichosporon cutaneum*. Measuring the total amount of oxygen uptake needs 15–20 minutes, and the recovery time between measurements is as much as 3–4 h [15]. In contrast, a fast method of determining oxygen consumption was developed by Riedel and Renneberg [16,17]. The prerequisite for the use of microorganisms for BOD sensors is a wide substrate spectrum. A new BOD sensor has been developed using a combination of two microorganisms that have different substrate spectra, the bacterium *Rhodococcus erythropolis* and the yeast *Issatchenkia orientalis*.

The combinationsensor with *Rhodococcus erythropolis* and *Issatchenkia orientalis* associates the specificity of both strains and can handle a wide range of wastewater substrates very precisely in minutes [13]. This combinationsensor was incorporated into a measuring tool to allow automatic measurement and marketed by Dr. Bruno Lange GmbH Berlin (Figure 2). This tool designated "ARAS" can be used for estimating BOD in various kinds of untreated and treated municipal and industrial wastewater.

The first successful application of the microbial combinationsensor is shown in Figure 3. It was used to measure the daily cycle of the organic load of a municipal sewage treatment plant in Germany in comparison to BOD_5 and COD (Chemical Oxygen demand). As shown, similar results were obtained for BOD values estimated by the microbial combinationsensor and determined by the conventional BOD_5 method. The range of sensor BOD values is related quite well to COD and BOD_5.

For testing under Asian conditions, the sensor was brought to the marshes of the Mai Po WWF Nature Reserve Area in Hong Kong. The aim was to investigate the

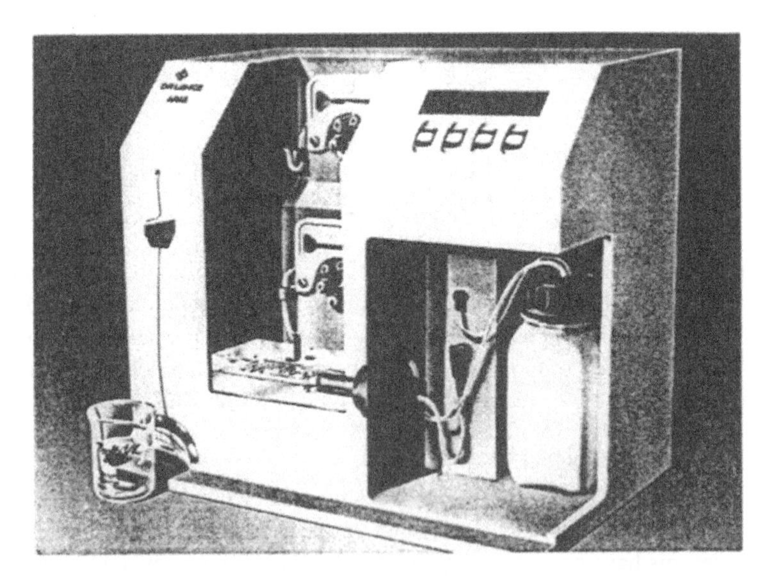

Figure 2. BOD-Biosensor system *ARAS* produced by Dr. Bruno Lange GmbH Berlin, Dusseldorf. Automatic pumping of buffer and samples are controlled by the program provided.

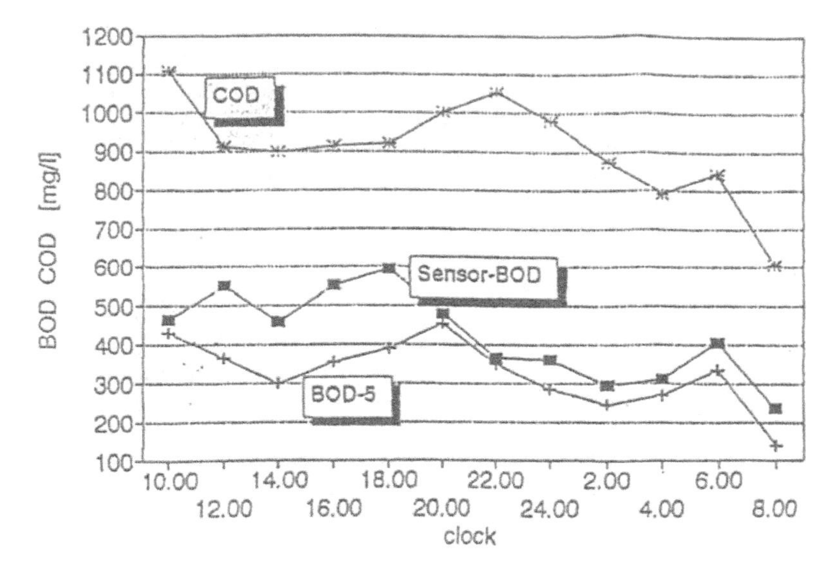

Figure 3. Day profile of a municipal wastewater plant in Berlin (Germany). Pollution levels increase from the start of the day and decrease during the night. All pollution parameters mentioned have the same trends and correlate well.

extent of water pollution caused by the rapid development of nearby new towns and the influence of the China Pearl River on an intact ecosystem.

Mai Po is located at the confluence of salt and freshwater and is composed of 40 pounds (see Figure 4), so-called "gei wai," for shrimp production along the seafront. In the first attempt, samples from all "gei wai" were collected and analyzed. The data show in general a very low sensor BOD value ranging from 1 to 33 mg/L. However, data did not clearly explain about where the pollution comes from. The influence of salt and freshwater was not clear either. Therefore, in the second attempt, samples were collected at the "bird-watching hide" outside the "gei wai," mangrove stand at the seafront and inside the "gei wai" in Mai Po. The sensor BOD values of the water samples were measured by the microbial sensor.

The preliminary results from these water samples show that the water samples outside the "gei wai" generally had a higher value of sensor BOD than the samples inside the "gei wai." Nevertheless, all the samples have lower sensor BOD values compared to the results of the five-day BOD test. The BOD_5 values are 2 to 10 times higher. Two conclusions were drawn from these observations:

1. Macromolecules (like starch and cellulose from pig farms) are present in the water samples which obviously cannot penetrate the outer protective mem-

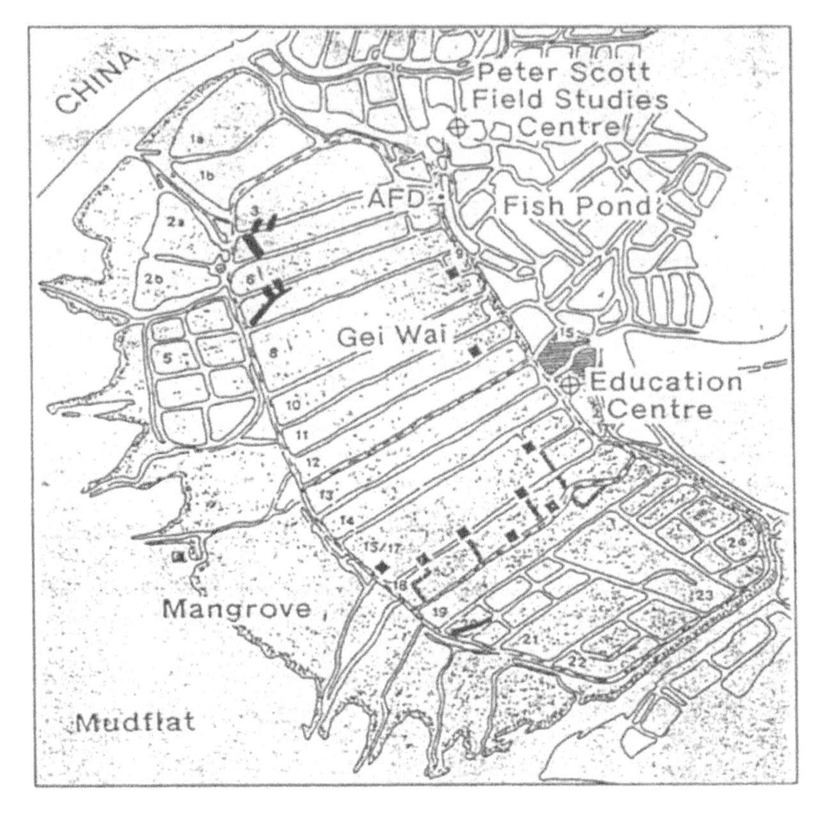

Figure 4. Map of Mai Po Marshes Nature Reserve in Hong Kong. Pollution coming in from Shenzen and from Chinese rivers was investigated.

brane of the sensor. Therefore, only small molecules enter the pores of the membrane and are converted by the microbes. In contrast, in the five-day BOD test, all biodegradable molecules, including macromolecules, are digested by the delocalized seeding microbes in the bottle.

2. The microbes in the sensor are definitely influenced by the salt content in the water samples. The salt content in the samples from inside the "gei wai" is due to less dilution by river and groundwater.

Thus, there are two problems to be solved.

2.1 Prehydrolysis of Macromolecules

Because large molecules cannot penetrate the outer semipermeable membrane of the microbial sensor, pretreating water samples was suggested to break the macromolecules (e.g., polysaccharides, proteins) into small molecules (e.g., monosaccharides, amino acids). These small molecules can penetrate the membrane and the microbes in the biosensor and convert them under O_2 consumption to create a signal. Three substances were used to prepare an "artificial waste water," starch powder, milk powder and cellulose, because they are the main constituents of domestic wastewater. The milk powder contains proteins and also fat. Therefore, we have a "high molecular weight cocktail." A first attempt using hydrolytic enzymes (proteases, amylases, and cellulases) was not successful. After this, acid hydrolysis was used. Different concentrations of hydrochloric acid were used and the duration of acidic hydrolysis was also adjusted to find an optimal condition for hydrolyzing artificial waste water. The concentration of hydrochloric acid was kept constant and the time for hydrolysis was varied. To make the future hydrolytic method simpler, room temperature was chosen first. Figure 5 shows that the sensor BOD value becomes steady in 3 to 10 minutes of hydrolysis. After this, we tried to find the lowest concentration of acid required for hydrolysis to make the experiment safer. Hydrolysis was performed using different concentrations of acid for 10 minutes. From Figure 6, we can see the trend that the higher the concentration of

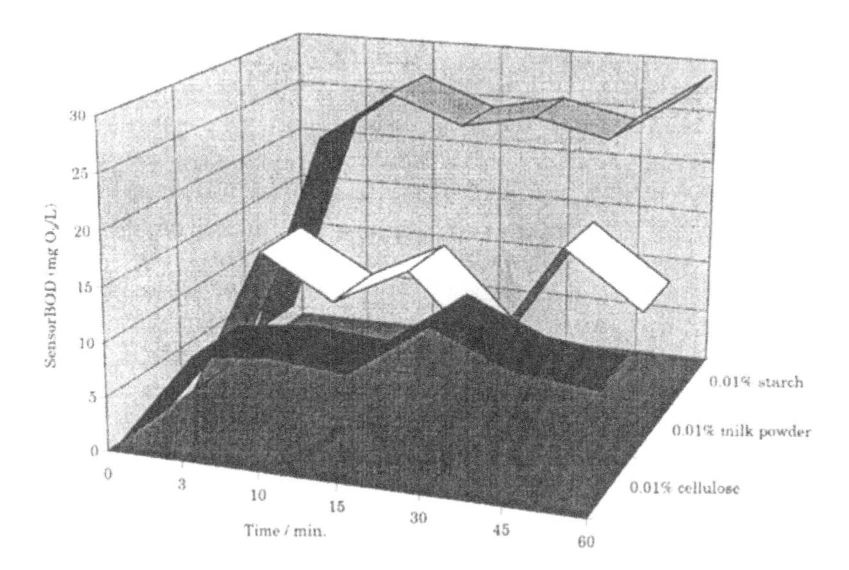

Figure 5. Hydrolysis of artificial wastewater by 6 M HCl at room temperature in different time frames. Three common substances in household wastewater (starch, milk powder, and cellulose) were hydrolyzed. With longer time of hydrolysis, the sensor BOD value increases because more macromolecules are cracked and taken up by the immobilized microbes. All macromolecules were broken down dramatically in three minutes. Starch is the easiest to hydrolyze into small molecules.

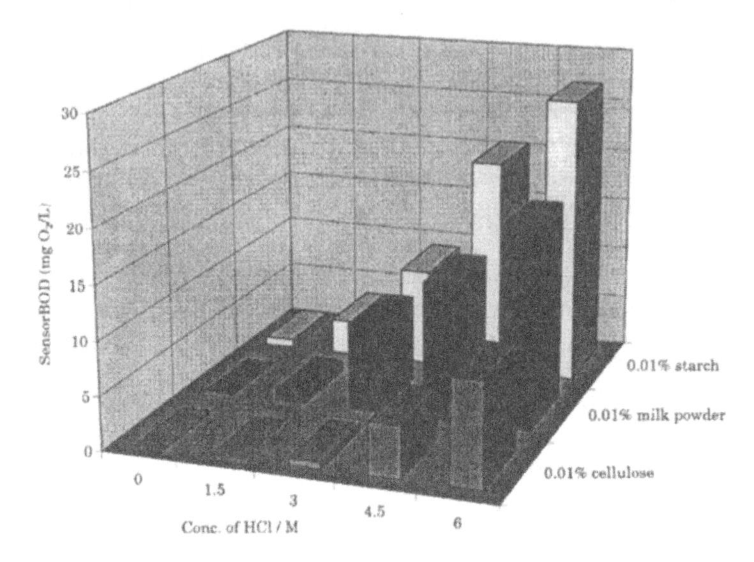

Figure 6. Hydrolysis of artificial wastewater at room temperature in 10 min with different acid concentrations. With increasing concentration of hydrochloric acid, the sensor BOD value increases because more macromolecules are broken down by the stronger acid and are taken up by the immobilized microbes.

acid used, the better the resulting signal of the sensor. By increasing the concentration from 3 M to 6 M, the sensor BOD values increased by nearly three times for starch and cellulose. Therefore, we concluded that it is better to use the highest concentration, 6 M hydrochloric acid.

The next step was to optimize the temperature of hydrolysis. We hydrolyzed a mixture of starch, milk powder, and cellulose—the "cocktail"—at different temperatures. Figure 7 shows that the higher the temperature, the higher the sensor BOD value obtained. Because we could not increase the water temperature beyond the boiling point at normal pressure, 100 °C was finally chosen as the optimal temperature for acid hydrolysis.

Knowing that 100 °C was the optimal temperature for hydrolysis at normal pressure, we returned to the incubation time. Originally it was assumed that 10 minutes would be enough for acid hydrolysis. However, it was found that a longer time is safer for complete acid hydrolysis (Figure 8). Thus, one hour hydrolysis with 6 M HCl at 100 °C was chosen finally as a safe way to hydrolyze most of the compounds.

Using this step, the sensor can also detect high molecular substrates with pretreated samples [18]. The water samples from Mai Po marshes, after prehydrolysis, gave results similar to the BOD_5 values.

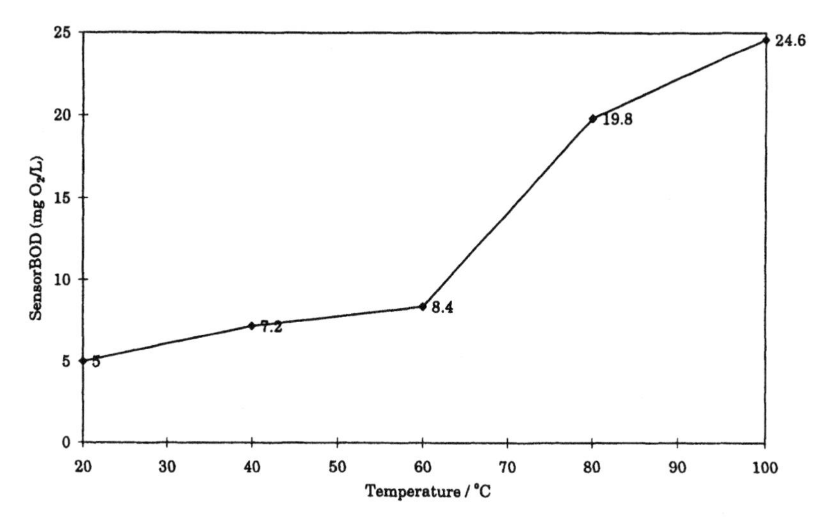

Figure 7. Hydrolysis of a substrate "cocktail" by 6 M HCl in 10 min at different temperatures. The mixture of starch, milk powder, and cellulose was hydrolyzed.

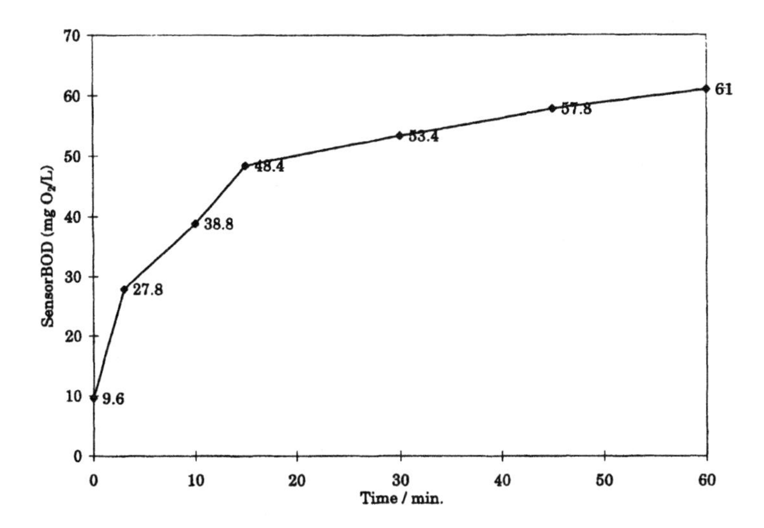

Figure 8. Hydrolysis of substrate "cocktail" by 6 M HCl at 100 °C in different times. With longer hydrolysis, the sensor BOD value increases because more macromolecules are broken down and taken up by the immobilized microbes. The sensor BOD increases by six times from no pretreatment to one hour of hydrolysis.

2.2 Effect of Salt on the Value of Sensor BOD

Different concentrations of sodium chloride were combined with the same amount of biodegradable substance (standard solution) and tested. Sodium chloride was chosen because it is the major component of seawater. It was suspected that the salt in seawater influences the microbes in the BOD sensor in small concentrations, because the microorganisms *Issatchenkia orientalis* and *Rhodococcus erythropolis* are extracted from freshwater media and neither are salt-tolerant. When salt is diffused into the immobilized microbes, the uptake rate of the biodegradable substances is increased, the metabolic rate of the microbes is raised from their normal condition, and thus will enhance the rate of oxygen uptake is enhanced and leads to a higher sensor BOD value. On the other hand, salt in higher concentrations may inhibit microbial metabolism. The investigation of the effect of salt on the biosensor is essential for further development of the sensor for particular areas of the world. For instance, seawater is used for flushing water in Hong Kong, whereas freshwater is used in other countries, such as Germany. In Hong Kong all wastewater is a mixture of salt and freshwater.

At 50 mM NaCl the combinationsensor showed about a 10% smaller signal compared to a solution containing no salt. Obviously, this problem can be solved only by using salt-tolerant microorganisms and is discussed in the latter section.

2.3 Analysis of Wastewater from a Hong Kong University

To test the ability of the microbial sensor to measure sewage water samples, we decided to measure the sensor BOD of sewage water of the Hong Kong University of Science and Technology (HKUST) located at Clear Water Bay in the east side of Kowloon, Hong Kong. Sewage water samples for elementary analysis were collected monthly from the wastewater pipe by the Safety and Environmental Protection Office (SEPO) of HKUST. The samples were evaluated by the sensor BOD test with and without pretreatment and by the BOD_5 test.

After a three-month trial, good correlation was found between sensor BOD and BOD_5 values (Figure 9). As can be clearly seen, better correlation can be obtained using prehydrolyzed samples. In addition, such correlation can be used to predict the result of the five-day BOD test from the five-min sensor BOD data.

For direct analysis of the polluted water, samples were collected freshly from the main wastewater pipe of HKUST. Both untreated and hydrolyzed samples were measured (Figure 10). The findings are summarized as follows:

- A first peak value of sensor BOD occurs at 8 A.M., when people get up, have breakfast, and bathe.
- In the afternoon, a peak occurs at 4 P.M. and in the evening at 9 P.M., a time when dishwashing in the HKUST restaurants is usually done.
- After 10 P.M., pollution levels go down.

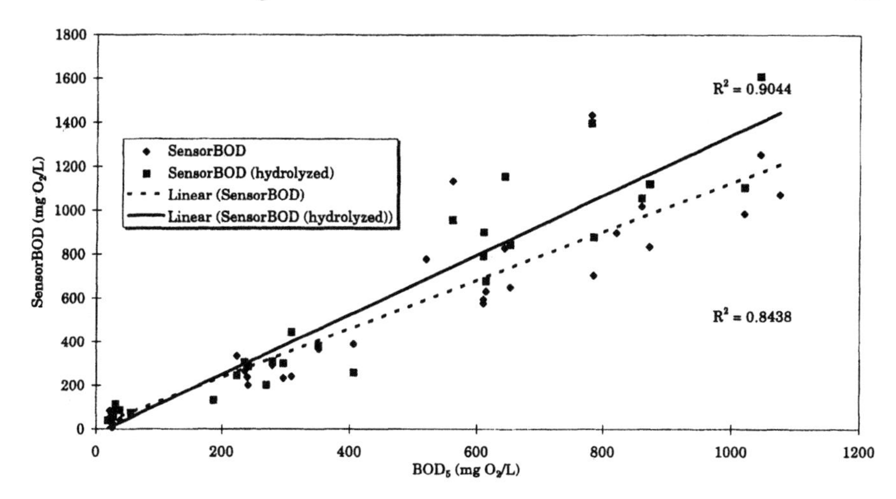

Figure 9. BOD_5 versus Sensor BOD. Both hydrolyzed and unhydrolyzed samples are compared. The best correlation is obtained by the pretreated samples (R^2 = 0.9044).

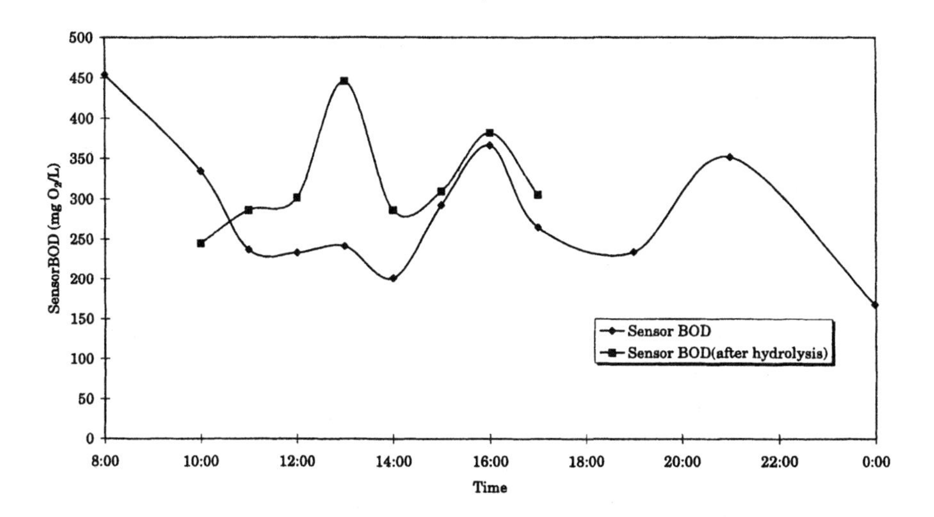

Figure 10. Day profile of wastewater at HKUST measured by ARAS Sensor BOD during the summer vacation. This preliminary test shows that macromolecules in sewage water samples were detected by sensor BOD after being pretreated with acid hydrolysis.

- The trends for hydrolyzed wastewater are quite similar to those for untreated wastewater except that there is a very high peak at 1 P.M. when large molecules (e.g., starch) are in the water. The starch may come mainly from cooked rice.
- The sensor values of the pretreated wastewater samples are always higher, because the samples include large molecules.

We compared the sensor BOD with the traditional five-day BOD test. Moreover, we also measured the sensor BOD of all samples after five days of storage at 4 °C to investigate any possible degradation of samples during storage. As shown in Figure 11, the results are quite satisfactory. The peaks of all three methods are reached at the same time. A recalculation factor between sensor BOD and BOD_5 can be found from the correlation. The remeasured sensor BOD after five days of storage was lower than that of the fresh directly measured samples. Because the remeasured sensor BOD and BOD_5 tests were performed on the same day, we used these two for direct comparison. Obviously, wastewater microbes in samples (even stored at 4 °C) can digest the organic substances in the samples. The largest difference is 140 mg/L. Most of them have differences within 50 mg/L and the percentage of error is less than 20% (Figure 11). These results are quite remarkable. It is also interesting to note that the sensor BOD value at lunchtime (12 P.M.) is increased by storage, indicating hydrolysis of macromolecules. At 9 P.M., however, the sensor value of the fresh sample was higher than that of the stored sample. One might speculate that fewer macromolecules are contained in the wastewater during the night.

The data shown were obtained at the end of a semester when the majority of students had already left the campus. However, we collected samples during the

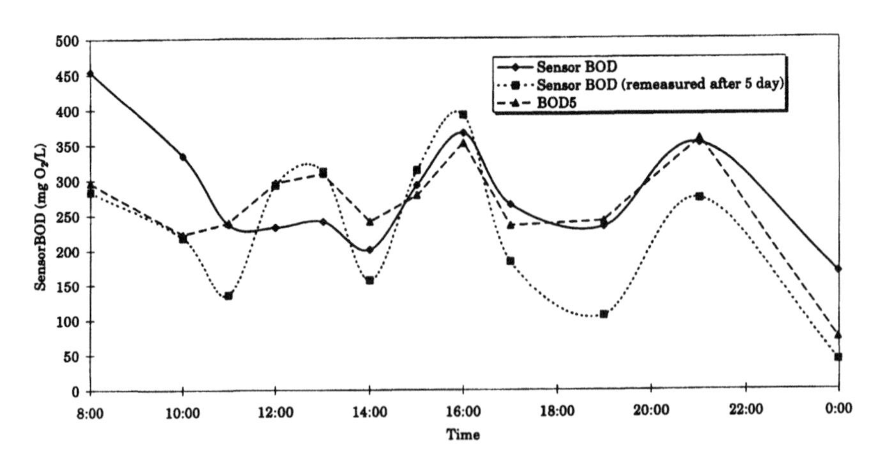

Figure 11. Day profile of wastewater at HKUST measured by ARAS Sensor BOD and BOD_5.

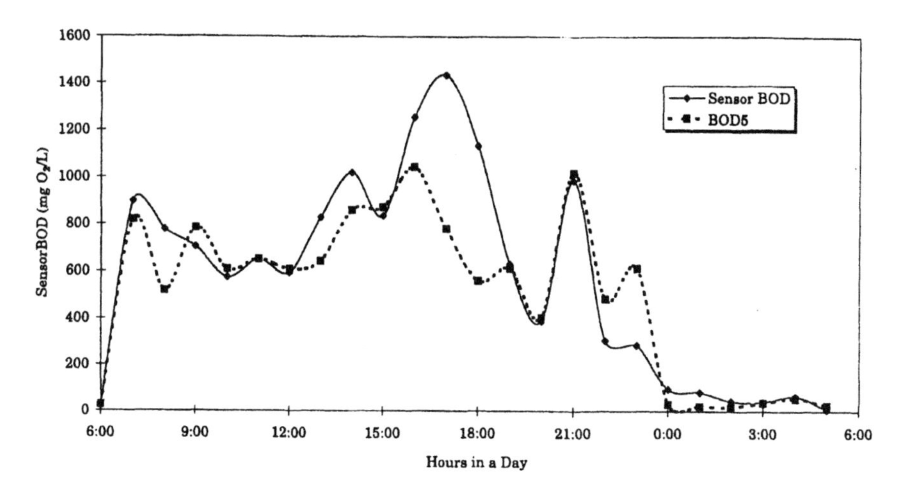

Figure 12. Pollution level measurement at HKUST during the semester.

active period at the university over a whole day on the 2nd and 3rd of October 1996 for 24 hours. It was a normal university day, and the sensor BOD values were expected to be much higher than for the previous tests when students were out for holidays. Indeed, from our wastewater profile (Figure 12), the sensor BOD value averaged four times higher than in the previous experiment. However, the trends followed the previous experiment. In this experiment, moreover, wastewater discharged after midnight was also checked and its sensor BOD value was very low (lower than 50 mg/L). At 6:00 the following morning, the sensor BOD value increased again.

3. A NOVEL APPROACH USING THE SALT-TOLERANT YEAST *ARXULA*

The microorganism used until now in commercial sensors (Nisshin Denki, Japan) has been mainly *Trichosporon cutaneum*. However, *Trichosporon* is pathogenic. In the commercial ARAS BOD sensor produced by Dr. Bruno Lange GmbH Berlin, two types of microbes are co-immobilized. One is bacteria (*Rhodococcus erythropolis*) and the other is a yeast strain (*Issatchenkia orientalis*). This combinationsensor uses bacteria and yeast to broaden the substrate specificity. However, the combinationsensor also has disadvantages.

Each microorganism is specific to certain substrates. For a broad substrate range, many different microorganisms should be mixed, as in activated sludge. However, we have found that with populations of mixed microorganisms, the sensor signal changes after several days of measurements and gives nonreproducible results. Therefore, it is better to use only one species of microorganism for microbial sensors. Our experience in Hong Kong tells us that this microorganism should also be salt-tolerant. A salt-tolerant yeast, *Arxula adeninivorans*, has an amazingly broad substrate range. It assimilates all sugars, polyalcohols, and organic acids except for ribose, lactose, lactic acid, and methanol [19]. As a result, an *Arxula*-based sensor should correlate with BOD_5 for many substrate and real samples [20].

Arxula adeninivorans isolated from wood hydrolysates in Siberia (Russia) was supplied from the yeast collection of the Institut fuer Pflanzengenetik und Kulturpflanzenforschung (IPK) in Gatersleben, Germany [21].

Poly(carbamoyl)sulfonate (PCS) was used to immobilize the microorganism. PCS was introduced and tested by Vorlop [22] and Muscat [23]. PCS is prepared from isocyanate-terminated polyurethane (PUR) prepolymer with bisulphite and is a better gel for entrapment than PUR and calcium alginate [22]. It combines the advantageous properties of calcium alginate (low toxicity) and the PUR matrix (high elasticity).

The *Arxula* sensor has very high stability and can be used for more than 40 days (Figure 13). A gradually increasing signal obtained during the first five days was assumed to be due to the growth of the cells. The gradually decreasing background current proves this assumption. The sensor shows a very stable signal up to 40 days. Five measurements were made for each set of data. The sensor was stable for more than 150 measurements. The percentage of error for each set of data was within

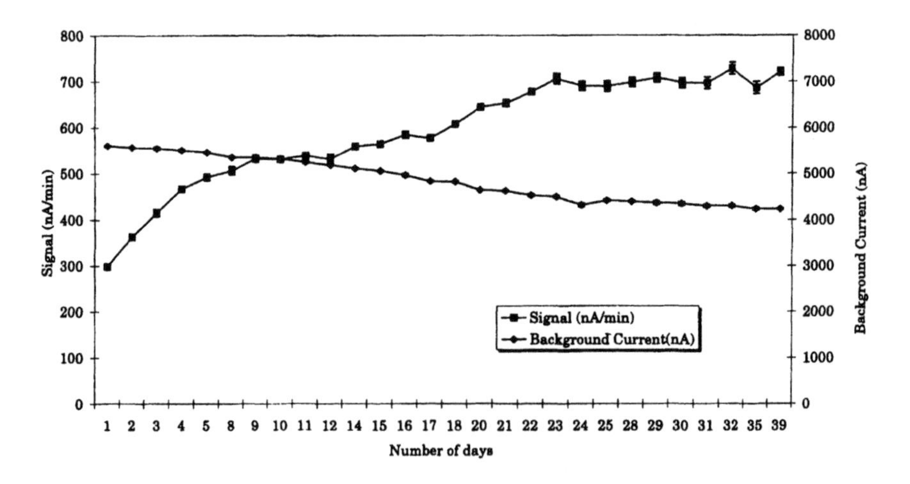

Figure 13. Changes in the signal and background current of the *Arxula* sensor. The signal is determined 70 s after sample injection. The signal increases from 300 nA/min up to a steady range of 680 nA/min after 40 days. The substrate for each measurement is 214.4 mg/L glucose and glutamic acid (GGA), and all data are based on five measurements. The background current indicates the saturated oxygen concentration in the gel layer in front of the oxygen sensor in the absence of the substrate.

10%. This is very important because we need a stable, accurate, and precise microbial sensor for long-term wastewater monitoring. Besides measurement frequency, *Arxula* has high storage stability. Two months of storage at 4 °C did not cause any problems.

We used 214.4 mg/L of glucose and glutamic acid as a calibration standard for a 275 mg O_2/L BOD value. The sensor has a linear response up to 550 mg/L (Figure 14). If we want to measure the samples larger than 550 mg/L with BOD, the sample must be diluted first. The linearity of this range is satisfactory ($r^2 = 0.9901$). The calculated limit of determination was 13 mg/L, and the limit of detection was 8 mg/L BOD.

For a strong signal, an optimal number of cells must be immobilized. However, a thicker cell layer hinders the diffusion of oxygen because diffusion is inversely proportional to the distance of diffusion. Then the sensor produces a lower signal instead. Because we want the substrate concentration to be the only parameter that causes a signal, the external oxygen concentration should not affect the signal and oxygen diffusion and should be at the maximal level. The background current at the maximal range is more than 4000 nA. For a gel layer without any cells, the background current is 5000 nA. For a 180 μg/mm^2 or higher concentration of *Arxula*, however, the background current drops to 2000 nA.

From our findings, the optimal cell loading for the *Arxula* sensor is between 75 and 100 μg/mm^2 (Figure 15). It corresponds actually to 0.5–0.75 g/mL wet weight of *Arxula* in a PCS gel mixture. We measured on the first day and the fifth day after

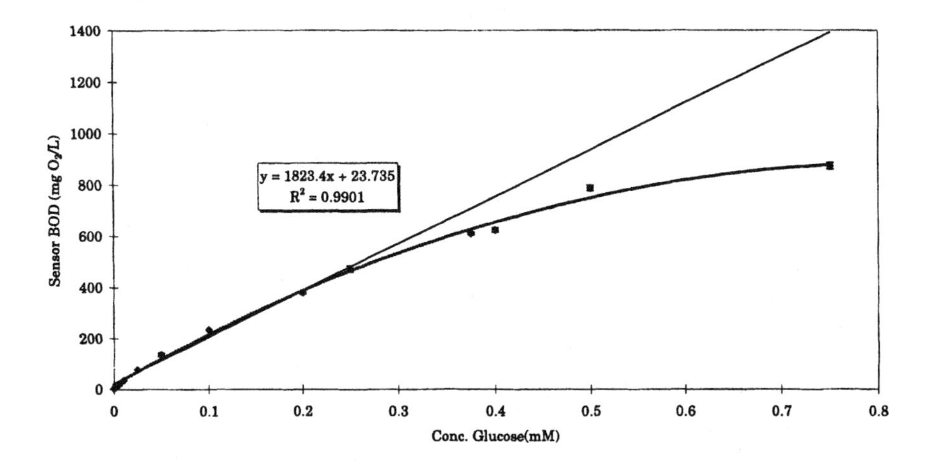

Figure 14. Calibration curve for the *Arxula* sensor. The sensor is calibrated by the GGA standard. The concentration of glucose is the final concentration in the measuring cell. The range from 0 to 0.25 mM of glucose is chosen for linear regression and the r^2 value is 0.9901.

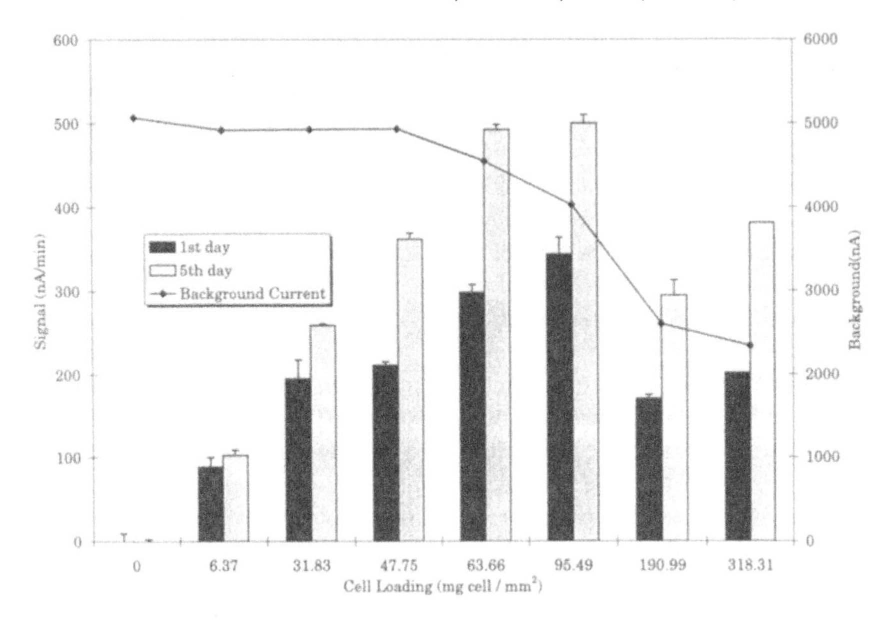

Figure 15. Profile of cell loading of the *Arxula* sensor. The dotted bar represents data from the first day of sensing. The solid bar represents data from the fifth day of sensing. The substrate for each measurement is 214.4 mg/L GGA, and each set of data is based on five measurements. The peak region is at 75–100 μg cell/mm^2. The diameter of the electrode is 5 mm. The cell loading was calculated from the volume of the cell and the gel mixture (20 μl), the cell concentration, and the area of electrode surface.

activation. The signal on the fifth day is much higher than that on the first day, and it is compatible with our previous result for stability measurement. We finally chose 100 μg/mm^2 for further measurements.

Figure 16 shows a comparison between the BOD_5 and sensor BOD value. The *Arxula* sensor BOD value correlates well with the BOD_5 below 550 mg/L. After reaching the linear range, the *Arxula* sensor is saturated. This problem can be overcome by diluting the concentrated sample back to the linear range. We also compared the result with the commercial BOD sensor that utilizes two microbial strains (*Rhodococcus erythropolis* and *Issatchenkia orientalis*). We found that the commercial sensor has a lower saturation level of glucose concentration. When we compared the error in 550 mg/L, the *Arxula* sensor has a 10% error but the commercial sensor has an error up to 40% compared to BOD_5. Therefore, the *Arxula* sensor has better correlation with BOD_5 than the commercial sensor. *Arxula* also has broader substrate specificity for various organic substances compared with the strains combined in the commercial sensor [20].

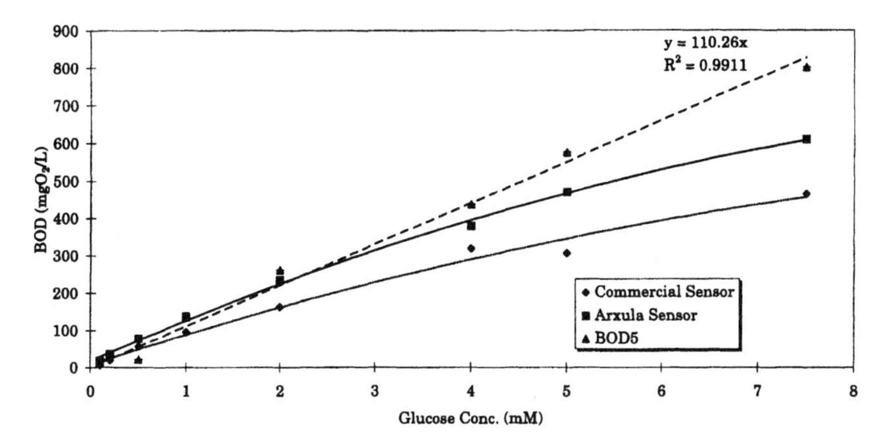

Figure 16. Comparison of accuracy between the novel *Arxula* sensor and the commercial BOD sensor ARAS. The sensor is calibrated by the GGA standard. The concentration of glucose is the final concentration in the measuring cell.

One of the main reasons for using *Arxula* was to overcome the salt-dependence of the commercial sensor ARAS. In many coastal or island places, seawater is used in toilet flushing. In wastewater measurement with microbial sensors, the salt concentration affects the BOD measurement. The commercial sensor ARAS is very sensitive to salt, and is activated by low concentrations of salt. At high salt concentration, it is deactivated. A salt-independent microbial sensor for BOD is needed. *Arxula* is salt-tolerant and stable to very high concentrations of salt—up to 120 mM of sodium chloride (Figure 17). However, the commercial sensor is 18% deactivated at this concentration. For sea salt concentration (~40 mM NaCl), the *Arxula* sensor is not affected by salt, but the commercial sensor suffers 10% deactivation.

Arxula has amazingly high stability. The *Arxula* sensor is stable for 40 days under working conditions without any decrease in activity. It has a linear range up to 550 mg/L. It shows better correlation with BOD_5 than the commercial sensor with *Rhodococcus erythropolis* and *Issatchenkia orientalis*. Unlike the commercial sensor, it is stable to salt concentration, which is very important for wastewater measurement in coastal areas and islands.

From our results, *Arxula* is a suitable strain for microbial sensor development [24]. After optimization, *Arxula* will certainly have better performance in wastewater measurement. From the analysis and experiments being done, the microbial wastewater sensor can be used as a method for rapid screen and direct determination of water pollution [25].

The fast response time of the microbial sensor—in minutes—can be of great help to wastewater treatment operations. When there is a large influx of wastewater into

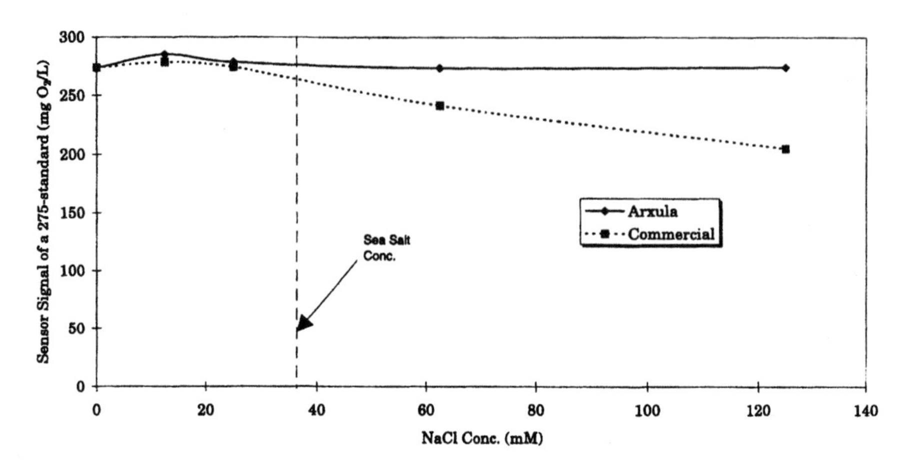

Figure 17. Comparison of salt effect between the *Arxula* sensor and the commercial sensor using *Issatchankia* and *Rhodococcus* cells. The sodium chloride concentration is the concentration in the measuring cell. The salt concentration of sea water is normally 3.8%. This corresponds roughly to 50 mM of NaCl.

the waste treatment plant, now the operator can monitor the extent of waste immediately. Action can be taken instantly, such as stepping up the process of biodegradation by activated sludge by increasing the air pumping rate in the treatment plant. On the other hand, energy-consuming air pumps can be switched off if clean water flows into the treatment plant.

Thus, microbial wastewater sensors can contribute a great deal to environmental protection and energy conservation.

REFERENCES

[1] Riedel, K., Renneberg, R., Wollenberger, U., Kaiser, G., Scheller, F. (1989). Microbial sensors: Fundamentals and application for process control. *J. Chem. Biotechnol.*, **31**, 502–504.

[2] Riedel, K. (1994). Whole-cell and tissue sensors. In *Food Biosensor Analysis*, Wagner, G., Guilbault, G.G. (Eds.). Marcel Dekker, New York, pp. 123–150.

[3] Riedel, K. (1996). Application of biosensors to environmental samples. In Commercial Biosensors: Application to Clinical, Bioprocess and Environmental Samples, Ramsay, G. (Ed.). Wiley, New York (in press).

[4] Riedel, K. (1994). Microbial sensors and their applications in environment. *Exp. Techn. Physics.*, **40**, 63–76.

[5] Karube, I., Nakahara, T., Matsunaga, T., Suzuki, S. (1982). *Salmonella* electrode for screening mutagens. *Anal. Chem.*, **54**, 1725–1727.

[6] Riedel, K. (1991). Biochemical fundamentals and improvement of selectivity of microbial sensors—a mini-review. *Bioelectrochem. Bioenerg.*, **25**, 19–30.

[7] Riedel, K., Renneberg, R., Scheller, F. (1989a). Studies in peptide utilization by microorganisms using biosensor techniques. *Bioelectrochem. Bioenerg.*, **22**, 113–125.

[8] Riedel, K., Lange, K.P., Stein, H.J., Kuehn, M., Ott, P., Scheller, F. (1990). A microbial sensor for BOD. *Water Res.*, **24**, 883–887.
[9] Riedel, K., Liebs, P., Renneberg, R., Scheller, F. (1988b). Characterization of the physiological state of microorganisms using the respiration electrode. *Anal. Lett.*, **21**, 1305–1322.
[10] Riedel, K., Renneberg, R., Liebs, P. (1985). An electrochemical method for determination of cell respiration. *J. Basic Microb. Biotechnol.*, **25**, 51–56.
[11] Standard Methods for the Examination of Waters and Wastewater, American Public Health Association, Washington, D.C., 16th ed., 1986, pp. 523–531.
[12] Karube, I., Matsuaga, T., Suzuki, S. (1977). A new microbial electrode for BOD estimation. *J. Solid Phase Biochem.*, **2**, 97–104.
[13] Hikuma, M., Suzuki, H., Yasuda, A., Karube, I., Suzuki, S. (1979). Amperometric estimation of BOD by using living immobilized yeasts. *Eur. J. Appl. Microbiol.*, **8**, 289–297.
[14] Tan, T.C., Li, F., Neoh, K.G. (1993). Measurement of BOD by initial rate of response of a microbial sensor. *Sensors and Actuators B*, **10**, 137–142.
[15] Tan, T.C., Li, F., Neoh, K.G., Lee, Y.K. (1992). Microbial membrane-modified dissolved oxygen probe for rapid biochemical oxygen demand measurement. *Sensors and Actuators B*, **8**, 167–172.
[16] Riedel, K., Renneberg, R., Kuehn, M., Scheller, F. (1988). A fast estimation of biochemical oxygen demand using microbial sensors. *Appl. Microbiol. Biotechnol.*, **28**, 316–318.
[17] Riedel, K. (1994). Microbial sensors and their applications in environment. *Exp. Tech. Phys.*, **40**(1), 63–76.
[18] Kwong, A.W.K., Chan, C., Renneberg, R. (1998). Monitoring biodegradable substances with high-molecular content with a microbial sensor, *Anal. Lett.*, **31**, 2309–2325.
[19] Middelhoven, W.J., de Jong, I.M., de Winter, M. (1991). *Arxula adeninivorans*, a yeast assimilating many nitrogenous and aromatic compounds. *Antonie Leeuwenhoek J. Microbiol.*, **59**, 129–137.
[20] Riedel, K., Lehmann, M., Renneberg, R., Kunze, G. (1998). *Arxula adeninivorans* based sensor for the estimation of BOD. *Anal. Lett.*, **31**, 1–12.
[21] Gienov, U., Kunze, G., Schauer, F., Bode, R., Hofemeister, J. The yeast genus *Trichosporon* Spec. LS3: Molecular characterization of genomic complexity. *Zbl. Microbiol.*, **145**, 3–12.
[22] Vorlop, K.-D., Muscat, A., Beyersdorf, J. (1992). Entrapment of microbial cells within polyurethane hydrogel beads with the advantage of low toxicity. *Biotechnol. Tech.*, **6**, 483.
[23] Muscat, A., Beyersdorf, J., Vorlop, K.-D. (1995). Poly(carbamoylsulphonate) hydrogel, a new polymer material for cell entrapment. *Biosensors & Bioelectronics*, **10**, 11–14.
[24] Chan, C., Lehmann, M., Tag, K., Lung, M.L.L., Kunze, G., Riedel, K., Renneberg, R. (1998). Measurement of biodegradable substances using the salt-tolerant yeast *Arxula adeninivorans* for a microbial sensor immobilized with Poly(carbamoyl)sulfonate (PCS). Part I: Construction and characterization of the microbial sensor. *Biosensors and Bioelectronics*. In press.
[25] Lehmann, M., Chan, C., Lo, A., Lung, M.L.L., Tag, K., Kunze, G., Riedel, K., Renneberg, R. (1998). Measurement of biodegradable substances using the salt-tolerant yeast *Arxula adeninivorans* for a microbial sensor immobilized with Poly(carbamoyl)sulfonate (PCS). Part II: Application of the novel biosensor to real samples from coastal and island regions. *Biosensors and Bioelectronics*. In press.

APPLICATION OF BIOSENSORS FOR DIAGNOSTIC ANALYSIS AND BIOPROCESS MONITORING

Gao Xiang Li and Jian Guo Liu

OUTLINE

Advances in Biosensors
Volume 4, pages 215–240.

ISBN: 0-7623-0073-6

ABSTRACT

Various kinds of biosensor systems have been developed and applied to medical diagnostics and monitoring and to controlling fermentative bioprocesses. Enzyme-electrode flow-injection analysis (FIA) systems used for determining glucose, uric acid, and free cholesterol in serum are described. A biosensor based on luminol chemiluminescence for serum uric acid determination and an optical biosensor based on an enzyme catalyzed-fluorescent reaction for detecting glucose are discussed. Enzyme field effect transistor (FET) systems were used for glucose and urea analysis in serum. Penicillinase and penicillin acylase-FET systems were used for monitoring the concentration of penicillin and penicillin G for process control in the penicillin fermentative industry. An automatic FIA biosensor system was developed for on-line monitoring of the glucose and glutamate concentration in the glutamate fermentative process. A piezoelectric immunosensor for the detection of C_2 type *Staphylococcus* enterotoxin is also discussed.

1. INTRODUCTION

Enzymes are becoming increasingly popular as analytical reagents because of their sensitivity, better selectivity, high degree of specificity, and rapid response. In many respects, microorganisms are a more practical source of enzymes for analytical use than animal tissue. In general, immobilized enzymes can improve the enzymatic stability and operative reuse. Immobilized enzyme reactors are combined with transducers to construct various biosensors, and they have attracted much attention in recent decades. Many investigations have been reported in the literature on developing and characterizing different types of biosensors. Presently biospecific sensors, including electrochemical detectors and biological macromolecules, are applied in medical diagnostics, the food industry, environmental analysis, and fermentative bioprocess monitoring and control. Biosensors would allow the direct, simple, continuous *in vivo* analysis of important metabolic parameters in body fluids. Recent advances in the field of biotechnology have greatly increased the need for new on-line biosensor technological development. The ideal biosensor system for bioprocess monitoring and control should be designed to monitor substrate uptake and product formation, and to provide a feedback system triggered by sensor response to adjust levels of substrates in the reaction broth. This system should be an automatic on-line and continuous FIA system so that optimal production of high-yield products are obtained. The aim of this chapter is to present results from our laboratory in developing various kinds of biosensors for diagnostic analysis and bioprocess monitoring and control.

2. ENZYME-ELECTRODE-FIA SYSTEMS

Electrochemical sensors that employ immobilized biocatalysts have definite advantages. In particular, an enzyme sensor has excellent selectivity for biological substrates, and the biosensor can directly determine a single compound in a complicated mixture without the need for a prior separation step. Most of the commercially available biosensors are based on enzyme electrode technology. In our laboratory, glucose oxidase (GOD), horseradish peroxidase (HRP), uricase, cholesterol esterase (CHE), and cholesterol oxidase (COD) were successfully immobilized on controlled pore glass by cross-linking with glutaraldehyde. GOD and glutamate oxidase (GLOD) were covalently coupled to nylon tubing. These immobilized enzyme reactors were combined with hydrogen peroxide (H_2O_2) or oxygen electrodes individually to construct the FIA systems for diagnostic analysis or fermentative monitoring and control.

2.1 Glucose Oxidase-Electrode System

A GOD-FIA system has been developed for clinically determining serum glucose [1]. GOD from *Penicillium nototum* was immobilized to alkylamine porous glass with glutaraldehyde. The immobilized biocatalyst contained about 25 mg/g glass protein with activity of 300–400 IU/g glass. The level of enzyme activity was sufficient for glucose analysis. The enzyme electrode FIA system response to glucose at concentrations below 500 mg/dL was perfectly linear. The flow rate of carrier solution has an important effect on the flow-injection analysis. In this system, the effect of various flow rates on the response value of the GOD electrode system was determined using samples with glucose concentrations of 100, 300, and 500 mg/dL. The response value of the GOD electrode system was highest at a liquid flow rate of 100 mL/h at least 2800 measurements were assayed for serum glucose determination. The response time of the GOD electrode system for each sample was 30 seconds and that for every assay cycle was about 1 minute. A comparative experiment on glucose concentration in 38 serum samples that had low, intermediate and high glucose concentrations was carried out by the GOD electrode method and by the autoanalytical method using a GOD reagent kit. Good correlation (r = 0.9977) was obtained. The within-batch relative standard deviation (SD) and coefficient of variation (CV%) were 0.036 and 0.65, respectively, for sets of 20 injections at various concentrations using a sample injection volume of 10 μL. The results of these experiments are shown in Table 1.

Table 1. Correlation and Precision for Serum Glucose Measured by the GOD Electrode

	Within-Batch		
Relative Coefficient[a], r	*Standard Deviation (SD)*	*Coefficient of Variation (CV%)*	*Linear Range Glucose, mg/dL*
0.9977	0.036	0.65	0–500

Note: [a]Against serum glucose assayed by GOD reagent kit method.

2.2 Uricase Electrode System

Uricase (E.C.1.7.3.3.), which catalyzes the oxidation of uric acid to allantoin, has been widely used as a clinical diagnostic enzyme to determine uric acid concentration in serum [2].

The principle of the assay is as follows:

$$\text{Uric acid} + H_2O + O_2 \xrightarrow{\text{Uricase}} \text{Allantoin} + H_2O_2 + CO_2$$

The serum uric acid concentration is an important index for clinically diagnosing gout, leukemia, toxemia of pregnancy, and severe renal impairment. A uricase electrode—FIA system has been developed for determining serum uric acid. We succeeded in screening out an uricase-producing rich strain from *Candida utilis* A.S.2.117. The optimum culture condition for uricase production was investigated (see ref. 30).

The uricase was purified and immobilized by covalent coupling to alkylamine controlled pore glass. The immobilized uricase (IMU) contained about 1.1% protein and its activity was about 5.0 IU/g glass. Some characteristics of the IMU and the native uricase have been compared [3]. A small IMU reactor (0.5 units IMU) was combined with an oxygen electrode to construct the FIA system and used for uric acid determination. The response value of the uricase electrode system was highest at a liquid flow rate of 360 mL/h. Using a sample injection volume of 20 μL, uric acid was linearly quantified in the range of 0–14 mg/dL. The response time

Table 2. Precision Data of FIA for Uric Acid Using Uricase Electrode System

Sample[a]	*$\overline{X}$ Uric Acid, mg/dL*	*n*	*SD*	*CV%*
Serum 1	3.7	5	1.51	3.20
2	6.3	5	2.40	3.40
Uric acid	6.0	11	1.20	1.20

Note: [a]Sample injection volume of 50 μL.

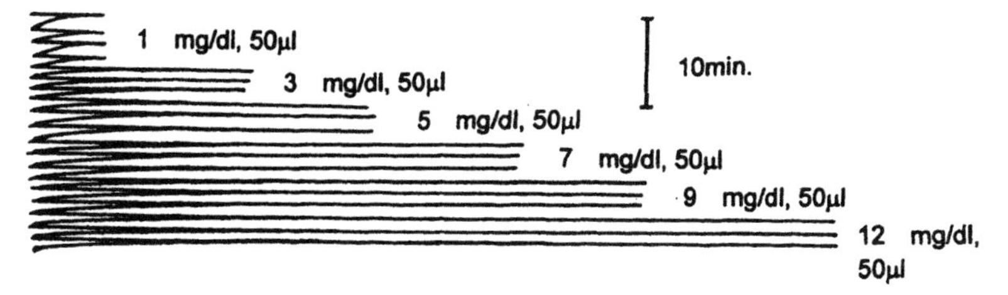

Figure 1. Recorder tracing of FIA for uric acid using a uricase electrode.

Table 3. Operational Stability of Immobilized Uricase Column-Electrode

Operational Time, h	*Activity of IMU*			*Half-life $t_{1/2}$, h*	*Decay Constant, h^{-1}*	*Assay Times*
	Initial, IU/g	*Remaining IU/g*	*Remaining %*			
201	0.50	0.132	26.4%	107	6.5×10^{-3}	3000

was within 1 minute. The within-run standard deviation (SD) and coefficients of variation (CV%) were 1.5 and 4.8, respectively (Table 2). Typical response peaks for the injection of uric acid samples are shown in Figure 1.

The enzyme column-oxygen electrode system was continuously operated at 25 °C using 10 μg/mL uric acid solution at pH 8.5 at a flow rate of 15 mL/h for 201 hours. The half-life of IMU was 107 hours under operational conditions. One enzyme reactor containing 0.5 U of IMU was used for more than 3000 assays (Table 3).

2.3 Cholesterol Oxidase-Electrode System

The concentration of total cholesterol and free cholesterol in serum are important indexes for diagnosing human heart and blood vessel disease. A cholesterol enzyme-electrode system has been developed for determining serum cholesterol [4]. The principle of the analysis is based on two consecutive reactions:

$$\text{Cholesterol ester} + H_2O \xrightarrow{\text{Cholesterol esterase}} \text{Cholesterol} + \text{Fatty acid}$$

$$\text{Cholesterol} + O_2 + H_2O \xrightarrow{\text{Cholesterol oxidase}} \text{Cholest-4-en-3-one} + H_2O_2$$

Cholesterol esterase (CHE) and cholesterol oxidase (COD) were also immobilized on alkylamine porous glass with glutaraldehyde. The coupled COD absolute activity reaches 13 U/g glass (dry weight), the relative activity was 30–45%, whereas the coupled CHE activity was 5.62 U/g glass and the relative activity was 20%. Some characteristics of the immobilized enzymes and native enzymes have

Table 4. Precision Data of COD-Electrode System for Determining Free Cholesterol in Serum

Serum Sample	*Free Cholesterol, $\bar{X}$ mg/dL*	*n*	*Within-Batch*	
			SD	*CV%*
1	30.9	20	1.88	4.7
2	70.3	20	4.31	4.2

also been comparatively investigated. There are not many differences except that the immobilized enzymes are more stable. Under continuous operation, the half-lives of immobilized COD and CHE were 53 hours and 27 hours, respectively. Immobilized COD was packed into a nylon tube 3 cm long by 0.4 cm in diameter to make an enzyme reactor. The COD reactor was combined with an oxygen electrode to construct a FIA system for determining free cholesterol in serum. The performance characteristics of the system were also discussed. The injection volume of serum sample for each assay was 50 μl. The COD electrode FIA system responses to cholesterol at concentrations below 50 mg/dL were perfectly linear. The response time of the COD electrode system was 23 seconds. In within-batch, the relative standard deviation (SD) and coefficient of variation (CV%) were 1.88–4.31 and 4.2–4.7, respectively for sets of 20 injections at two cholesterol concentrations (Table 4).

2.4 On-Line Monitoring of Glucose and Glutamate Concentrations in Fermentative Broth Using an Enzyme Electrode-Automatic FIA System

Recent advances in the field of biotechnology have greatly increased the need for new on-line electrochemical biosensor technological development. The ideal biosensors for bioprocess control should be designed to monitor substrate uptake, growth rate, and product formation continuously. In addition, they can be automated to trigger a feedback mechanism that adjusts the levels of substrates in the fermentation broth, so that the system holds promise for optimizing the production of high-yield products, such as glutamate, organic acids, and pharmaceuticals. Many investigations have been reported in the literature in developing various types of biosensors for monitoring and controlling biotechnological processes [5,6].

We have successfully developed an automatic multichannel FIA enzyme electrode system for on-line monitoring and control of glucose and glutamate concentrations in the glutamate fermentation process [7]. The principle of measurement is summarized in the following reactions:

$$\text{Glucose} + O_2 + H_2O \xrightarrow{\text{GOD}} \text{Gluconolactone} + H_2O_2$$

$$\text{2-L-glutamate} + O_2 + H_2O \xrightarrow{\text{Glutamate oxidase}} \text{2-Oxoglutarate} + 2NH_3 + H_2O_2$$

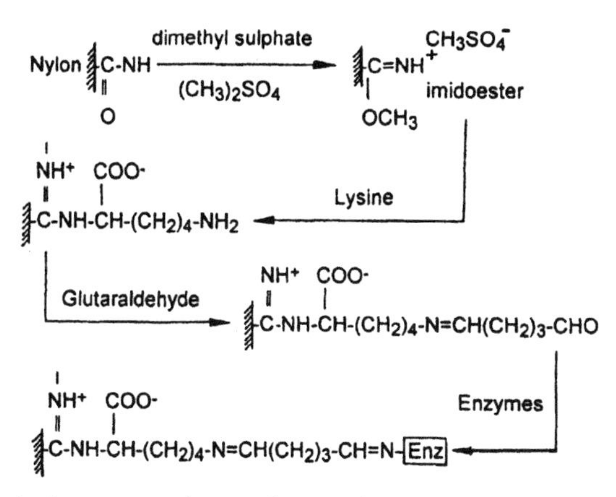

Figure 2. Chemical reaction schemes for attaching enzymes to an *o*-Alkyated nylon.

The direct electrochemical oxidation of hydrogen peroxide results in a current proportional to the glucose and glutamate concentrations in the follow chamber:

$$H_2O_2 \rightarrow O_2 + 2H^+ + 2e^-$$

Glucose oxidase (GOD) and glutamate oxidase (GLOD) have been covalently coupled on the inside of the nylon tube with glutaraldehyde. The procedure of immobilized enzyme is as follows (Figure 2).

A system for analysis had the following design. The automatic FIA system consists of filter equipment, pumps, sample dilution equipment, an autoinjection valve, immobilized GOD and GLOD nylon tube reactors, hydrogen peroxide electrodes, a detector, a recorder, and a feedback controlled system. A diagram of the automatic on-line FIA biosensor system is shown in Figure 3. The system can also act as a multisensor by using different enzyme tubes.

To determine glucose using this automatic biosensor system, the flow rate of carrier buffer solution has an important influence on the FIA system. The response value increases with decreasing flow rate, but the response time is longer at lower flow rates. So the optimum flow rate was set at 1.5 mL/min. The response time was within 1 minute. The linear range for glucose determination increased with increasing sample dilution. The response of the FIA biosensor system was linear up to 600 mg/dL of glucose concentration. The optimum pH of the carrier buffer was pH 5–6. The optimum temperature of the GOD tube reactor was 40 °C. The continuous measurement of metabolic parameters is an important part of developing electrochemical sensors, and stability, selectivity, and sensitivity are essential requirements. For this automatic on-line FIA GOD biosensor system, the variation coefficients (CV%) for 20 responses to 100 mg/dL, 400 mg/dL, and 1000 mg/dL of glucose concentrations were 2.96%, 2.4%, and 3.2%, respectively, (Table 5). The response sensitivity of the GOD sensor was 4.2–4.4 mv/mg/dL.

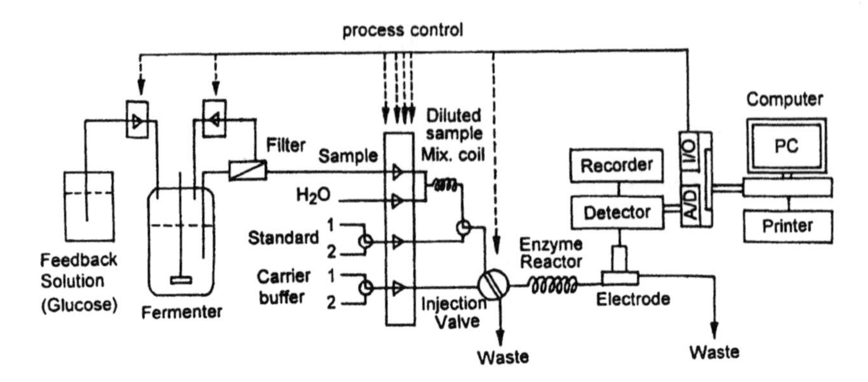

Figure 3. Diagram of automatic on-line FIA biosensor system.

Continuous monitoring and feedback control of the glucose levels in the fermenter finished with the system in glutamate fermentation. When the glucose concentration was set at 300 mg/dL and a computer was used to control the glucose level in the fermenter, the maximum error was less than 5% (Figures 4 and 5).

The automatic biosensor system was used for continuously measuring the glucose concentration in the fermentation liquor during the entire glutamate production process using a 200-liter fermenter. The glucose concentration in the fermentation broth was determined by comparing the results of the automatic biosensor method (Y) with those determined by the GOD reagent kit method (X) and a linear correlation was obtained. The regression equation and correlation coefficient are $Y = 0.348 + 0.9617\ X$ and $r = 0.9794$, respectively (Table 6).

The percentage of recovery by the biosensor method ranged from 97.9–107% (Table 7) which is adequate for routinely determining the glucose concentration in the fermentation process.

We also used the GLOD nylon tube sensor and the automatic FIA system for on-line monitoring of glutamate. The glutamate oxidase nylon tube biosensor

Table 5. Precision Data of GOD Nylon Tube Sensor for Glucose Analysis[a]

Concentration of Glucose, mg/dL	Response Value $\bar{X}$ mv	Assay Times n	Within-Batch SD(mv)	Within-Batch CV%
100	119	20	3.55	2.96
400	432	20	10.40	2.40
1000	801	20	25.90	3.20

Note: [a]Length of GOD-nylon tube:52.5cm.

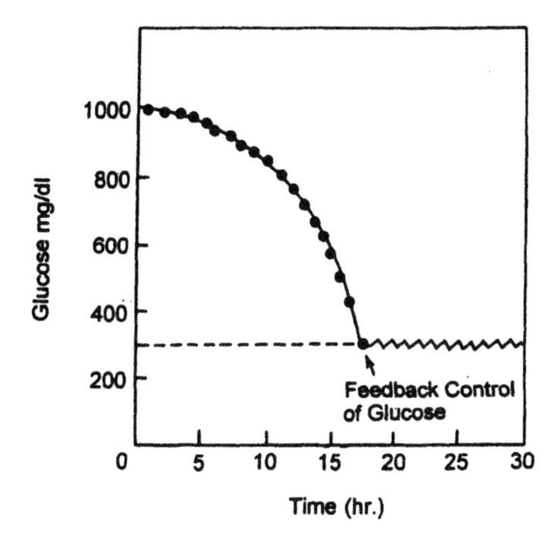

Figure 4. Automatic FIA GOD sensor system used for on-line feedback control of glucose in glutamate fermentation.

responds linearly to glutamate over the concentration range of 10–60 mg/dL. The response time is 30 seconds. The within-batch CV% for 20 performances that responded to 10 mg/dL, 40 mg/dL, and 100 mg/dL of glutamate concentration were 6.27%, 2.34%, and 2.62%, respectively. One nylon tube reactor of immobilized GLOD was used for more than 2000 assays (Table 8).

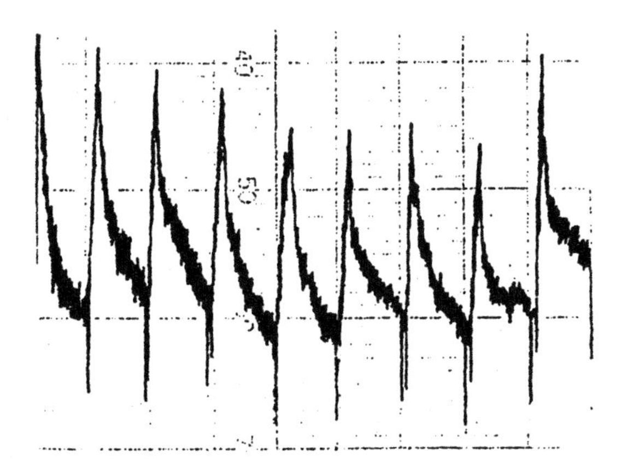

Figure 5. Recorder tracing of on-line glucose in feedback control for fermentation.

Table 6. On-line Monitoring of Glucose Concentration in Glutamate Fermentation Process

Fermentative Time, h	*Biomass O.D. 600 nm, 25-fold Dilution*	*pH*	*Biosensor Method (Y), Glucose %*	*GOD Kit Method (X), Glucose %*
0	0.059	7.60	11.04	11.44
12	0.112	7.95	10.98	11.08
16	0.134	8.19	11.10	11.09
20	0.331	7.52	11.16	11.49
24	0.625	6.47	10.50	10.85
28	0.980	5.66	9.17	10.13
32	–	5.86	7.90	8.01
36	1.604	5.37	5.99	6.76
40	1.796	5.17	6.06	6.28
48	1.808	5.02	5.30	4.81

Table 7. Recovery of Glucose Added in Fermentation Broth

Fermentation Sample	*Original, %*	*Added, %*	*Measured, %*	*Recovery, %*
1	3.742	2	5.77	101.4
2	3.742	4	8.04	107.0
3	3.742	8	11.57	97.9

Table 8. Precision Data of GLOD Nylon Tube Sensor for Glutamate Analysis

Concentration of Glutamate, mg/dL	*$\overline{X}$, mv*	*n*	*Within-Batch*	
			SD(mv)	*CV%*
10	88.4	20	5.5	6.27
40	354.4	20	8.3	2.34
100	557.1	20	14.6	2.62

3. H^+-ISFET BASED BIOSENSORS

The Enzyme Field Effect Transistor (ENFET) is composed of a hydrogen ion selective field effect transistor (H^+-ISFET) and an immobilized enzyme membrane covered on the gate region of the ISFET. ENFET has several advantages over the enzyme electrode including small volume, rapid response, and low output impedance. Furthermore, it makes the applications of differential measurement and integration technology possible [8].

Differential measurement has many advantages compared with single measurement. First the differential model measurement compensates automatically for the effects caused by some external disturbances such as pH variation of the bulk solution and temperature variation of the environment. Drift of output voltage with time is also decreased by using differential measurement. Comparisons between single model output and differential model output are shown in Table 9. Figure 6 shows the time response curves of the single model and the differential model when 1 mL of 0.01 M, pH 6.5 phosphate buffer is added to 4 mL of 0.01M, pH 7.0 phosphate buffer. In Figure 6, $t = 0$ refers to the time when 1 mL phosphate buffer is added. From Figure 6 it can be seen that the output of differential model FET returns to zero after about 10 s, i.e., the differential model FET is insensitive to bulk pH variation. The transitional process during the initial 10 s is caused by the difference in the pH response speed between the BSA membranes covering on the gate regions.

Gold or platinum "pseudo-reference electrodes" can be used for the differential model ENFET. In this way, the reference electrode can be integrated in the FET chip, and thus the sensor is miniaturized. When a gold electrode is immersed in solution, its potential is unstable due to the effects of various kinds of oxidation and reduction at the gold-solution interface. Therefore, a gold electrode cannot be used as the reference electrode for the single model FET. However, it can be used as the reference electrode for the differential model FET, because the potential instability can be compensated for. Table 10 compares single FET output with differential model FET output based on different types of reference electrodes. Data shown in the Table 10 are drifts of output voltage within 3 min of the preparation of the BSA membrane on the gate region of an H^+-ISFET.

Table 9. Comparisons of the Output of Single Model FET with Differential Model FET

Parameter Variation	*Single Output*	*Differential Output*
Temperature variation, mV/°C	1.3–1.6	≤ 0.2
Bulk pH variation, mV/pH	50–53	0.2–0.3
Drift with time, mV/h	0.5–1.0	< 0.5

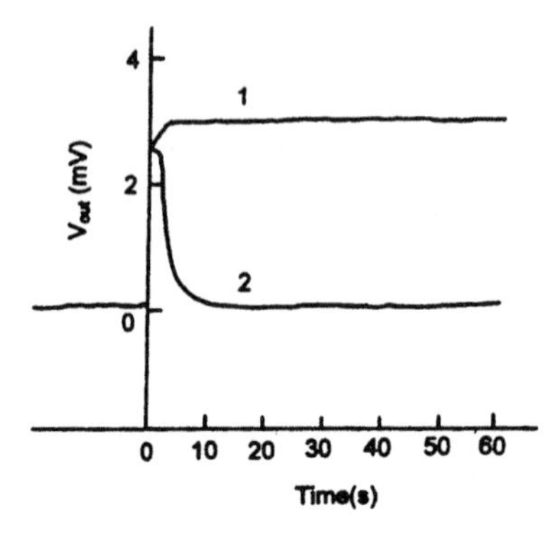

Figure 6. Relationships of differential output and single output of ENFET with time under bulk pH variation.

The typical circuit principle for the differential model ENFET is shown in Figure 7. FET1 and FET2 are two H^+-ISFETs that have the same characteristics. The gate region of FET was covered with a membrane of cross-linked bovine serum albumin(BSA) and contained enzyme, while the other gate region of FET was only a cross-linked BSA membrane. The difference between the source voltages of the two FETs was output through a differential amplifier.

To band the cross-linked BSA and enzyme membrane chemically to the gate region of FET, the amino groups were introduced by treating the surface silicon nitride sensitive film of FET with γ-aminopropyl-triethoxysilane (APTES). Using different analytic enzymes to catalyze the reaction involving the change in H^+ ion concentration, the ENFET can be used to analyze different substrates. In the sections that follow, some ENFETs based on this technology are introduced.

3.1 Penicillinase-FET for Determining Penicillin in Fermentation Broth

Penicillin G content in fermentation broth is an important parameter in the fermentation process. Traditional methods were the chemical iodometric method, which is time-consuming and has low accuracy or, more recently, the high pressure liquid chromatography (HPLC) method,which is costly and requires toxic analytic reagents. The ENFET method has several attractive features, including rapid

Table 10. Comparisons of the Output of Single Model FET with Differential Model FET under Different Types of Reference Electrodes

Reference Electrode Type	*Single Output*	*Differential Output*
Gold electrode	3–4 mV	0.2–0.3 mV
Calomel electrode	0.5–0.6 mV	0.2–0.3 mV

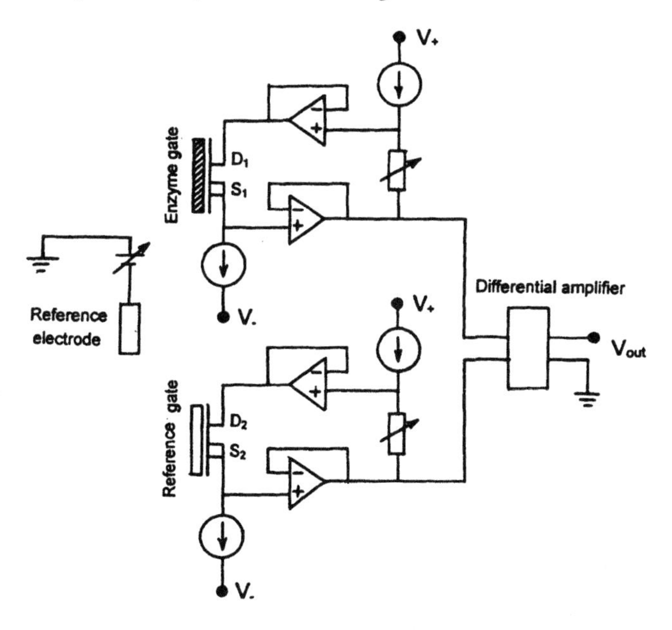

Figure 7. Circuit diagram for differential measurement of ENFET probe.

response, low cost, and greater accuracy and compatibility with differential model measurement and integration circuit technology. Therefore, a miniaturized penicillinase-FET can be fabricated [9,31].

The ENFET assay is based on the following chemical equation:

$$\text{Penicillins} \xrightarrow{\text{Penicillinase}} \text{Penicilloate} + H^+$$

The change in hydrogen ion concentration in the reaction system can be detected by H^+-ISFET. The response time of penicillinase-FET to various concentrations of penicillin is short, and the signal change reaches a steady-state value after about 30 s. The response reproducibility of penicillinase-FET was tested for assaying 10 mmol/L penicillin in a 0.01 mol/L phosphate buffer, pH 7.0. The average response for eight performances was 78 ± 2 mV, the standard deviation was 1.67 mV, and the coefficient of variation 2.1%, respectively. When penicillinase-FET was immersed in a 0.01 mol/L phosphate buffer and stored in a refrigerator, the penicillinase-FET had a lifetime of five months and only a slight decline in output.

The penicillinase-FET was used for measuring the penicillin concentration in the fermentation liquor of 23 batches in a seven-day batch fermentation process for penicillin production. A comparison of the penicillin concentration in the fermentation broth determined by the penicillinase-FET method with those determined by the iodometric method gave a linear correlation $Y = 1.007\ X\text{-}57$. The correlation coefficient for the two methods was 0.9991, which satisfies the requirements for penicillin analysis in fermentation broth. The probe was used for more than 1000 assays during the fermentation process.

3.2 Urease-FET for Clinical Analysis of Urea Content in Human Serum

Quantitative analysis of urea in blood and urine of patients is one of the important means for diagnosing nephritic disease, which is a criterion for evaluating kidney function. Conventional analysis of urea in blood and urine can be performed by spectrophotometric methods, based either on an enzyme-catalyzed specific reaction or on a nonenzyme-catalyzed reaction. However, these methods generally involve rather complicated and delicate procedures. Combining the urease with H^+-ISFET to fabricate the ENFET can be used to determine the urea concentration in blood and urine [32]. The assay is based on the following equation:

$$\text{urea} + 2H_2O + H^+ \xrightleftharpoons{\text{Urease}} HCO_3^- + 2NH_4^-$$

The change in hydrogen ion concentration can be detected by H^+-ISFET. The time-response curves of the urease-FET to different concentrations of urea indicate that the response reaches 95% of its steady-state value within 1 min. Twenty repetitive assays were made on 100 mg/dL urea in buffer solution. The standard deviation and coefficient of variation for 20 measurement were 1.39mV and 1.44%, respectively. Table 11 shows the effect of stabilizer on the operational stability of the urease-FET. In the absence of stabilizer, no response was detected after storing the urease-FET overnight in a refrigerator. Because the response sensitivity decreases by about 10% after 15 assays in the presence of stabilizer 1 and after 250 assays within 1.5 months in the presence of stabilizer 1 and stabilizer 2, urea in both the buffer and the serum were assayed to test the operation stability. The urease-FET was stored at 4 °C between measurements.

A serum sample containing urea was divided into two parts of equal volume, a definite volume of urea solution of known concentration was added to one of them, and buffer was added to the other. Then, the urea concentrations of these two parts were determined. The ratio of the covered amount of urea to that of the added amount was defined as the recovery of the determination of urea by the urease-FET. An average recovery of 99.77% was obtained from the recoveries for three added

Table 11. Effect of Stabilizer on Operational Stability of Urease-FET

Added Stabilizer	*Assayed Number*	*Decrease in Response Sensitivity, %*
None	Overnight at 4 °C	No response
1	15	10
1+ 2	250	10

Table 12. Data on Blood Urea Nitrogen Determined by Both Enzymatic Kit and ENFET Methods

Sample No.	*Enzymatic Kit, mg/dL*	*ENFET, mg/dL*	*Sample No.*	*Enzymatic Kit, mg/dL*	*ENFET, mg/dL*
1	16.8	16.5	26	14.0	13.8
2	11.0	10.5	27	29.0	26.0
3	21.0	21.0	28	10.0	10.0
4	15.5	15.2	29	20.0	20.5
5	14.3	14.0	30	12.0	11.4
6	9.0	9.4	31	15.0	15.2
7	13.8	13.8	32	9.9	9.8
8	10.3	10.5	33	4.4	4.3
9	13.8	12.9	24	10.0	11.1
10	44.1	39.7	35	15.0	12.9
11	51.7	51.4	36	7.7	7.0
12	14.0	12.7	37	18.0	19.4
13	13.0	11.5	38	10.0	11.2
14	15.0	13.8	39	8.4	7.8
15	19.0	17.8	40	11.0	12.5
16	23.0	20.3	41	16.0	14.6
17	13.0	13.6	42	27.0	28.2
18	19.4	17.2	43	18.0	16.0
19	21.0	18.4	44	46.0	46.3
20	11.0	1.9	45	25.0	26.6
21	14.0	13.3	46	10.0	16.0
22	6.7	6.9	47	17.0	16.0
23	61.0	59.7	48	16.0	16.5
24	14.0	13.8	49	17.0	17.9
25	19.0	18.4	50	10.0	9.5

amounts. Table 12 shows the results of serum analysis for 50 blood samples taken from 50 patients at a hospital. The same samples were analyzed spectrophotometrically by the hospital laboratory. The results give a regression equation, $Y = -0.1272 + 0.9695\,X$ and a correlation coefficient of 0.9912, which indicates good agreement between the two methods.

3.3 Penicillin Acylase-FET for Determining Penicillin G

Penicillinase can be used to develop the ENFET for determining penicillin G in fermentation broth. However, because of the specificity of the enzyme, it was impossible to determine the real concentration of penicillin G. Compared with

Table 13. Reproducibility of the Penicillin Acylase-FET

Penicillin G Concentration, mM	*Output, mV (X)*	*Assayed Times, n*	*SD, mV*	*CV, %*
0.5	2.205	20	0.1050	4.76
2.0	7.245	20	0.0999	1.38
8.0	26.59	20	0.1373	0.52

penicillinase, penicillin acylase possesses a higher specificity to penicillin G. The assay is based on the following reaction [10]:

$$\text{Penicillin G} + H_2O \xrightleftharpoons[\text{Acylase}]{\text{Penicillin}} \text{6-Aminopenicillanic acid} + \text{Phenylacetic acid}$$

The change in H^+-ion concentration in the reaction solution can be detected by H^+-ISFET. The response of the penicillin acylase-FET to different concentrations of penicillin G was fast. Usually the output voltage reached the maximum value within 30 seconds after measurement is initiated regardless of the concentration of substrate. In 20 mmol/L phosphate buffer at pH 7.0, the linear range of the calibration curve for measuring penicillin G was from 0.5 mmol/L to 8 mmol/L. The penicillin acylase-FET had good reproducibility when measured at pH 7.0, in a 20 mmol/L phosphate buffer, (Table 13). The values of the coefficient of variation of the output at three different concentrations were all below 5%.

When stored in a 20 mmol/L phosphate buffer at pH 7.0 at 4 °C and periodically measuring the 5 mmol/L penicillin G solution at room temperature, the lifetime of the three ENFETs was six months, and more than 1000 runs were performed without a decrease in the output voltage value. The penicillin acylase FET was used to determine the penicillin G content during the fermentation process in the North China Pharmaceutical Factory. Comparing the values determined by the high pressure liquid chromatography (HPLC) method with the values assayed by the sensor method, these two methods had good coincidence (Figure 8). The correlation coefficient was $r = 0.9944$ and the regression equation was $Y = 1.034\,X\text{-}2083.7$ (X values were assayed by HPLC, and Y values were assayed by the penicillin acylase-FET method).

3.4 Glucose Oxidase-FET for Determining Serum Glucose

Determining glucose is essential in clinical analysis in diagnosing diabetic patients and in the microbiological and food industries for process control. A number of methods for glucose determination are well known, of which the enzyme-based assays are considered to be the most specific. By combining enzyme

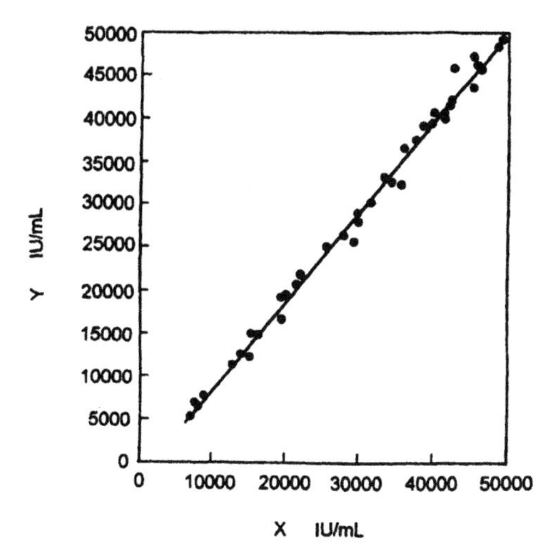

Figure 8. Comparison of the values assayed by the sensor with the values assayed by HPLC. The fermentation broth was assayed in a 20 mM, pH 7.0 phosphate buffer system. X-values were assayed by HPLC, and Y-values assayed by the sensor. $Y = 1.034$ X-2083.7 ($r = 0.9944$).

immobilization technology with H^+-ISFET, a miniature biosensor ENFET can be fabricated [11]. Glucose oxidase catalyzes the oxidation of glucose according to the following equation:

$$\text{D-glucose} + O_2 \xrightleftharpoons{\text{GOD}} \text{D-gluconolactone} + H_2O_2$$

Glucose can be determined by measuring the production of gluconic acid, into which D-gluconolactone is converted. Gluconic acid is a weak acid. Therefore, the pH of the analyte strongly affects the sensor characteristics. Responses of the ENFET to 20 mg/dL glucose at pH 5.5–8.0 are shown in Figure 9. Above pH 6.5, output increases with rising pH. This can be attributed to the ionization of gluconic acid at high pH.

The time curve of glucose oxidase-FET for various concentrations of glucose showed that the signal change reached 96% of the steady-state value within 2 min. The calibration graphs are influenced strongly by buffer capacity. At higher concentrations, more hydrogen ions generated enzymatically are neutralized, and thus the sensitivity of the ENFET is reduced. Meanwhile, the upper limit of the linear range shifts to high glucose concentrations. At low buffer concentrations, as expected, the sensitivity is higher, and the limit of the linear range is lower. When measured in a 5.0 mmol/L pH 7.0 phosphate buffer, the response sensitivity, linear range, and linear correlation coefficient of calibration curve were 0.23 mV/mg/dL, 2–25 mg/dL, and 0.9924, respectively. The response reproducibility of glucose

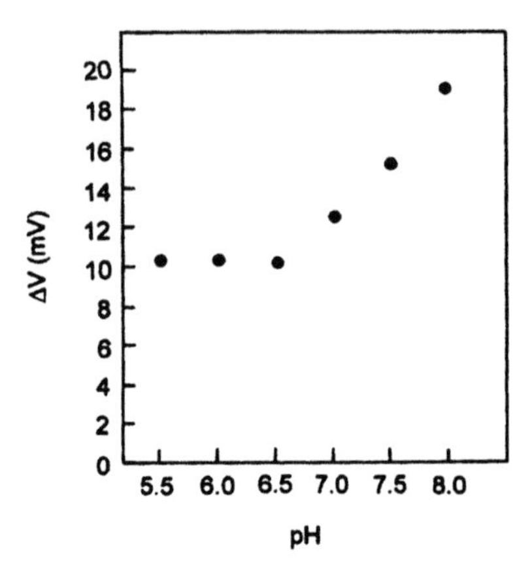

Figure 9. Effect of pH on the response of glucose-ENFET. Conditions: 1 mM phosphate buffer, glucose concn 20 mg/dL, room temperature.

oxidase-FET to 20 mg/dL glucose in a 1 mmol/L, phosphate buffer pH 7.0 showed that the average response for 20 measurements was 10.78 mV, the standard deviation was 0.25 mV, and the coefficient of variation was 2.3%, respectively. The glucose oxidase-FET was used to determine glucose either in buffer solution or in serum. After 250 assays within two months, the response decreased to 90% of the initial value. Between measurements, the ENFET was stored at 4 °C in a phosphate buffer pH 7.0. A comparison of the glucose concentration in serum determined by the glucose oxidase-FET method with those determined by the enzymatic kit method gave a linear regression $Y = -2.049 + 1.0539\,X$ (X values were assayed by the enzymatic kit method, and Y values were determined by the glucose oxidase-FET method). The correlation coefficient for the two methods was 0.9737, which satisfied the requirements for clinical analysis.

4. CHEMILUMINESCENT AND FLUORESCENT BIOSENSORS

In recent years, the optical enzyme biosensor has been widely used as an advanced analytical method because of its higher sensitivity, higher selectivity, rapid responsiveness, higher stability, wider measurement range, and smaller sample size. Furthermore, chemiluminescence or fluorescent detection does not require a reference electrode and is not subject to electrical interference. Therefore, it is inherently safer than electrical devices and is convenient for miniaturizing of the system for continuous *in vivo* biomedical measurement because there is no danger of electrical shock. The optical biosensor could be used in immunoassay, biochemical analysis, microbiology, pharmacology, food science, and other related fields. Recently a large number of both chemiluminescent and fluorescent reactions have been reported [12–15]. This section introduces biosensors based on chemiluminescent and fluorescent-FIA systems developed in our laboratory.

4.1 Chemiluminescent Biosensor for Determining Serum Uric Acid

Serum uric acid concentration is an important index for clinically diagnosing severe renal disease, gout, leukemia, and toxemia. Biosensors enable quick, simple, and convenient detection of serum uric acid concentration. In our enzyme electrode research [3] we found that the sensitivity of the oxygen electrode did not meet the demands of serum uric acid measurement, especially when the concentration of uric acid was below 3 mg/dL. The high sensitivity of luminol chemiluminescence overcomes this drawback. The analytical limit can reach about a 10^{-6} M level [16]. Following is the reaction sequence of the luminol chemiluminescence-biosensor for uric acid measurement:

$$\text{Uric acid} + O_2 \xrightarrow{\text{Uricase}} \text{Allantoin} + H_2O_2 + CO_2$$

$$H_2O_2 + \text{Luminol} \xrightarrow{\text{Ferricyanide}} \text{Aminophthalate anion} + \text{Photon} + N_2 + H_2O$$

The first reaction is highly specific for uric acid. However, the second reaction is catalyzed by enzyme (peroxidase) or metal ions such as $Fe(CN)_6^{-3}$, Fe^{+3}, Co^{+2} and so on that have an oxidation state [17]. Provided that all the conditions are constant, the emitted photon intensity is proportional to the concentration of the serum uric acid. The uricase was immobilized by covalent coupling to the alkylamine controlled pore glass with glutaraldehyde. The immobilized uricase was packed in a small plastic tube, 2.0 mm i.d. and 30 mm long to build the enzyme column. Figure 10 presents the flow-through diagram of the biosensor.

The tubing used in the FIA system was 1.0 mm i.d.. A considerable amount of light was emitted by admixing only luminol and ferricyanide solution. Therefore,

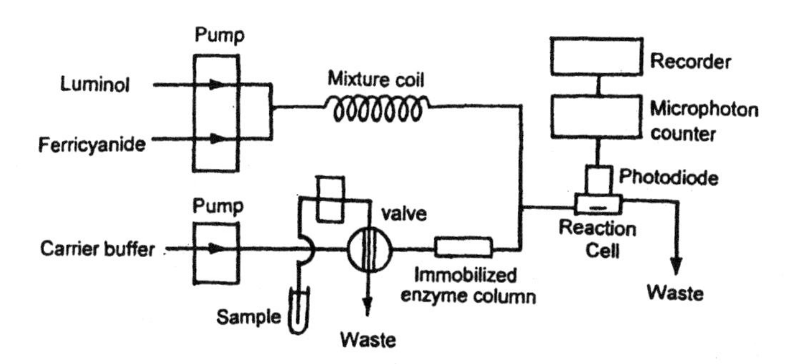

Figure 10. Diagram of FIA for Determining uric acid using a uricase-optical biosensor system.

a delay coil of suitable length was introduced to ensure the decay of such transient emissions before the mixture came into contact with hydrogen peroxide. A certain volume of serum sample met the carrier buffer through the valve injector, and then it flowed into the immobilized enzyme column. After enzymatic reaction, the mixture joined the luminol and ferricyanide solution in the reaction cell. The photons generated by the chemiluminescent reaction were detected by a photodiode attached to the face of the reaction cell. The photon intensity was shown on the window of the photoelectric generator.

The sensitivity of the biosensor was conspicuously affected by the concentrations of luminol and ferricyanide and the of reagent and carrier flow rates. Each factor was optimized for serum uric acid detection. It was also found that a reasonable combination of factors decreased the noise during measurement. The respective optimal conditions finally used for the sensor were as follows: 1.4 mmol/L luminol, 6.7 mmol/L ferricyanide, carrier flow rate 0.6 mL/min, and reagent flow rate 1.35 mL/min. The enzyme-optical biosensor FIA system can successfully be used for measuring the concentration of serum uric acid [18].

The response time of the biosensor to various concentrations of serum uric acid in a 17 μL sample volume was short. The photon intensity of the reaction reached a peak value several seconds after injection. Each sample required about 1.5 minutes for measurement.

Typical linearity of the working curve was from 1 mg/dL up to 20 mg/dL (r = 0.9999). This linear range was wider than that obtained by the electrochemical electrode method [3] and the results reported by Tabata [19].

Pooled serum samples that had low and high contents of uric acid were repeatedly analyzed 20 times (within a day) or 20 times for 10 days (day to day). The imprecision within a day was below 5%, and the day-to-day imprecision was below 8%, both of which are sufficiently precise and reproducible for clinical analysis.

Pooled samples of known uric acid concentration were supplemented with uric acid concentrations of 6 mg/dL, 8 mg/dL, and 9 mg/dL, respectively. The final concentrations measured showed that the recovery of this method was 93% to 109% (Table 14).

The operational stability of this biosensor depends mainly on the stability of the immobilized uricase column. Used at room temperature (25–32 °C) and stored at 4 °C, the immobilized uricase column retained 94% of its original activity even after 2000 runs for five and half months of continual usage. No decrease in photon intensity was observed.

Sixty-one serum samples of high and low uric acid content were measured by both the uricase optical sensor method and the standard colorimetric method routinely used in clinical labs, employing the enzymatic kit (Shanghai, China) in conjunction with a Hitachi 7150 autoanalyzer. As shown in Figure 11 excellent agreement between the two methods was obtained. The calculated linear regression and correlation coefficient were $Y = 0.4 + 0.938\ X$ (X data were measured by the

Table 14. Recovery of the Uricase Optical Sensor Method

Sample Initial Concn. mg/dL	*Added Concn. mg/dL*	*Measured Concn. mg/dL*	*Recovery, %*
3.9	6	9.74	96.8
	9	12.30	93.0
3.4	6	9.93	109
	9	12.89	105
2.5	8	10.47	99.6

enzymatic kit, and *Y* data were measured by the uricase optical sensor method) and $r = 0.9909$, respectively.

Ascorbic acid gave a large negative interference in the luminol chemiluminescent reaction catalyzed by peroxidase, but it gave a smaller negative interference in the same reaction catalyzed by ferricyanide [19]. This biosensor employs ferricyanide to catalyze the luminol chemiluminescent reaction and to decrease the interference of ascorbic acid in serum, which is quite suitable in clinical diagnosis. Meanwhile, the biosensor employing the FIA system compared with the batch analysis system, has the advantage of smaller sample volume and shorter analytical time. In addition, the quantity of immobilized enzyme in the membrane-type biosensor is limited by

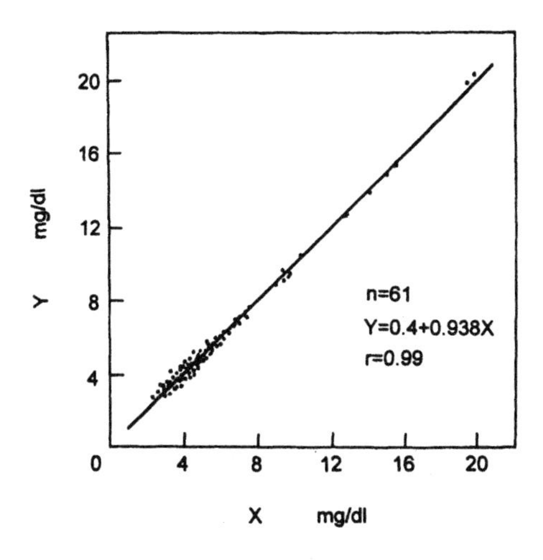

Figure 11. Correlation and regression line for serum uric acid measured by the uricase optical sensor method against serum uric acid measured by the enzymatic kit method. *X*: used enzyme kit method. *Y*: used enzyme optical sensor method.

the size of the reaction cell, whereas the quantity of immobilized enzyme in the column-type biosensor can be increased as the length of column increases to extend the lifetime of the biosensor. The chemiluminescent biosensor, which uses an immobilized enzyme column, showed no decline of output, even after 2000 runs over five and half months of continual usage. It is much more convenient, stable, and less expensive than the enzymatic kit with a soluble enzyme. The precision, reproducibility, and linear range of the biosensor satisfies the requirement of serum uric acid determination for clinical diagnosis.

4.2 Enzyme Fluorescent Sensor for Measuring Glucose

An optical biosensor based on an enzyme that catalyzes a fluorescent reaction for detecting glucose is presented because fluorescent procedures are several orders of magnitude more sensitive than electrochemical methods (10^{-12} M concentrations are determinable compared to 10^{-3}– 10^{-6} M by the electrochemical method). Fluorometric analysis depends on producing a fluorogenic compound as a result of an enzyme-catalyzed reaction. The rate of production of the fluorescent compound is related to both the enzyme and the substrate concentrations. Most of these procedures have been adapted to the autoanalyzer for rapid analysis. A fluorometric assay procedure for amino acid oxidase based on converting the nonfluorescent homovanillic acid to the highly fluorescent-2′-dihydroxy-3,3′-dimethoxydiphenyl-5,5′-diacetic acid has been described [20]. The principle of the fluorometric measurement of glucose is summarized by the following reactions:

$$\text{Glucose} + O_2 + H_2O \xrightarrow{\text{GOD}} \text{Gluconolactone} + H_2O_2$$

H_2O_2 + 2 [homovanillic acid: CH_2COOH, H_3CO, OH] $\xrightarrow{\text{POD}}$ [dimer: CH_2COOH, CH_2COOH, H_3CO, OCH_3, OH, HO]

Nonfluorescent

Fluorescent

λ_{ex}, 315 nm, λ_{em}, 425 nm

GOD and peroxidase (POD) were covalently immobilized on alkylamine porous glass with glutaraldehyde [21]. In the present procedure, 860 U/g of immobilized GOD and 300 U/g immobilized POD can be obtained. The GOD-POD column was combined with a fluorescent spectrophotometer (Hitachi) to construct the optical FIA biosensor system, and it was used for glucose determination [22]. The fluorescent compound produced in the enzyme coupled reaction described can be assayed by fluorescence at $\lambda_{excitation}$ of 315 nm and $\lambda_{emission}$ of 425 nm. The concentration of the homovanillic acid (HVA) solution is an important factor in the FIA system.

Table 15. Precision Data for Fluorescent Analysis of Glucose

Glucose Concn. mg/dL	*Glucose Added µL, µg*	*n*	*SD*	*CV%*
10	2 µl (0.2 µg)	10	0.012	1.25
200	2 µl (0.4 µg)	20	0.026	3.66

The effect of various concentrations of HVA on the fluorescent intensity of the optical biosensor system was determined using HVA concentrations of 1, 2.5, 5, and 12.5 mg/mL with an injection volume of HVA of 100 µL. The responsive fluorescent intensity of the optical biosensor system was highest at an HVA concentration of 12.5 mg/mL. The response time of the fluorescent biosensor was 1 minute for each sample. Using a sample injection volume of 2 µL, glucose was linearly quantified in the range of 50–600 mg/dL. The SD and CV% for within-batch responded to glucose concentrations of 10 mg/dL and 200 mg/dL, using a sample injection volume of 2 µL of glucose solution, were 0.012–0.026 and 1.25–3.66, respectively. The within-batch CV% for 10 measurements responding to glucose of 0.2 µg was 1.25%. Therefore, the fluorescent biosensor has excellent sensitivity for glucose assays (Table 15). Furthermore, the GOD-POD fluorescent biosensor had good operational stability.

5. PIEZOELECTRIC IMMUNOSENSORS

Quartz crystal microbalances are suitable transducers for biochemical sensing in general [23]. The first application of a piezoelectric (PZ) detector was reported by King in 1964 [24]. The detection of gaseous substrates by PZ immunosensors was first reported by Hlavay and Guilbault [25]. However, recent advances in PZ research have shown that quartz crystals can oscillate in contact with solutions and several pioneering studies have been reported [26–28]. It has been shown that the resonant frequency of quartz crystal changes depends on the biochemical concentration of the liquid. The following equation describes the frequency-to-mass relationship:

$$\Delta F = -2.3 \times 10^{-6} F^2 \frac{\Delta M}{A}$$

ΔF is the change in fundamental frequency of the coated crystal, F is the resonant frequency of the crystal, A is the area coated and ΔM is the mass deposited. PZ immunosensors are developed for monitoring biospecific interactions between antibodies and antigens, and this technique is more sensitive than the conventional method.

Table 16. Immobilization of SEC Antibody by Various Methods

Immobilized Method	*Protein Added, mg*	*Protein Uncoupled, mg*	*Protein Coupled, mg*	*Percentage of Immobilized Protein, %*
APTES	24.12	5.40	18.72	77.6
Protein A	24.12	4.37	19.75	81.8
PEI	24.12	4.81	19.31	80.0

5.1 A Piezoelectric Immunosensor for Detecting *Staphylococcus* Enterotoxin C_2

We applied the PZ immunosensor to determining *Staphylococcus* enterotoxin C_2 (SEC), a representative biological warfare agent [29]. In the present study, the antibody of SEC was immobilized on AT-cut PZ quartz crystals of 3.58 MHz resonant frequency, and gold electrodes were used to develop a microgravimetric biosensor for detecting SEC concentration in food. The crystal's electrode surface was chemically modified by acid treatment to generate a hydroxylated carrier surface which, on further reaction with 3-aminopropyltriethoxy silane (APTES) and then with cross-linker glutaraldehyde, provided a surface for binding an antibody covalently. The second method for immobilizing of the SEC antibody involved coating the quartz crystal with protein A on a gold electrode. In the third method for immobilizing the SEC antibody, the quartz crystal was coated with polythylenimine (PEI) and then subjected to further reaction using a glutaraldehyde linkage. The crystal surface coated with a concentration of 5 mg/mL of SEC antibody solution was optimal. The percentage of immobilized protein on the crystal surface was 77.6–81.8% (Table 16).

The PZ immunosensor was used to directly determine SEC in food. A frequency change of approximately 10 Hz occurred for each added 2.5 ng/mL of SEC. The calibration curve was linear over an SEC concentration range of 10–2000 ng/mL. The within-batch standard deviation for six measurements responding to SEC of

Table 17. Precision Data for a PZ Immunosensor for SEC Analysis

Immobilized Method	*Between-Batch*		*Within-Batch*	
	ΔF (n = 6)	*SD*	*ΔF (n = 6)*	*SD*
Protein A	124	±7.61	133	±3.56
PEI	118	±8.21	110	±3.45
APTES	104	±10.02	106	±5.65

25 ng/mL was 3.56–5.65 and the between-batch SD for six days ranged from 7.6–10.0 (Table 17).

The biosensor responds only to SEC and does not respond to other SE types, for example, enterotoxin A and enterotoxin B. The PZ immunosensor was stable for three months when stored at 4 °C or at room temperature in dry conditions. Each antibody membrane was used for more than six assays.

6. CONCLUSIONS

Many immobilized enzymes and proteins can successfully be used for constructing various biosensors with high efficiency, enhanced stability, high sensitivity, low sensitivity to interference, and a wide linear range of analyte concentrations. The biosensor devices can be easily automated or combined with flow-injection systems to extend their capability for continuous and repeated assays. Biospecific sensors should become multifunctional microdevices for optimizing fermentation processes. The application of biosensor systems promises new possibilities for rapid analysis and control of human diseases and environmental pollutants.

REFERENCES

[1] Li, G.X., Yin, M., Liang, X.X., Liu, J.G. (1988). In *Annual Report of National Labs of Transducer Technology*, pp. 34–38.

[2] Li, G.X., Shao, Y., Liu, J.G., Wang, G.S. (1988). In *Annual Report of National Labs of Transducer Technology*, pp. 21–27.

[3] Li, G.X., Shao, Y., Liu, J.G., Wang, G.S. (1990). In *Proceedings of Biosensors '90*, Singapore, p. 297.

[4] Zhong, L.C., Li, G.X. (1990). In *Annual Report of National Labs of Transducer Technology*, pp. 56–59.

[5] Geise, R.J., Yacynych, A.M. (1990). In *Sensors in Bioprocess Control*, Twork, J.V., Yacynych, A.M., Eds., Marcel Dekker: New York and Basel, pp. 173–191.

[6] Reilly, M.T., Charles, M. (1990). In *Sensors in Bioprocess Control*, Twork, J.V., Yacynych, A.M. Eds., Marcel Dekker, New York and Basel, pp. 243–291.

[7] Li, G.X., Liu, J.G., Xie, S.Z. (1995). In *Proceedings of The Second East Asia Conference on Chemical Sensors*, Xian, China, pp. 267–269.

[8] Wang, Z.X., Li, S., Zhong, L.C., Li, G.X. (1990). *Chin. J. Biotechnol.*, **6**(2), 149–156.

[9] Zhong, L.C., Li, G.X. (1992). In *Proceedings of The Fourth International Meeting on Chemical Sensors Tokyo (Sensors and Actuators B. 13–14, 1993)*, Yamazoe, N., Ed., pp. 570–571.

[10] Liu, J.G., Liang, L., Li, G.X., Han, R.S., Chen, K.M. (1996). In *Proceedings of The Fourth China-Japan Joint Symposium on Enzyme Engineering*, Guilin, China, pp. 47–48.

[11] Zhong, L.C., Han, J.H., Li, G.X., Cui, D.F., Fan, J. (1991). In *Annual Report of State Key Lab. of Transducer Technology*, pp. 34–40.

[12] Mccapra, F. (1987). In *Biosensors: Fundamentals and Applications*, Turner, A.P.F., Karube, I., Wilson, G.S. Eds., Oxford Science, pp. 617–637.

[13] Deacon, J.K., Thomson, A.M., Page, A.L., Roberts, P.R., Stops, J.E., Whiteley, S.W., Attridge, J.W., Love, C.A., Robinson, G.A, Davidson, G.P. (1990). *Biosensors & Bioelectronics*, **6**, 193–199.

[14] Attridge, J.W., Daniels, P.B., Deacon, J.K., Robinson, G.A., Davidson, G.P. (1991). *Biosensors & Bioelectronics*, **6**(3), 201–214.
[15] Wilson, R., KremeskÖtter, J., Schiffrin, D.J., Wilkinson, J.S. (1996). *Biosensors and Bioelectronics*, **11**(8), 805–810.
[16] Karube, I., Yokoyama, K. (1992). In *Proceedings of the Fourth International Meeting on Chemical Sensors Tokyo (Sensors and Actuators B. 13–14, 1993)*, Yamazoe, N. et al. Ed., pp. 12–15.
[17] Bostick, D.T., Hercules, D.M. (1975). *Anal. Chem.*, **47**(3), 447–452.
[18] Liu, J.G., Guo, J., Li, G.X. (1995). *Chin. J. Biotechnol.*, **11**(3), 177–183.
[19] Tabata, M., Fukunaga, C., Ohyabu, M., Murachi, T. (1984). *J. Appl. Biochem.*, **6**, 251–258.
[20] Guilbault, G.G. (1976). In *Handbook of Enzymatic Methods of Analysis: Clinical & Biochemical Analysis*, Marcel Dekker, Vol. 4, pp. 68–70.
[21] Guo, J., Mo, P.S., Li, G.X. (1990). *Appl. Biochem. Biotechnol.*, **23**, 15–24.
[22] Li, G.X., Zhong, L.C. (1990). *Proceedings of The First China-Japan Joint Symposium on Enzyme Engineering*, Wuxi, China, pp. 61–63.
[23] Suleiman, A.A., Guilbault, G. G. (1992). In *Biosensors: Fundamentals, Technologies and Applications*, Scheller, F., Schmid, R.D., Eds., VCH, New York, pp. 491–500.
[24] King, W.H. (1964). *Anal. Chem.*, **36**, 1735–1738.
[25] Hiavay, J., Guilbault, G.G. (1977). *Anal. Chem.*, **49**, 1890–1898.
[26] Muramatsu, H., Dicks, J.M., Tamiya, E., Karube, I. (1987). *Anal.Chem.*, **59**, 2760–2763.
[27] Muramatsu, H., Tamiya, E., Karube, I. (1988). *Anal.Chem.*, **60**, 2142–2146.
[28] Muramatsu, H., Kimura, K., Ataka, T., Homma, H., Miura, Y., Karube, I. (1991). *Biosensors & Bioelectronics*, **6**(4), 353–358.
[29] Gao, Z.X., Tao, G.Q., Li, G.X. (1996). In *Proceedings of The Fourth China-Japan Joint Symposium on Enzyme Engineering*, Guilin, China, p. 51.
[30] Liu, J.G., Li, G.X. (1989). *Acta Microbiol. Sin.*, **29**(1), 45–50.
[31] Zhong, L.C., Li, G.X., Wang, Z.X., Liu, L.N. (1990). *Chin. J. Biotechnol.*, **6**(2), 145–150.
[32] Zhong, L.C., Han, J.H., Li, G.X., Cui, D.F., Fan, J., Yang, X.L. (1992). *Chin. J. Biotechnol.*, **8**(1), 57–65.

NOVEL IMMUNOSENSORS FOR RAPID DIAGNOSIS OF ACUTE MYOCARDIAL INFARCTION:
A CASE REPORT

Reinhard Renneberg, Shong Cheng,
Wilhelmina A. Kaptein, Calum J. McNeil,
Judith Rishpon, Bernd Gründig, John Sanderson,
Albert Chu, and Jan F.C. Glatz

OUTLINE

Advances in Biosensors
Volume 4, pages 241–272.

ISBN: 0-7623-0073-6

ABSTRACT

In contrast to enzyme sensors, immunosensors are still in the prototype stage. The reasons for this are the more complex recognition and signal generation and also the lack of demand for an important and profitable analyte (like glucose for enzyme sensors) in pushing the business. Instead there may be clusters of analytes which make sense to detect, like myocardial infarction markers.

This case report analyzes the example of immunosensors for the early infarction marker, fatty acid-binding protein (FABP) and the strategies to reach the final goal of a fast, reliable, rugged, and sensitive immunosensor which can save millions of lives. The steps to develop immunosensors are analyzed, starting from antibody production, ELISA development, random and site-directed immobilization of antibodies, enzyme immuno electrochemical sensors, surface plasmon resonance sensors, and end up with optical dipstick type sensors.

1. INTRODUCTION

In China, Southeast Asia, and the Pacific Rim countries, the quality of life for all people has become highly valued. In addition to improvements in the availability of shelter and food, the most important impacts on the quality of life have been in the areas of health, physical fitness, and protection from diseases. Research in medicine has increasingly depended on technological advances. Asian and Pacific Rim countries must not be left behind as technology advances. Development of bioanalytical high technology in the region is essential. Biosensors are such high-tech bioanalytical devices. They use a biological recognition element (e.g., enzyme, antibody, receptor, cell) in contact with a transducer (e.g., electrochemical, optical, mass-sensitive sensors) to detect analytes (mostly reagentless) in complex samples like blood, plasma, urine, sweat, or wastewater.

Currently, the market for biosensor-based diagnostics is comparatively small. In U.S. dollars, worldwide sales in 1994 for biosensors have been estimated by Taylor [1] at $400 million. More than 50% of all biosensor sales were in the medical field, primarily due to sales of glucose sensors. Taylor predicts a $720 million market in 1999 and by the year 2004 nearly $1 billion in sales. These estimates rely just on the "narrow" definition of biosensors used by *Biosensors & Bioelectronics*, i.e., they do not include other bioanalytical tests and equipment such as dipsticks, ELISAs, and DNA probes.

The "ever more cost-effective delivery of health care" mind-set that has gripped the health-care delivery industry in the 90s has created an opportunity for diagnostic instruments that can both help lower overall costs and improve the quality of health care. Many attributes now sought in health-care diagnostics are found in biosensor-based systems. Specifically, sophisticated and sensitive assays that detect low-concentration analytes without sample pretreatment and are reagentless in small samples are desired elements in good diagnostic systems.

2. THE NEED FOR RAPID SENSORS OF ACUTE MYOCARDIAL INFARCTION

In Hong Kong, heart disease followed by acute myocardial infarction (AMI) is currently the second most common cause of death after cancer (Figure 1). Many myocardial infarction deaths typically occur suddenly before access to medical care. Therefore, the need for rapid diagnosis of myocardial infarction is crucial.

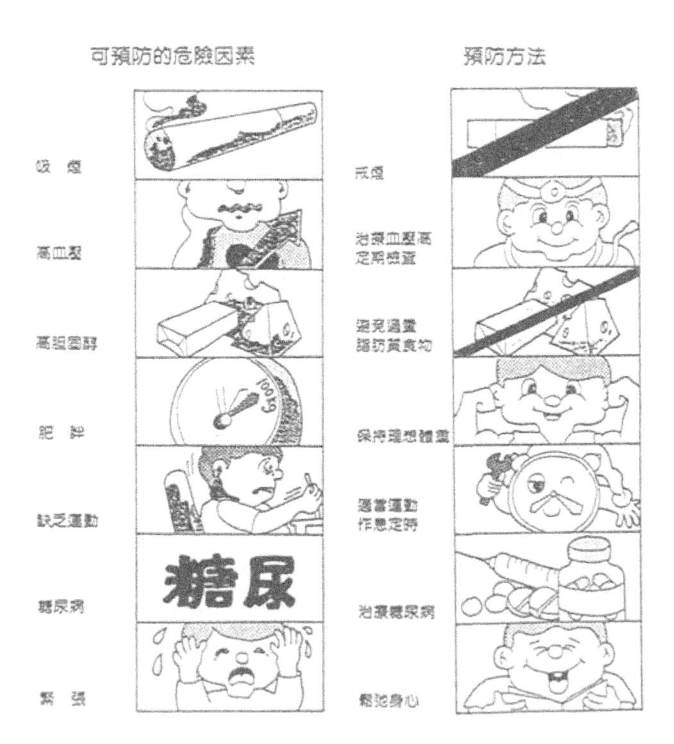

Figure 1. Campaign poster for prevention of heart disease in Hong Kong. The risk factors are shown at the left-hand side (including diabetes, written in Chinese).

The current diagnosis of AMI in human is based on three criteria: the presence of a typical pain syndrome, typical electrocardiographic changes, and the release of proteins from the heart into plasma. The first two criteria can readily be assessed on admission of a patient, but they lack sufficient sensitivity and specificity in diagnosis. Therefore, conclusive evidence that AMI (and of minor myocardial injury) has occurred is usually obtained from the rise and fall of plasma concentrations of cardiac proteins, such as creatine kinase MB (CK-MB), myoglobin (MYO), and troponin T (TnT).

Of all patients entering the hospital emergency room with suspected AMI, about 40% have clear symptoms (clinical history and ECG changes), so that appropriate (usually thrombolytic) treatment can be started without delay. In these patients, biochemical markers are used merely to confirm AMI and estimate infarction size. In contrast, AMI cannot readily be recognized in the remaining 60% of patients. Moreover, the prevalence of AMI in this group is only about 10%. This means that the majority of patients in this group, if properly diagnosed, could be sent home soon.

A more rapid and appropriate diagnosis of AMI is strongly indicated for

1. successful patient treatment,
2. cost reduction,
3. legal liability of practitioners.

Thus, there is an urgent need for strategies to improve diagnostic rapidity and accuracy to identify those patients who present with acute coronary symptoms but for whom early discharge is a safe option.

Consequently, the primary goal is to develop a biosensor system for a rapid and accurate AMI test

- which can be performed in minutes instead of hours and days after infarction,
- which is portable and easy to use (even for the layman),
- which is applicable directly at the patient's location, and
- which gives a quantitative signal.

In Asian countries the government and private hospitals and the medical doctors will be the users of infarction biosensors. They all need a direct on-site (in the presence of the patient) test to obtain a rapid diagnosis (in minutes) in emergencies or for decision making. These customers do not have or do not want to use an expensive and time-consuming centralized laboratory. In recent years, several companies active in the field of cardiac markers have introduced specific sticks (whole blood assays based on chromatographic tests) to test for infarction. Examples are the systems of Boehringer Mannheim (Germany) for troponin T and Spectral (Canada) for different markers [2]. However, such sticks give only qualitative readings (i.e., below or above a certain discriminator level) and suffer from appreciable interobserver differences.

A rapid quantitative AMI-biosensor should establish the occurrence of a myocardial infarction within 1–3 h after the onset of symptoms even away from the hospital. Without exaggeration, it is clear that such a biosensor could save millions of human lives.

In the opinion of leading specialists in cardiology, an even more important goal for biosensor diagnostics would be the *exclusion* (ruling out) rather than *inclusion* (establishment) of AMI because the proper exclusion would save health-care providers time and effort and patients millions of dollars. In the United States, there are about 2,700,000 hospital admissions per year of patients complaining of chest pain who eventually turn out *not* to have had AMI and who could have been sent home the same day *if* the diagnosis of AMI had been properly excluded at an earlier time. It can be estimated that if we were able to discern AMI from non-AMI effectively in patients at an earlier time, this would save up to 40–50% of bed occupancy in the coronary care units [equivalent to a saving of $60, 000 (U.S.) per average hospital per year].

Thrombolytic therapy is indicated in patients with AMI *only* (i.e., infarction must be confirmed) and should be started as early as possible so as to limit heart muscle injury (the doctors rule is: "time saves muscle"). Thrombolytics are expensive and can, if wrongly administered to a healthy person—even cause some serious complications, e.g., brain damage.

In Germany, which has a population of 80 million, 600,000 people suffer from AMI each year of whom 200,000 die immediately or within several days after the infarction. In 1995, around 400,000 heart infarction dipsticks (based on the late cardiac marker troponin T) were sold by Boehringer Mannheim. This German-based company expects an "explosion in sales" in the next several years.

It can be roughly estimated that at least one test is needed per myocardial infarction. The specialists assume, however, a future demand of at least three tests per AMI:

- one immediately upon admission to the hospital,
- another 1 h later, and
- the last test 3 h after admission.

Obtaining real data about heart infarctions in Asian countries is very difficult. Even a quite rare recent study [3] gives direct numbers only for hospital admissions per 10,000 population in Hong Kong and the United States (1986/88). For heart disease, there are 59.8 admissions/10,000 in Hong Kong compared to 155.9/10,000 in the United States. Taking into account a population of 6 million in Hong Kong, a total of around 36,000 hospital admissions per year were made in 1988 for heart problems.

At present, the rate of AMI in Hong Kong is about one-third to one-fourth of that in Europe. Chinese people do not develop vascular disease as often as Caucasians, even given the same risk factors. Data produced at the Prince of Wales Hospital in Hong Kong (J. Sanderson, personal communication) clearly show, for instance, that the endothelial function in the Chinese is preserved even in smokers and in old age. In part, this may be genetic but it also may be related to the diet, especially green tea, which is rich in antioxidants. The increase in the AMI rate in Hong Kong has resulted mainly from the general increase in the length of life. The mean age of AMI patients in Hong Kong is about 10 years older than in Europe or the US. Clearly, the number of AMIs will rise in Hong Kong from increasing Westernization, especially from changes in diet, but this number will fortunately never reach European levels.

Even so, in China which has a population of 1.3 billion and an AMI rate which is one-third of the European level (taking Germany as a standard) the estimated number of acute myocardial infarctions annually would be three million.

The Health Departments of the Asian governments are interested in *minimizing the costs* of AMI treatment by limiting the number of people actually admitted to hospitals to those who actually are suffering from AMI and then *maximizing the level of heath care* for these patients.

The 1987 mean hospital cost per bed day was $537 (U.S.) in the United States [3]. In Hong Kong (HK), the average was only $572 (HK; i.e., seven times less). The average hospital length of stay in Hong Kong (1985/86) was 8.6 days per admission. This means a total of $177 million (HK) of costs for hospitalization of people with heart problems. Assuming that 20% of the patients do not actually have AMI and could be safely sent home, this would mean annual savings of $35 million (HK) if a reliable early infarction test were available! For China, which has 1.3 billion people, a rough calculation of savings is simply overwhelming.

The compelling need for accurate, reliable, and fast AMI tests cannot be ignored by scientists or the government. This is an essential development task. Seven years ago, we began the development of a rapid AMI detection system based on appropriate biosensors. Biosensors using immunorecognition principles (immunosensors) promise sensitive and selective reaction, rapid response, on-site operation, and cost-effective production and operation.

A breakthrough in this process of developing a sensor for rapid AMI detection was the recent finding by Jan Glatz et al. [4–9] that heart-type fatty acid-binding protein (FABP) appears in plasma as quickly as 1.5–3 hours after the onset of the first clinical symptoms of AMI. Except for myoglobin, all of the other established infarction markers (creatine kinase, lactate dehydrogenase, and even the new marker troponin T) are *late markers*, that is, they appear only 6–8 hours after myocardial infarction (Figure 2).

Therefore, FABP is an outstanding new *early* infarction marker. By monitoring FABP in plasma, the assessment or exclusion of an infarction is possible as early as 1.5 hours after the onset of the first symptoms of AMI. Early detection is important for starting early successful treatment.

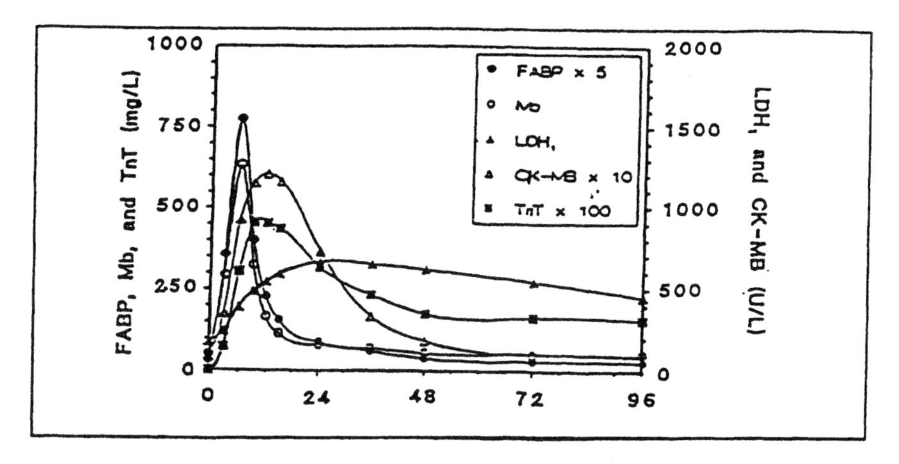

Figure 2. Mean plasma concentrations of FABP, myoglobin (Mb), creatine kinase CK-MB, troponin T (TnT), lactate dehydrogenase isoenzyme 1 (LDH1) as a function of time after onset of AMI in 49 patients [6].

3. PHYSIOLOGICAL BACKGROUND

Acute myocardial infarction (AMI) in humans is usually assessed or excluded by measuring the activities of cardiac enzymes in plasma, such as creatine kinase isoenzyme MB (CK-MB) and an isoenzyme of lactate dehydrogenase (LDH1), or of the concentrations of cardiac proteins, such as myoglobin (MYO) and the troponins T and I (TnT and TnI).

The differences among the various marker proteins in the time at which they appear in plasma after AMI relate to the molecular size of the proteins. Smaller proteins generally appear in plasma more rapidly than larger proteins. Because MYO is a small protein (MW 17.2 kDa), it appears in substantial quantities in plasma within 2–3 h after the onset of AMI, whereas other proteins take at least four hours to appear. It has been suggested that MYO may be a biochemical marker especially suitable for the early assessment of AMI. Myoglobin, however, is also released if skeletal muscles are injured.

Heart-type fatty acid-binding protein (FABP) is a small (MW 15 kDa) cytoplasmic protein that is abundant in cardiomyocytes and is thought to be involved in myocardial lipid homeostasis [4,8,9]. It has been found that FABP is released in substantial amounts from human hearts after AMI and that, like MYO, plasma FABP concentrations increase considerably within three hours of AMI and return to normal within 24 hours. Because their small molecular size, both FABP and MYO are (in contrast to CK-MB, troponins and LDH1) eliminated by the kidneys. Therefore, the ratio of FABP to MYO can be used to discriminate cardiac from skeletal muscle injury [7]. Glatz et al. have compared the release of FABP, MYO,

CK-MB, and LDH1, troponin T and I into plasma after AMI and estimated infarction size from the corresponding plasma curve areas (see Figure 2).

As a plasma marker for ischemic myocardial injury, FABP shares several characteristics with other cardiac proteins (myoglobin, troponin T and I) and with cardiospecific enzymes (CK-MB, LDH1), but it also has some unique qualities that enhance the detection and evaluation of AMI. Like myoglobin, the rapid appearance of FABP in plasma after tissue damage permits the early assessment or exclusion of AMI and the immunohistochemical confirmation of very recent myocardial infarction. This application is also enhanced by the relatively rapid elimination of FABP from plasma because this keeps the steady state plasma concentration of this protein at a low level. The consequence of such low concentrations in normal plasma is that the release of only minute amounts of FABP from myocardial cells significantly raises its concentration in plasma, making FABP a diagnostic marker with high sensitivity. Furthermore, the rapid clearance of FABP also allows identifying recurrent infarctions more easily. Recent single-center and multicenter studies have shown in larger groups of patients that the performance of plasma FABP in detecting AMI in patients admitted to the hospital who have chest pain is significantly better than that of MYO.

Normally, the FABP concentration in human plasma is 1–3 ng/mL [10]. The discriminator value is usually set between 5 and 10 ng FABP/mL. A typical AMI (e.g., loss of 10 grams of healthy myocardium) leads to plasma FABP up to 200 ng/mL.

4. THE DEVELOPMENT OF FABP IMMUNOSENSORS

4.1 The First Steps: Random and Site-Directed Immobilization of Catcher Antibodies

Immunosensors use antibodies or antigens in (more or less intimate) contact with transducers. Electrochemical, optical and mass-sensitive transducers have been used to manufacture immunosensors. Different electrochemical approaches have been undertaken by our groups in Berlin-Buch (until 1991), Muenster (Germany), Newcastle-upon-Tyne (United Kingdom), and since 1995, in Hong Kong [11].

The concept of a membrane enzyme-immunosensor for detecting proteins was developed [12,13] before the German reunification in Berlin-Buch by transferring the ELISA sandwich principle onto a sensor head covered with a cellulose membrane (Figure 3). To detect the therapeutic protein interferon alpha 1 (IFNa1), the membrane was modified to bind the capture antibodies randomly and by site-directed immobilization (Figure 4).

The site-directed immobilization on regenerated cellulose membranes (RCM) takes advantage of the carbohydrate moieties of antibodies that are located on the heavy chains in the C_H2 domain, distal to the antigen-binding site. Mild oxidation of these carbohydrates with sodium metaperiodate results in formyl groups, which

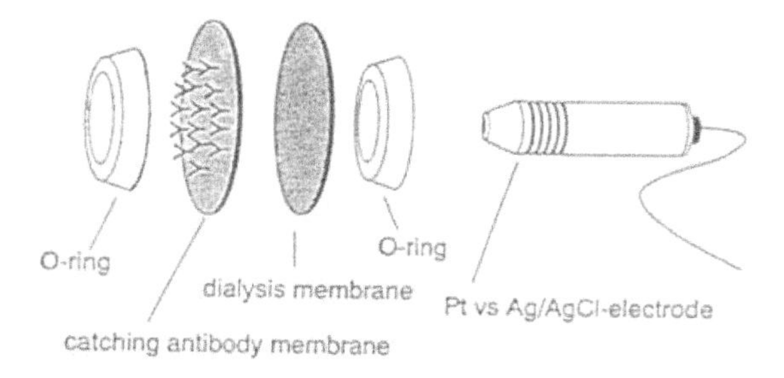

Figure 3. General setup of an amperometric immunosensor.

Random immobilization of capture antibodies
on regenerated cellulose membrane

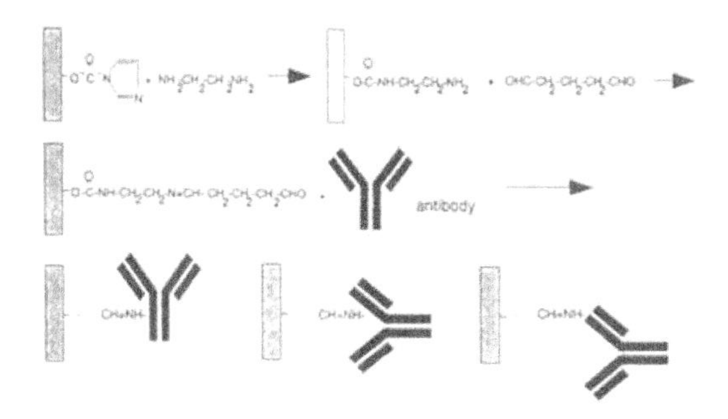

Site-directed immobilization of capture antibodies
on regenerated cellulose membrane

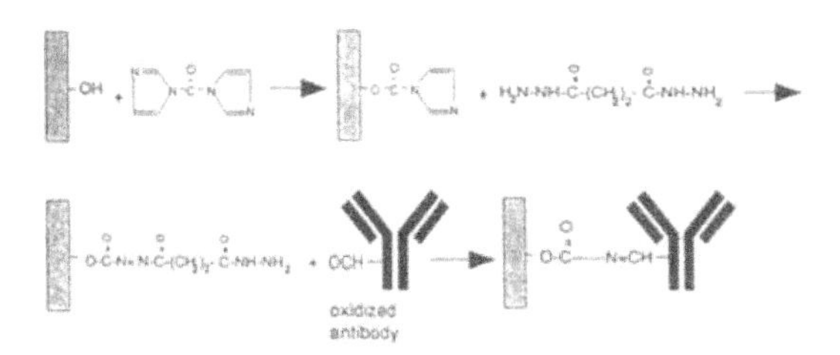

Figure 4. Methods for immobilizing capture (catcher) antibodies: random (upper scheme) and site-directed immobilization on regenerated cellulose membrane.

can be coupled specifically to hydrazide-activated supports. During random immobilization on RCM, antibody molecules are coupled to membranes via amine-reactive groups [13].

The term "site-directed immobilization of an antibody" describes coupling a molecule to known positions within the three-dimensional structure of the immunoglobulin. Thus the attachment of terminal hydrazide groups of the membrane to carbohydrate moieties of the monoclonal antibodies avoids impairing antigen-binding capacity. This is obvious by comparing the effects of immobilization procedures for the capture antibody with the performance of the immunosensor [12,13]. When the immunosensor uses random immobilization, a measuring range of 100–1000 × 10^3 U/mL was reached for interferon. For site-directed immobilization, from 3–12.5 × 10^3 U/mL, i.e., a much more sensitive but smaller measuring range, was obtained (Figure 5). The measuring time is shortened from 50 min (random

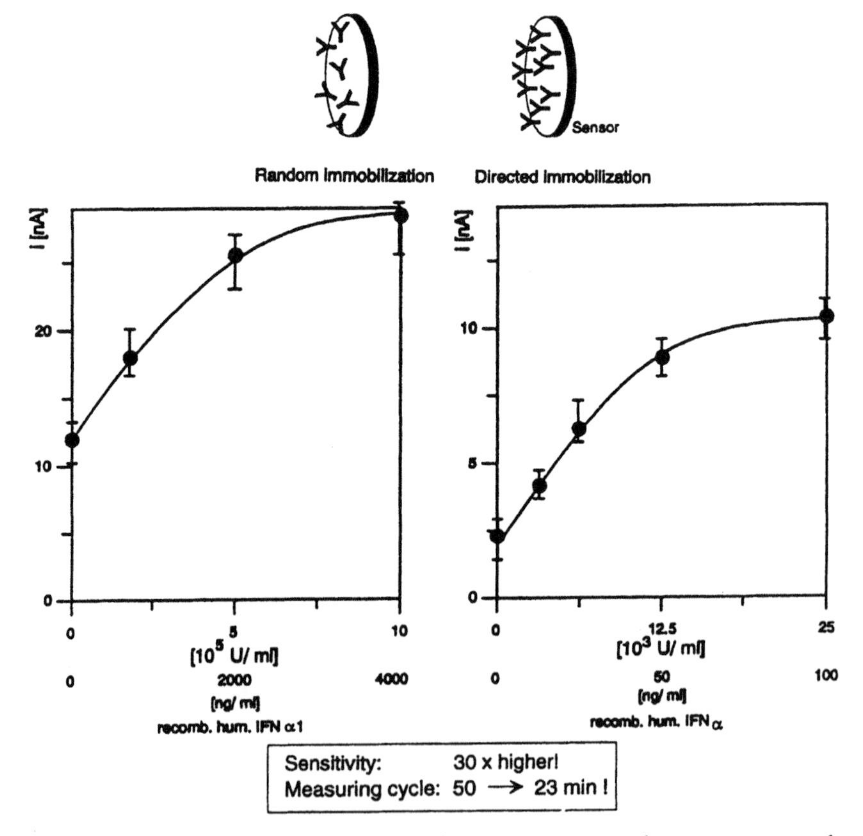

Figure 5. Comparison of sensitivities and measuring time of immunosensors for detecting interferon alpha using randomly and site-directed antibodies [13].

immobilization) for one cycle (with regeneration) to 23 min (site-directed immobilization). The "random" sensor is used 15 times, whereas the "site-directed" sensor is reusable only nine times. The capture antibodies were regenerated (removal of the bound antigen) with glycine-HCl buffer (pH 2.2).

To examine the influence of capture antibody orientation, different immobilization techniques were compared in detail after Renneberg and Warsinke moved from the biosensor group in Berlin-Buch to ICB Muenster following German reunification in 1991. Using another antigen (apolipoprotein E), the Ph.D. student Markus Meusel [14] obtained improved results, again with site-directed immobilization, by coupling the carbohydrate moieties of antibodies to terminal hydrazide groups on the membrane surface. In contrast, the random derivatization of primary amino groups of the monoclonal antibodies led to a total loss of antigen-binding capacity in this assay configuration.

4.2 The First FABP Sensor

Coincidentally, Jan Glatz and colleagues in Maastricht discovered the exciting properties of FABP. It was natural that we should try together to develop new biosensor technology to measure the new infarction marker FABP.

The first electrochemical FABP sensor was developed by the Muenster immunosensor [Fraunhofer Institut fuer Chemo- und Biosensorik, (ICB)] group during the Ph.D. work of Cordula Siegmannn-Thoss [15]. Using bovine FABP as a model, two approaches based on the classical enzyme immunoassay were tested: (1) the competition type, in which defined amounts of glucose oxidase (GOD)-labeled FABP compete with free FABP to bind to the immobilized capture antibodies; (2) the sandwich type, where the free FABP binds to the immobilized capture antibodies and forms a sandwich with secondary GOD-labeled (detector) antibodies (Figure 6). In both approaches, the resulting decrease in oxygen concentration after adding glucose was determined by using an amperometric oxygen electrode.

A sensor calibration curve for bovine FABP using the competitive principle, nitrocellulose as the membrane, and GOD as the marker was linear within the range up to 100 ng FABP/mL buffer. In principle, this revealed the possibility of attaining a measuring range for FABP useful in clinical diagnosis. However, the nonspecific binding of the conjugate to the blocked membrane limited the ultimate sensitivity of the measurement. For this reason and because of the difficult optimization of assays of the competitive type, we turned to the sandwich principle. Moreover, the nitrocellulose membranes displayed a significant amount of leaching out of the noncovalently immobilized antibodies. Taking into consideration the low binding capacity of filter papers for antibodies, we selected the more effective "Immunodyne" membrane for all further experiments. This ready-to-use cellulose membrane is chemically preactivated and binds proteins covalently after incubation.

Because the antibodies covalently couple to "Immunodyne," regeneration of the antibody membrane became possible. Indeed, the use of the glycine/HCl buffer, pH

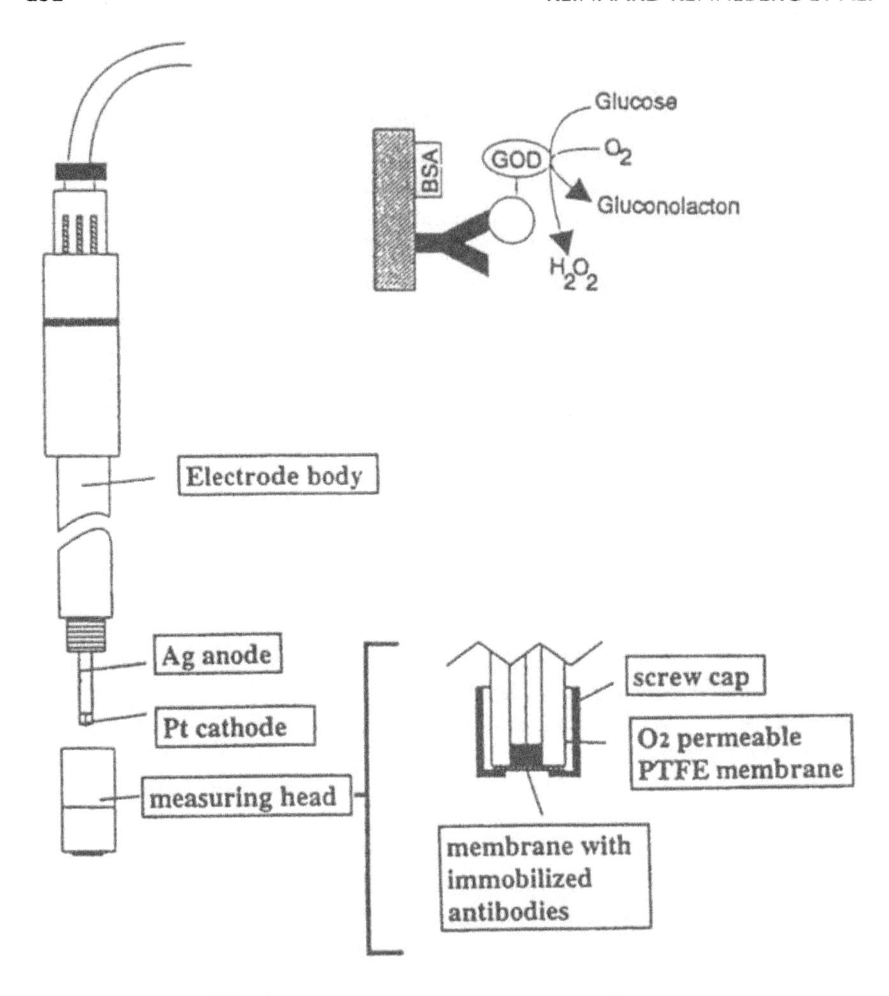

Figure 6. Setup of the first amperometric immunosensor for FABP using an oxygen electrode and the sandwich approach [15]. The computer simulation (courtesy Dietmar Schomburg, University of Cologne) shows the spatial arrangement of catcher antibody, FABP, detector antibody, and marker enzyme (here: GOD subunit).

2.7, allowed the rapid dissociation of the antibody/antigen/antibody-enzyme conjugate complex. The overall signal obtained initially increased with the number of regenerations. However, after the fourth regeneration, it scattered within a small mean. Taking into account an acceptable relative error of ±10%, this renewable immunosensor was usable for 10 measurements.

The sensitivity of this immunosensor permitted determining FABP in the concentration range 5–100 ng FABP/mL buffer. Because all of the samples had to be diluted with the buffer at least 10 times, the overall sensitivity was not sufficient to distinguish low physiological values of FABP.

The sandwich principle for determining human FABP by using monoclonal antibovine-FABP antibodies as the capture and GOD-labeled polyclonal goat-antihuman-FABP antibodies as the detector gave better results by far. The monoclonal antibodies were immobilized on the "Immunodyne" membrane, which was mounted in front of the Clark-type oxygen electrode. The sensor was operated with a single use of the membrane. In clinical application, the regeneration technique could not be used anyway for safety reasons, i.e., risk of infection.

The calibration curve was linear ($r = 0.98$) within the range 5–80 ng FABP/mL buffer and had a standard deviation of about 10%. Recovery experiments using normal plasma spiked with purified human FABP yielded an average recovery of 93 ± 5%, which compared well with the recovery of the sandwich ELISA (93.5%) [15]. Thus, the development of this enzyme immunosensor made measuring human FABP in plasma samples of cardiac patients feasible for the first time. Figure 7 shows data for the release of FABP into the plasma of a cardiac patient who had diagnosed anteroseptal myocardial infarction and a normal kidney function measured with the immunosensor. Although this immunosensor reduced the assay time to only 30 min, the preparation of the immuno membranes was very laborious. One must also mention the insufficient sensitivity for discriminating normal patients from patients with a very early AMI.

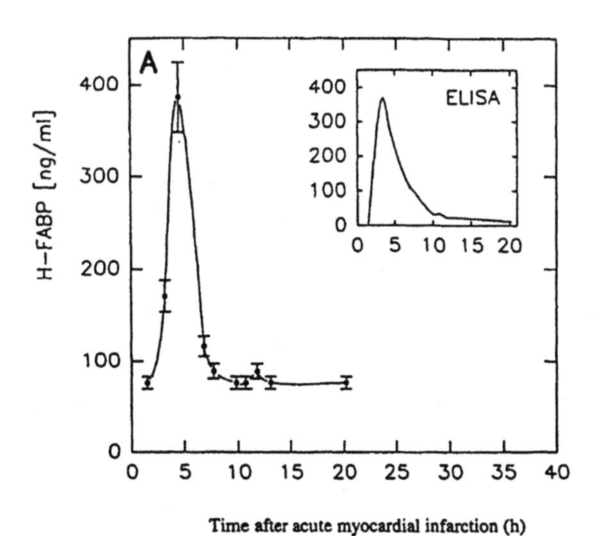

Figure 7. FABP values detected with the immunosensor shown in Figure 6 and an ELISA test in plasma samples of a patient who had diagnosed anteroseptal myocardial infarction after [15].

4.3 Monoclonal Antibodies and ELISA Development

ELISA methods developed in 1992 in Maastricht and in 1995 in Japan for estimating FABP take 4 h [6] and 1.5 h [16], respectively. The preparation of monoclonal antibodies against human FABP was accomplished by Werner Roos in Switzerland [17] at Hoffmann-La Roche, Basel. Antibodies were purified from crude culture supernatant by protein G chromatography following standard procedures.

Both Roos in Basel and in parallel the Ph.D. student Wendy Shong Cheng in Hong Kong have used the BIAcore instrument (Pharmacia) to characterize the monoclonals. In Hong Kong we tried to find the antibodies best suited for constructing a sandwich-type immunosensor. The immunoassay binding studies of monoclonal antibodies to FABP were performed with six hybridoma-derived antibodies supplied by Werner Roos. The novel technique of Surface Plasmon Resonance (SPR) detection was applied to study the affinity and epitopes of heart FABP. In contrast to most applied techniques for intermacromolecular detection, SPR detection does not require labeled interactants. It also allows one to visualize the macromolecular interaction in real time.

Rabbit antimouse (RAM) Fc polyclonal antibody was used as the primary capturing reagent and bound to the dextran-covered surface of the sensor chip. Then, it bound the Fc portion of anti-FABP monoclonals for further kinetic and epitope investigation. Therefore, the antigen-binding properties of the monoclonals have not been influenced by immobilization. The RAM Fc surface used was very stable even after treatment with a regeneration buffer of 100 mM HCl. There was no loss of ligands after the regeneration cycle. The stable RAM Fc surface allowed us to be confident about the kinetic analysis.

The quantitative analysis of the interaction between heart FABP and monoclonals revealed information about rate and affinity constants (Figure 8). The model used for interpreting binding data assumed Langmuirian binding in the proportions of one to one. The association rate constants for heart FABP to monoclonal antibodies were established in a range from 3.13×10^4 to 2.38×10^5 $M^{-1}s^{-1}$ and dissociation rate constants fall within 10^{-3} s^{-1}. From the association and dissociation rate constants, the calculated affinity constants were between 1.26×10^7 to 1.27×10^8 M^{-1}. This shows that the monoclonals have appropriate binding power for the antigen. The calculated dissociation constants have the value of 7.97×10^{-9} to 7.94×10^{-8} M.

The epitope specificity for heart FABP against monoclonal antibodies was studied and an epitope map was constructed (Figure 9). Monoclonals were bound sequentially to heart FABP. The results provide further evidence about the specific binding sites of the monoclonal antibodies. Three separated epitopes of heart FABP were defined for the monoclonals. It was concluded that the tested antibodies were very well suited for binding FABP. The epitope map predicted what pairs of Mab

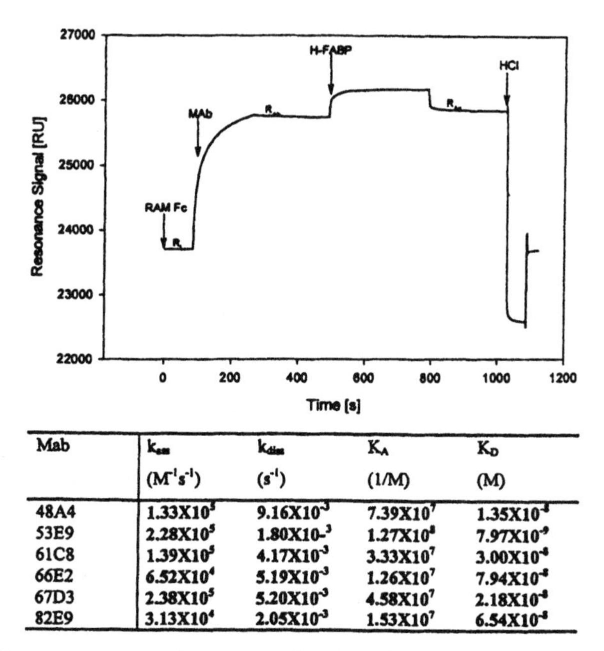

Mab	k_{ass} $(M^{-1}s^{-1})$	k_{diss} (s^{-1})	K_A (1/M)	K_D (M)
48A4	$1.33X10^{5}$	$9.16X10^{-3}$	$7.39X10^{7}$	$1.35X10^{-8}$
53E9	$2.28X10^{5}$	$1.80X10\text{-}^{3}$	$1.27X10^{8}$	$7.97X10^{-9}$
61C8	$1.39X10^{5}$	$4.17X10^{-3}$	$3.33X10^{7}$	$3.00X10^{-8}$
66E2	$6.52X10^{4}$	$5.19X10^{-3}$	$1.26X10^{7}$	$7.94X10^{-8}$
67D3	$2.38X10^{5}$	$5.20X10^{-3}$	$4.58X10^{7}$	$2.18X10^{-8}$
82E9	$3.13X10^{4}$	$2.05X10^{-3}$	$1.53X10^{7}$	$6.54X10^{-8}$

Figure 8. Characterization of six monoclonal antibodies against human heart FABP using Surface Plasmon Resonance (BIAcore device, Pharmacia). The established apparent kinetic constants are shown in the table.

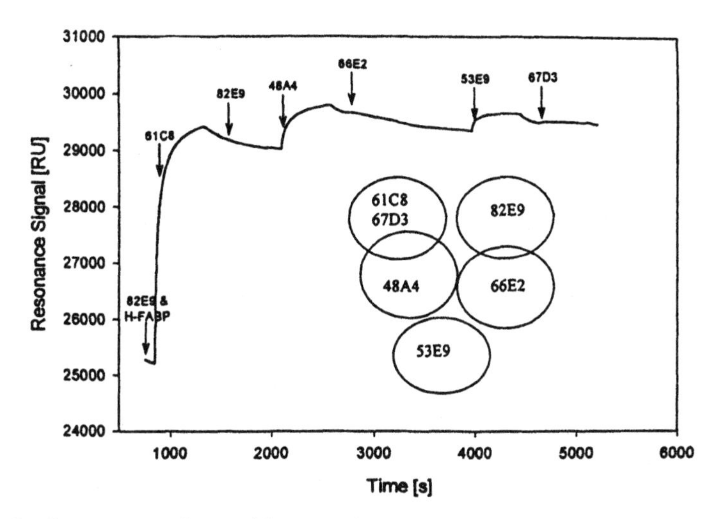

Figure 9. Sensorgram obtained by SPR demonstrating the sequential binding of six monoclonal antibodies to FABP. The antibody 82E9 was bound to rabbit antimouse Fc. Insert: Two-dimensional representation of the binding pattern of the six monoclonal antibodies to the three different epitopes of the heart FABP.

should be used to construct an ideal sandwich-type immunosensor for early detection of acute myocardial infarction.

In all of the following experiments both in Maastricht and Hong Kong, the monoclonal antibody 67 D3 was used as the capture antibody and 66 E2 as the detector (labeled with an enzyme). Sensitive enzyme-linked immunoassays (ELISAs) have been developed. Figure 10 shows the test by Glatz et al. using these monoclonal antibodies and horseradish peroxidase (HRP) as the marker enzyme.

In 1994/95, Ulrich Kunz and Andreas Katerkamp at ICB Muenster tried to use SPR to detect FABP (Figure 11). They constructed planar and fiber optic SPR devices [18,19]. For a competitive FABP assay, they covered the silver layer of the sensor surface with human FABP and blocked the surface with a neutral protein (BSA). The sample containing FABP was preincubated for 30 min with a fixed amount of polyclonal anti-FABP antibodies. When injected into the measuring cell, sample FABP and surface-bound FABP did compete for binding with the antibodies. The result of this competition was clearly seen with SPR. The detection limits for both sensors are 200 ng FABP/mL. The sensor surface was generated 20 times with glycine/HCl buffer (pH 2.2). Despite its low sensitivity, the fiber optic device could be useful in giving real-time data. It could be miniaturized for remote sensing, maybe even for *in vivo* use [19].

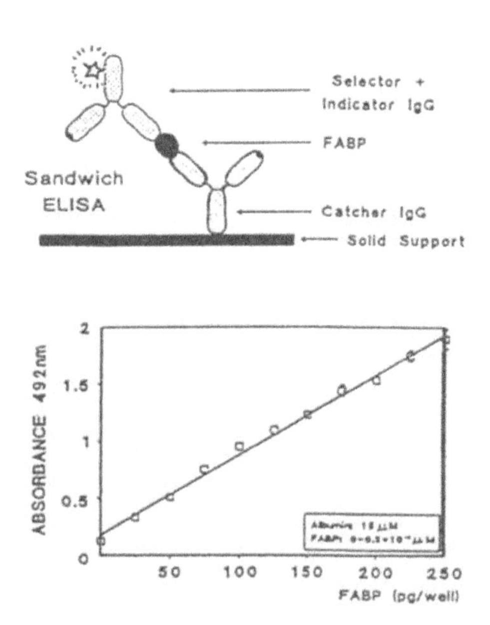

Figure 10. Sandwich ELISA developed by Wodzig et al. [10] using two monoclonal antibodies as catcher and detector antibody and horseradish peroxidase as an enzyme label.

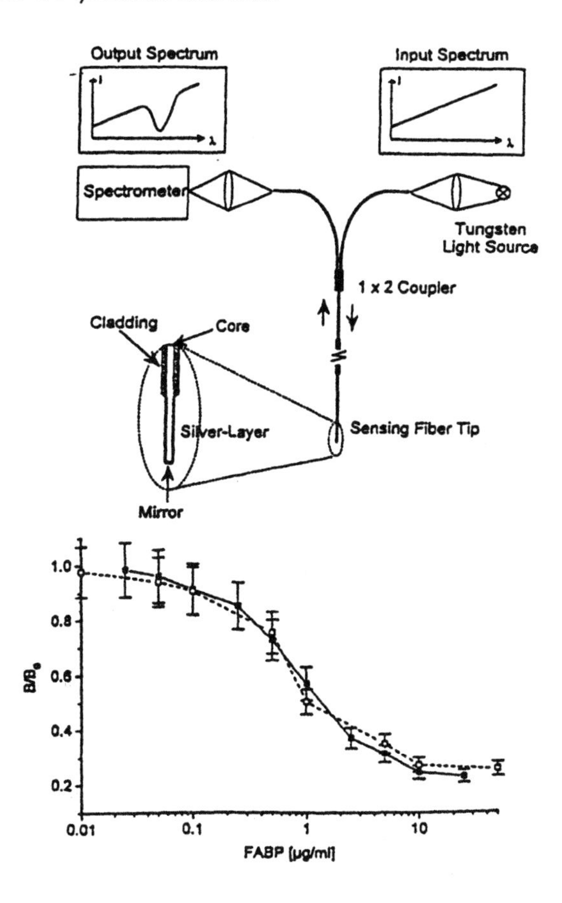

Figure 11. Experimental setup of the fiber optical Surface Plasmon Resonance device [18] and calibration curves for planar and fiber optic SPR sensors (competition principle).

4.4 The EUROCARDI Trial and the FABP Sensor

The FABP immunosensor development was pushed forward greatly by a European Union project, "Eurocardi," during 1994–1996 under the leadership of Jan F.C. Glatz at CARIM (Maastricht, The Netherlands). A practical prototype immunosensor was developed with assistance from a "European Union Concerted Action" grant and was tested in hospitals in Germany, The Netherlands, U.K., and Denmark. The Concerted Action group also investigated the value of FABP as an early infarction marker [20].

For this immunosensor, alkaline phosphatase (AP), a marker enzyme that higher specific activity than GOD was used. William Heineman's group in Cincinnati was the first to develop a special "electrochemical substrate," *p*-aminophenylphosphate (pAPP), for AP [21]. A stable reference potential at different currents and a reproducible potential plateau of the electrochemical active product is the basis of electrochemical determinations. A cyclic voltammogram for the substrate pAPP, the product, *p*-aminophenol (pAP), and carbonate buffer, pH 9.6, revealed oxidation waves of pAPP and pAP with peak potentials at about +200 and +800 mV versus Ag/AgCl, respectively (Figure 12) [22]. Thus, the product is easily detected by electrochemical oxidation without interference from the substrate and buffer. The hydrodynamic voltammogram for pAP in a carbonate buffer showed that the limiting oxidation current starts at potentials higher than +300 mV versus Ag/AgCl. Thus, a potential of +350 mV was chosen to detect *p*-aminophenol in subsequent studies [22] by Andrea Schreiber et al.

The development of the EUROCARDI sensor started with reusable carbon working electrodes (graphite rod) and disposable nitrocellulose membranes covering the electrodes. A semiautomatic device was constructed by Rainer Feldbruegge and the Dept. of Immunosensors at the Institute CB in Muenster for automated incubation and rinsing steps (Figures 13, 14).

Four prototype instruments were built to be supplied to all partners of the "Eurocardi" project. The first measurements were promising (Figure 15). Four immunosensors, one reference sensor and three sensors for real samples, detected FABP in parallel. However, it was soon clear that the procedure of changing the

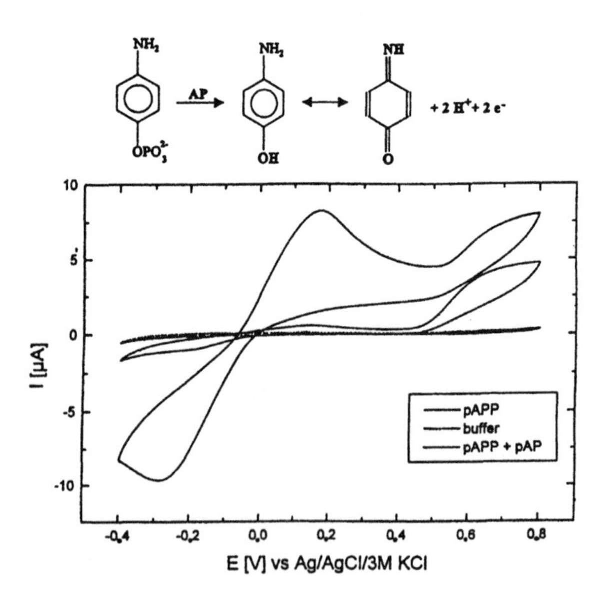

Figure 12. Reaction sequence and cyclic voltammogram for substrate and product of alkaline phosphatase and for the carbonate buffer (pH 9.6) [22].

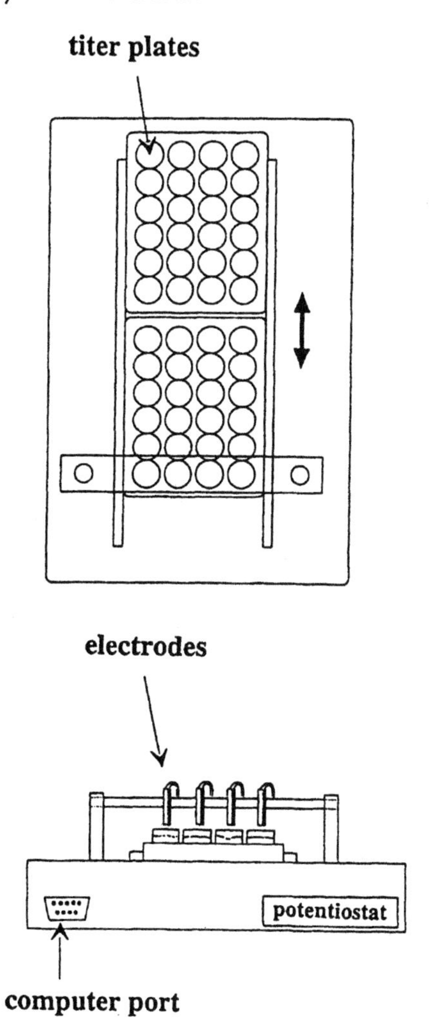

Figure 13. Setup of the semiautomatic EUROCARDI device for measuring FABP.

membranes for each measurement was tedious and also inadvisable. When real blood samples are used, laboratory personnel should avoid any additional contact with them.

This practical fact also solved the question of whether to design a regenerating sensor or a one-use diagnostic. A one-use sensor was finally designed. The reusable graphite rod sensor was replaced by a screen printed graphite sensor (Figure 16). Because now there was no longer a need for a stable multiple-use immunosensor, the question of random versus site-directed immobilization could also be solved. In favor of the simple (and cheaper) approach, adsorption of antibodies on graphite

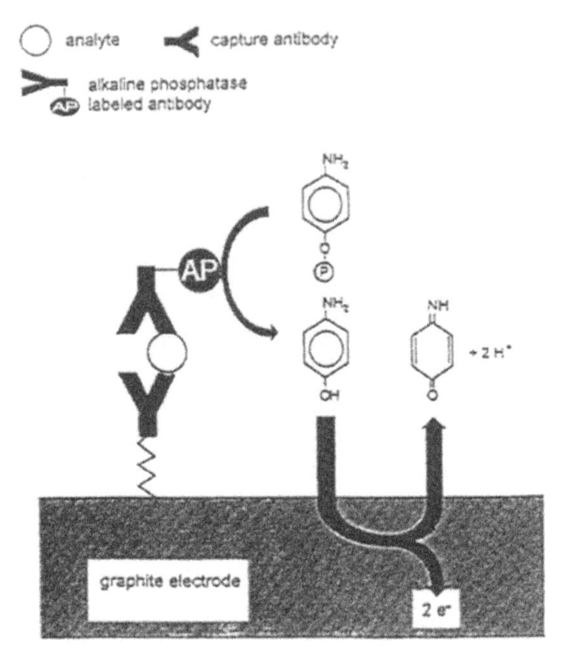

Figure 14. Principle of the amperometric EUROCARDI immunosensor.

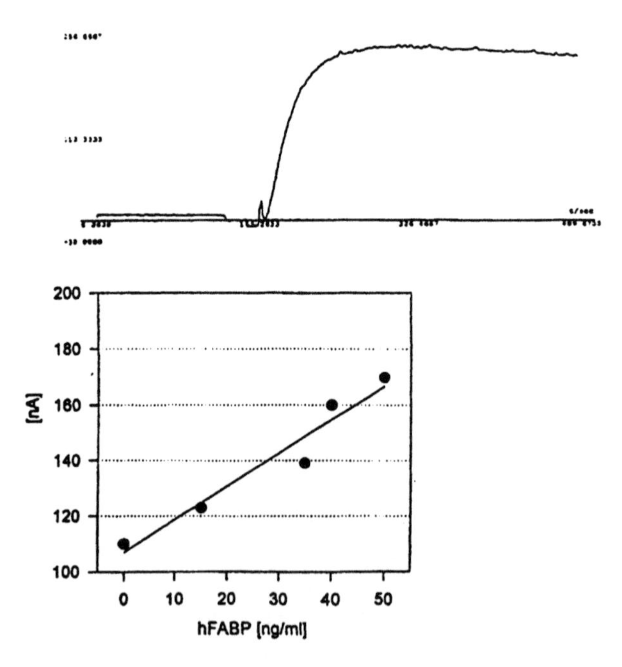

Figure 15. First results obtained with the semiautomatic EUROCARDI device.

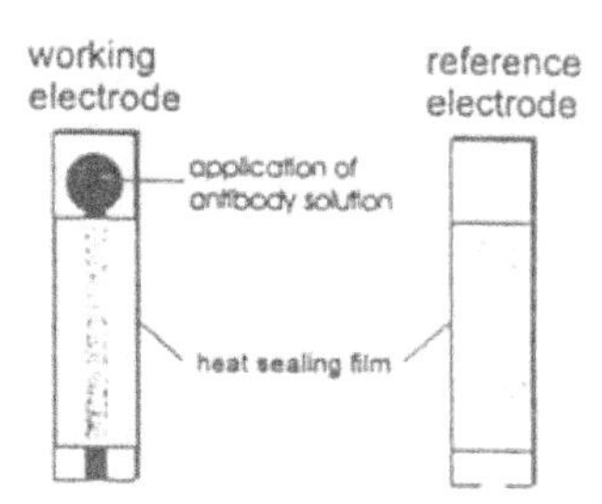

Figure 16. Preparation of the working electrode by screen printing [22].

was chosen instead of sophisticated site-directed immobilization. Because the sensor was not intended for reuse, there was no need to stabilize the catcher. This approach somewhat follows the development of disposable enzyme sensors for glucose.

The screen printed immunosensor development was led by Rainer Feldbruegge and the immunosensor group with Andrea Schreiber and Goeran Key at the ICB Muenster [22] in collaboration with Jan F.C. Glatz's group in Maastricht.

The working electrodes were designed as follows. A graphite paste was mixed with standard screen printing oil and the mixture added to a heat sealing film with the help of a screen printing pattern and kept at 333 K for 1 h. The connecting link and the working electrode were separated by a heat sealing film covering part of the electrode (Figure 16). Then, the electrodes were incubated for 10 min at room temperature with antibody solution. After passive adsorption of the antibodies, the electrodes were rinsed once with buffer, and the remaining protein binding sites were blocked for 20 min with tris buffer containing 2% bovine serum albumin, followed by a final rinse with buffer. The reference electrodes were produced in the same way using an Ag/AgCl paste instead of graphite.

To estimate the reproducibility of the working electrodes without adsorbed antibodies, repeated measurements were carried out with different concentrations of *p*-aminophenol (pAP). Five pAP concentrations were measured with six electrodes each. Standard deviations of measured currents in all cases were below 10% and demonstrated similar performance of the electrodes in signal generation without the biological components. This was a prerequisite for a correct measurement with the immuno electrodes containing the adsorbed monoclonal capture antibodies for detecting FABP in real plasma samples.

A series of immuno electrodes was prepared as described before and subsequently air dried and stored at either 277 K or at room temperature. After various storage times and temperatures, the performance of these electrodes was compared to that of the freshly prepared immuno electrodes. The electrodes stored at 277 K were stable for at least three months, whereas those stored at room temperature produced significantly reduced signals after three months. Thus, it is possible to prepare antibody-coated working electrodes in batches and store them in a refrigerator until use. A recovery test was performed using normal human plasma spiked with eight different concentrations of recombinant FABP in the range of 0–350 ng/mL. The mean values calculated from eight measurements of each sample revealed recovery rates between 80 and 100%. To determine the assay precision, four patient plasma samples with different FABP concentrations ranging from 0–300 ng/mL FABP, as assessed by ELISA, were measured eight times. The intra- and interassay coefficients of variation were between 10 to 15% and 9 to 16%, respectively. Although, in principle, the disposable electrodes give one data point only per sample the precision of the interassay, whose coefficients of variation were between 9 and 16%, was fairly good.

To calibrate the immunosensor, human plasma samples collected from healthy donors were spiked with different amounts of FABP. Although not linear over the whole range, the resulting standard curve allowed calculating FABP concentrations in real plasma samples in the range from 10 to 350 ng/mL (Figure 17). The background due to nonspecific binding was <5 ng/mL in all measurements. Then, the performance of the immunosensor was compared with a new reference ELISA [10]. This novel one-step fast (30 min) ELISA for FABP developed by Wodzig et al. in the Maastricht group employed the same monoclonal antibodies in the sandwich configuration as the immunosensor. Six plasma samples, taken from a

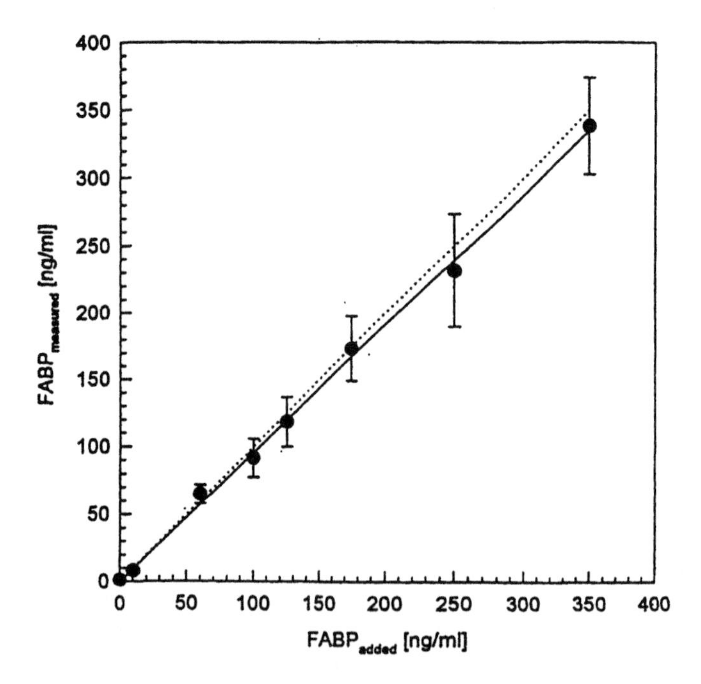

Figure 17. Determination of the recovery rate by eight independent measurements of plasma samples containing different amounts of FABP [22].

patient during the first hours after AMI, were measured. As shown in Figure 18, the FABP concentrations determined by the two methods (ELISA and biosensor) and by two different devices correspond quite well. The *sensitivity* of the immunosensor was a main problem. The sensor was too insensitive to distinguish low values of FABP (just between normal and pathological values at the borderline).

The immuno electrodes were prepared by a screen printing technique that can easily be scaled up and automated for cost-effective mass production. Because the graphite electrodes that have the immobilized capture antibody can be stored in the refrigerator for at least three months, they can be produced in large batches in advance. At only 20 min, the assay time is short compared to conventional ELISA assays. However, because of the urgency of an infarction situation, one should look for even faster and more simple detection methods.

Using the EUROCARDI laboratory prototype, it can be shown that FABP measurement for blood plasma is feasible. This sensor, however, had the disadvantage of several separation (incubation and washing) steps. The entire measurement was restricted to quite a complicated semiautomatic laboratory device and needed at least 20 min to complete. This sensor could be used only by well-trained personnel.

In the following section we discuss the most promising approaches to producing simple, reliable, sensitive FABP sensors for practical medical use.

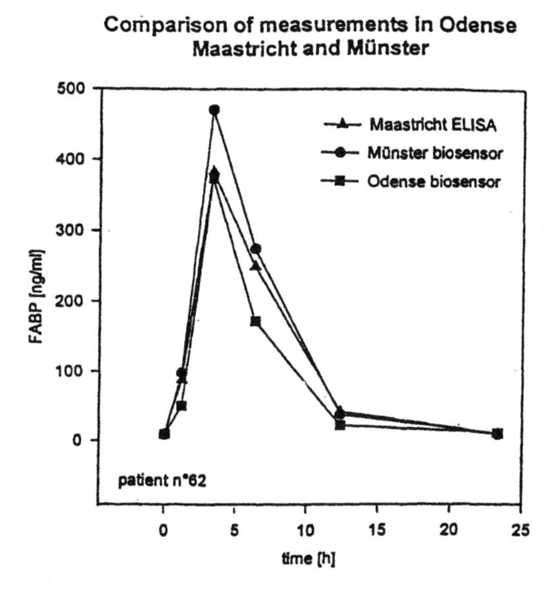

Figure 18. FABP in plasma samples of a patient who had AMI. Comparison of measurements with two immunosensors and with the reference ELISA.

4.5 New Ideas in Immunosensing

4.5.1 Nonseparation Thick-Film Immuno Sensors

To develop nonseparation electrochemical sensors, all interferences from the real sample should be eliminated. This can be done with novel mediator-modified redox electrodes (working free of interferences near 0 mV vs. Ag/AgCl) using imprinted polymer thick-film pastes. The working electrode can be combined together with reference and auxiliary electrodes on a plastic chip. The working electrode can also be printed as a microelectrode array (similar to thin-film technology). Therefore, high current densities can be obtained. The antibodies will be directly immobilized on the working electrode, as in the case of the screen printed immunosensor.

For nonseparation assays, a principle developed by Calum McNeil's group in Newcastle-upon-Tyne [23] can also be used for FABP detection. Screen printed electrodes carry immobilized catcher antibodies and horseradish peroxidase (HRP) as the indicator enzyme. HRP detects H_2O_2 with high sensitivity. H_2O_2 is generated from a special substrate by alkaline phosphatase, the marker enzyme of the indicator antibody. H_2O_2 is detected only if it is generated near the electrode surface (Figure 19). H_2O_2 in bulk (by unbound marker enzyme) is destroyed by catalase and does not interfere.

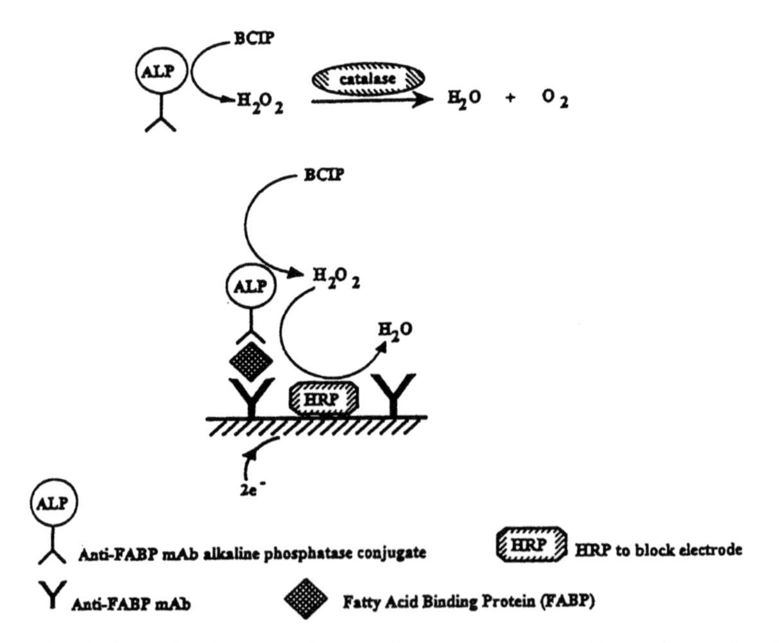

Figure 19. Schematic diagram of separation-free noncompetitive electrochemical immunosensor approach incorporating screen printed electrodes with immobilized antibodies and horseradish peroxidase as indicator enzyme [23].

In Judith Rishpon's group in Tel Aviv, in collaboration with the Hong Kong and Maastricht groups, another approach, an enzyme channeling immunoassay [24], was successfully used to measure FABP. Glucose oxidase (GOD) generates H_2O_2 at the sensor surface after the addition of glucose. The HRP reaction (iodide to iodine) takes place only if secondary HRP-labeled antibodies were bound to FABP at the catcher antibody (Figure 20). As the figure shows, this promising principle works elegantly in absence and presence of FABP. No separation step is needed.

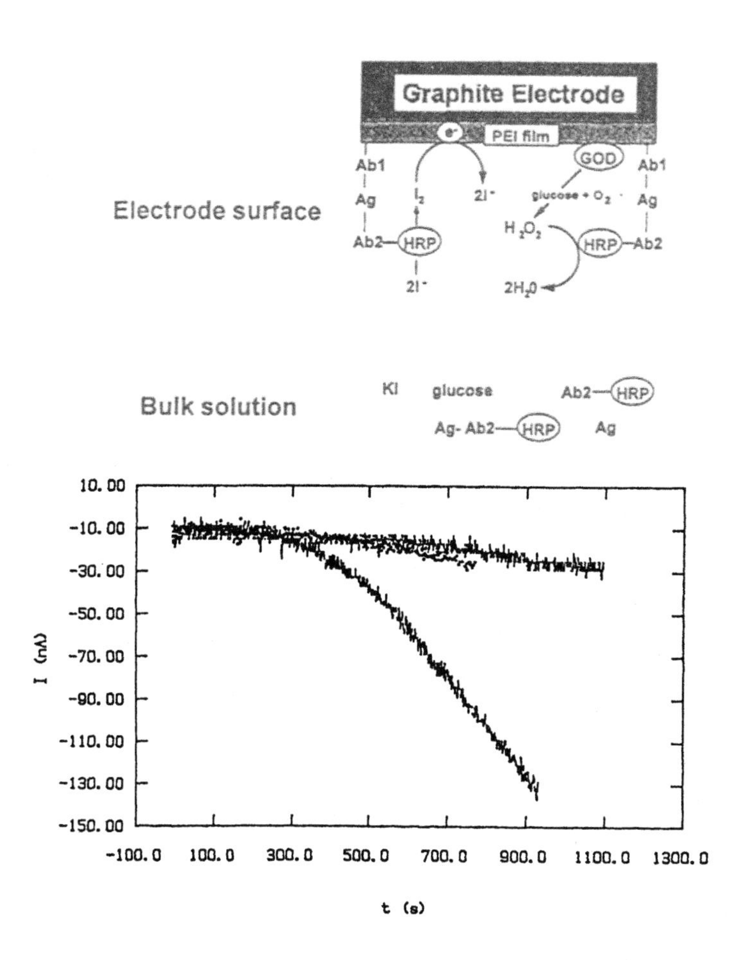

Figure 20. Principle of separation-free, amperometric enzyme channeling immunoassay that has immobilized GOD and antibody on the surface of a polyethylene-imine (PEI) modified graphite electrode (after Ivnitsky and Rishpon [24]). Experimental data show the feasibility for FABP detection: The horizontal curves show the reaction at zero FABP concentration, the declining curve after addition of FABP.

4.5.2. *Simultaneous Measurement of Two Analytes*

To correctly diagnose AMI, the simultaneous detection of an early and a late marker would be desirable to avoid problems if, for example, the AMI occurred 24 h before and FABP is already eliminated from the plasma. There is no single "ideal infarction marker protein." Therefore, we combine an early, yet not totally specific, marker (FABP) with an absolutely cardiospecific but late marker (e.g., troponin I). Ideally, we want to obtain a diagnostic window of 1–24 (or more) hours.

Renneberg and McNeil developed an idea for doing this with an electrochemical immunosensor (Figure 21). FABP and troponin from plasma are bound by the

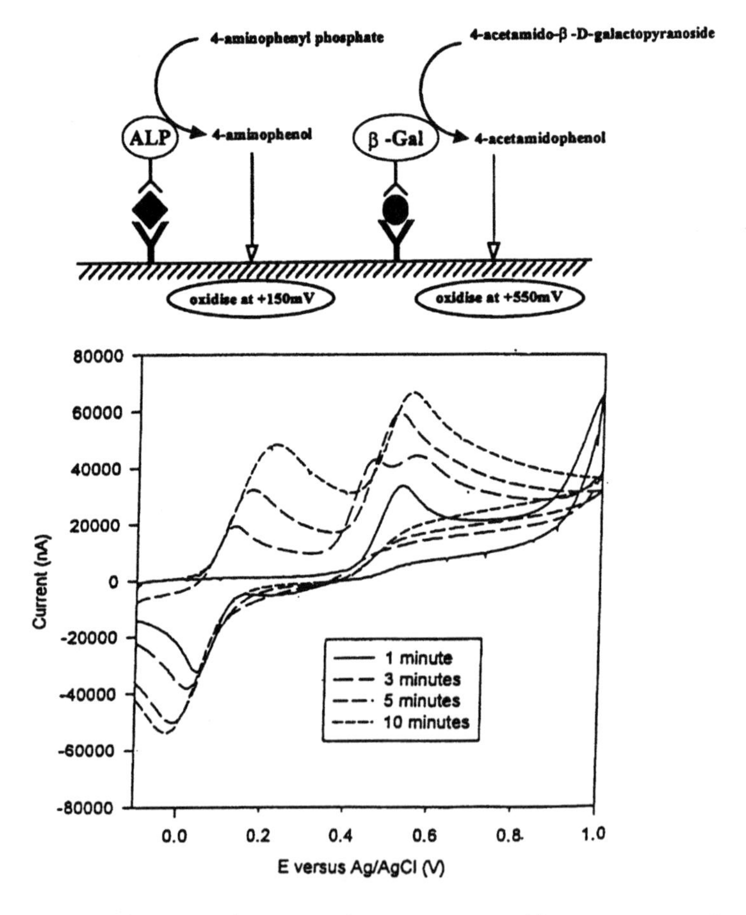

Figure 21. Simultaneous detection of two antigens with one sensor using two different monoclonals and two different marker enzymes. Cyclic voltammograms of the mixture (Experiment: R.Keay, McNeil's group in Newcastle-upon-Tyne).

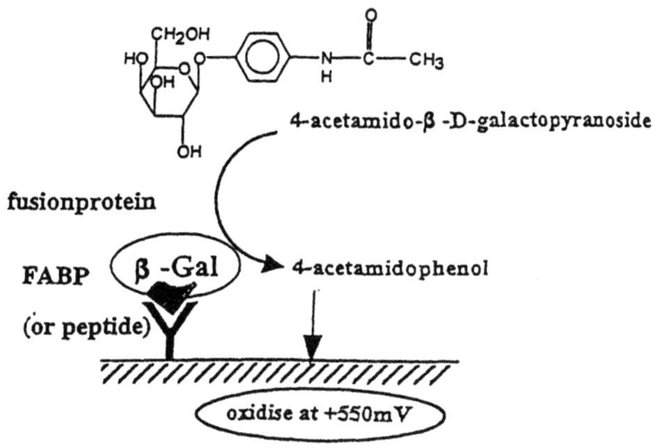

Figure 22. Principle of a simplified FABP test using a fusion protein FABP-galactosidase and a novel substrate yielding the electrode-active product acetaminophen.

evenly mixed corresponding catcher antibodies at the sensor surface and then sandwiched with a secondary detector antibody (directed against another epitope on the FABP or troponin molecules). This secondary antibody against FABP is conjugated with the enzyme alkaline phosphatase, which converts the available substrate *p*-aminophenyl phosphate directly to the electrode-active product *p*-aminophenol (at +150 mV vs. Ag/AgCl). To make a simultaneous troponin measurement, the antitroponin detector antibody is labeled with a different marker enzyme, beta-galactosidase. In realizing an idea of Ulla Wollenberger (Frieder W. Scheller's group, University of Potsdam), a new substrate, 4-acetamido-D-galactopyranoside, was synthesized by the Hong Kong biosensor group. It yielded, with the galactosidase, a product that is electrode-active at a different potential (+550 mV vs. Ag/AgCl). Hence, the two parallel sensor signals are directly proportional to the two marker enzyme activities and the corresponding FABP and troponin concentrations (Renneberg and McNeil, unpublished).

Another idea is also quite obvious for simplifying the FABP assay, "fusion proteins" (FABP with enzymatic activity), would be of great help. Because the human FABP gene has been isolated, FABP (or a part of its molecule) can be fused into one protein with the hydrolytic enzyme galactosidase. Then, 4-acetamido-D-galactopyranoside can be used as a substrate. It yields the product acetaminophen (paracetamol) which is highly electrode-active (Figure 22). The drug acetaminophen is a major interferent in glucose sensors.

4.5.3 Future On-Line FABP Detection

Because FABP is eliminated by the kidneys, the FABP concentration in blood decreases after reaching a maximum (see Figure 2). This allows (in contrast to the high molecular weight CK-MB, LDH, and troponins) an estimation of recurrent infarctions. An alarm system was designed in the Hong Kong biosensor group by Helma Kaptein in collaboration with Jakob Korf (University of Groningen, The Netherlands) [25]. The system monitors on-line the increase in FABP in plasma. This means that a flow-through immunosensor is required. The system should be operated automatically, i.e., there is no need to take samples.

In standard displacement flow immunoassays, an analyte in the sample creates an active dissociation of labeled antigens (or antigen homologues) from an antigen-binding site of an immobilized antibody, whereafter the labeled substance is measured downstream. Such systems have been described for molecules up to 1 kDa (mainly pesticides and drugs). We could demonstrate for the first time in the literature displacement for detecting a small protein, heart FABP (15 kDa). The displacement system applies an inverse setup. Enzyme-labeled monoclonal antibodies are associated with an antigen (FABP) immobilized on *N*-hydroxysuccinimide (NHS) activated Sepharose and displaced by analyte FABP in the sample. It allows detecting both physiological (2–12 μg/L) and pathological concentrations (12–2000 μg/L) of FABP in an on-line flow system (Figure 23). A protein equal in size to FABP, lysozyme, did not show a replacement.

This system will be now equipped with a flow-through electrochemical sensor for detecting the enzymatic activity of the label. Our plan is to combine the FABP sensor with enzyme sensors for glucose and lactate and make the future alarm system even more useful for Emergency Units in the hospitals.

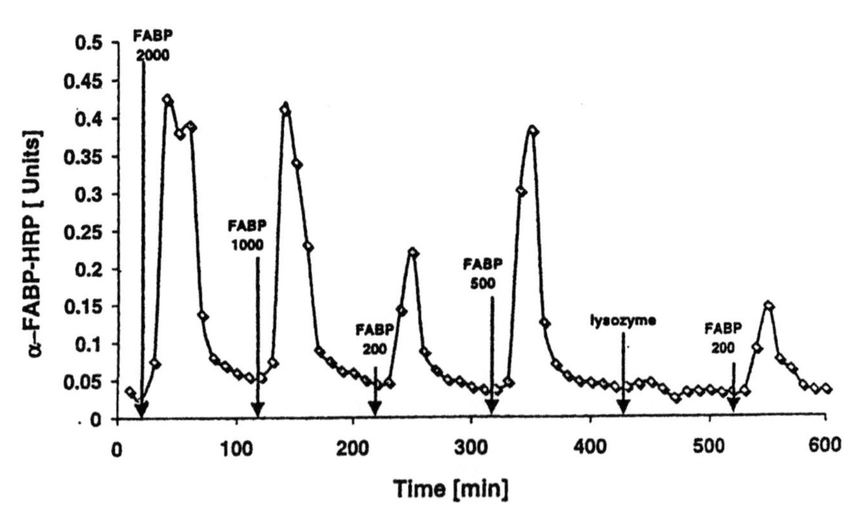

Figure 23. On-line detection of FABP with a flow-through system using an immunoreactor that has immobilized FABP and the displacement of the bound antibodies by sample FABP [25]. Different amounts of FABP are injected on-line into the system causing displacement of the bound labeled antibody. In contrast, lysozyme injection does not cause a displacement.

4.5.4 Immuno Affinity-Filtration Chromatography

A challenge to electrochemical immunosensors are optical assays. A research collaboration between Albert Chu (EY Laboratories, Hong Kong and San Mateo, California), the HKUST Biosensor & Bioelectronics Group, and Jan Glatz (CARIM, Maastricht, The Netherlands) made a first attempt to overcome the difficulties of electrochemical assays.

In the past, EY Labs has successfully developed immuno devices for HIV tests, for hepatitis B and C, malaria, *Helicobacter pylori* and others using colloidal gold as a label. An obvious advantage of colloidal gold as a label is that it gives an instant signal (instead of a time-dependent reaction with enzymes). Enzymes, on the other hand, further amplify the signal just by adding incubation time.

The EY Laboratories' principle, which is patented, avoids the problem of all of the dipsticks. It does not use a time-consuming horizontal chromatographic separation like the Spectral [2] or Boehringer dipsticks for cardiomarkers. This horizontal separation leads to a 7–15 min overall measuring time. Instead in the EY Laboratories' approach, the chromatography is vertically designed in an ingenious way which speeds up the antibody-antigen binding kinetics. The capture antibodies were bound at the surface of a chemically modified cellulose pad in a sophisticated way to avoid unspecific binding of detector antibodies. The deeper cellulose layers suck the fluid of the sample through within less than one minute. First of all a drop of the sample is added to the surface and sucked in immediately. In a second step, a "developer," the gold-labeled detector antibody, is added. A washing solution sweeps off all unbound or unspecifically bound gold-labeled detector antibodies. Now, a clear red spot is visible.

The FABP assay was tried on the basis of this so-called "immuno affinity-filtration chromatography" (IAFC; Figure 24) principle and uses a sandwich technique in which a mixture of goat polyclonal and mouse monoclonal (66E2) capture antibodies to human FABP are bound to the membrane and colloidal gold-labeled mouse monoclonal (67D3) antibodies are used as detector. The result can be seen as a red spot of colloidal gold if FABP is present. If no antigen is in the sample, the membrane remains white. One to two drops of blood plasma were used, and the overall performance time was 1–2 min. This amazingly short time contrasts with the long time needed for migration type dipsticks.

In Maastricht we studied 242 serial blood samples obtained from 83 patients admitted to the hospital with chest pain (72 AMI, 9 unstable angina, 2 other causes) and validated the test against a quantitative FABP immunoassay (established sandwich ELISA). Using an upper reference level of 12 ng FABP/mL (83 samples below, 159 samples above the cutoff), the specificity of the rapid test was 96% and the sensitivity 86% [26].

This was an amazing first result for a rather unoptimized attempt using no additional device but the experimenter's eyes. For a decision: "yes, AMI! " or "no AMI" this would be a great step forward. We concluded from the beginning that

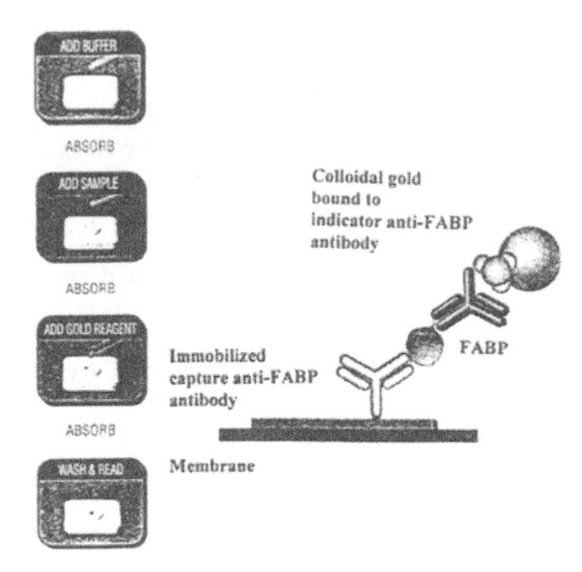

Figure 24. Principle of immuno affinity-filtration chromatography (IAFC) for detecting FABP (modified after Albert Chu, EY Laboratories, Hong Kong and San Mateo).

the rapid FABP test has good diagnostic performance and thus should be of significant value in deciding treatment strategy early after myocardial infarction.

With a grant from the Hong Kong Government's Industry Department, a quantitative FABP sensor will be developed now. It will be complemented by a troponin I sensor to detect in parallel a late marker for AMI and to broaden the diagnostic window. A novel CCD camera will be used in future as the transducer to make quantitative detection possible.

Both the qualitative "yes/no" sensor and the quantitative FABP/troponin I sensor will hopefully help make a practical breakthrough in AMI diagnostics in Hong Kong, China, later Asia, and the whole world.

5. WHAT CAN BE LEARNED FROM THIS CASE?

The principles and strategies outlined in this case report are general, i.e. by exchanging the antibody the biosensor can in principle be adapted to *any other analyte*, such as other cardiac markers (like troponine I), toxins, hormones, or pesticides.

We have learned several lessons:

- Begin the development in the closest collaboration possible with the future user. The final aim to save human life is a very strong motivating impulse for all collaboration partners. The permanent critical feedback from future users is crucial.

- Make the frame of technology as broad as the situation permits and do not stick to your own approaches. Be open to innovation, even though long years of research seem to be "wasted." This was true for the site-directed immobilization of the antibodies, for the principle of reusing the immunosensor, and partly for the electrochemical approach.
- "The better is the enemy of the good." Make the sensor simple, reliable, sensitive, and last but not least, cost-effective!
- Our case study shows clearly that only interdisciplinary teamwork can master the problems of modern biosensor research.

It is a clear illustration of Louis Pasteur's remark "Chance favours the prepared mind..."

ACKNOWLEDGMENTS

Parts of our work has been financially supported by the European Union (grant EC CT93.1692), by a British Council grant, a HK University of Science and Technology Research Infrastructure Grant (RIG), and the Netherlands Heart Foundation.

REFERENCES

[1] Taylor, R.F. (1996). Chemical and biological sensors: Markets and commercialization. In *Handbook of Chemical and Biological Sensors* (Taylor, R.F., Schultz, J.S., eds.) Institute of Physics, Bristol and Philadelphia.

[2] Jackowski, G. (1998). Method and device for diagnosing and distinguishing chest pain in early onset thereof. US Patent 5, 710, 008.

[3] Hay, J.W. (1995). *Health Care in Hong Kong*. The Chinese University Press, Hong Kong.

[4] Kleine, A.H., Glatz, J.F.C., Van Nieuwenhoven, F.A., Van der Vusse, G.J. (1992). Release of heart fatty acid binding protein into plasma after acute myocardial infarction in man. *Mol. Cell. Biochem.*, **116**, 155–162.

[5] Glatz, J.F.C., Van Nieuwenhoven, F.A., Van Dieijen-Visser, M.P., Hermens, W. Th., Van der Vusse, G. J. (1993). Fatty acid-binding protein and myoglobin as plasma marker for the early assessment of acute myocardial infarction in man. *Tijdschr. Ned. Ver. Klin. Chem.*, **18**, 144–150.

[6] Glatz, J.F.C., Kleine, A.H., Van Nieuwenhoven, F.A., Hermens, W.Th., Van DieijenVisser, M.P., Van der Vusse, G.J. (1994). Fatty-acid-binding protein as a plasma marker for the estimation of myocardial infarction size in humans. *Brit. Heart J.*, **71**, 135–140.

[7] Van Nieuwenhoven, F.A., Kleine, A.H., Wodzig, K.W.H., Hermens, W.T., Kragten, H.A., Maessen, J.G., Punt, C.D., Van Dieijen, M.P., Van der Vusse, G.J., Glatz, J.F.C. (1995). Discrimination between myocardial and skeletal muscle injury by assessment of the plasma ratio of myoglobin over fatty acid-binding protein. *Circulation*, **92**, 2848–2854.

[8] Glatz, J.F.C., Van der Vusse, G.J., Simoons, M.L., Kragten, J.A., Van Dieijen-Visser, M.P., Hermens, W. T. (1998). Fatty acid-binding protein and the early detection of acute myocardial infarction. *Clin. Chim. Acta*, in press.

[9] Glatz, J.F.C. (1998). Fatty acid-binding protein as plasma marker for the early detection of myocardial injury. In *Myocardial Damage: Early Detection by Novel Biochemical Markers*, Kaski, J.C., Holt, D.W., eds. Kluwer, Boston, pp. 73–84.

[10] Wodzig, K.W.H., Pelsers, M.M.A.L., van der Vusse, G.J., Roos, W., Glatz, J.F.C. (1997). One-step enzyme-linked immunosorbent assay (ELISA) for plasma fatty acid-binding protein. *Ann. Clin. Biochem.*, **34**, 263–268.

[11] Renneberg, R. (1996). Amperometric enzyme-immunosensors for medical and environmental application. *Invited Plenary Lecture, The Fourth World Biosensor Congress*, Bangkok (to be published in *Biosensors & Bioelectronics*).

[12] Renneberg, R. (1992). Amperometric immunosensors for process control. In *Biosensors: Fundamentals and Applications. GBF Monographs*, Vol. 17, VCH, Weinheim.

[13] Warsinke, A. (1992). Development of amperometric enzyme immunosensors for bioprocess and environmental analytics (in German), Ph.D. Monograph, Humboldt University, Berlin.

[14] Meusel, M., Renneberg, R., Schmitz, G., Spener, F. (1995). Development of a heterogeneous amperometric immunosensor for the determination of apolipoprotein E in serum. *Biosensors Bioelectron.*, **10**, 577–586.

[15] Siegmann-Thoss, C., Renneberg, R., Glatz, J.F.C., Spener, F. (1996). Enzyme-immunosensor for diagnosis of myocardial infarction. *Sensors & Actuators*, **B30**, 71–76.

[16] Ohkaru, Y., Asayama, K., Ishii, H., Nishimura, S., Sunahara, N., Tanaka, T., Kawamura, K. (1995). Development of a sandwich enzyme-linked immunosorbent assay for the determination of human heart type fatty acid-binding protein in plasma and urine by using two different monoclonal antibodies specific for human heart fatty acid binding protein. *J. Immunol. Methods*, **178**, 99–111.

[17] Roos, W., Eyman, E., Symannek, M., Duppenthaler, J., Wodzig, K. W., Pelsers, M., Glatz, J. F. C. (1995). Monoclonal antibodies to human heart fatty acid binding protein. *J. Immunol. Methods*, **183**, 149–153.

[18] Kunz, U., Katerkamp, A., Renneberg, R., Spener, F., Cammann, K. (1996). Sensing fatty acid-binding protein with planar and fiber-optical surface plasmon resonance spectroscopy devices. *Sensors & Actuators*, **B32**, 149–155.

[19] Glatz, J. F. C., Renneberg, R., McNeil, C.J., Spener, F. (1995). Electrochemical and integrated-optical immunosensors for heart-type fatty acid-binding protein—a new plasma marker for acute myocardial infarction. Abstract *3rd Eur. Conference on Engineering and Medicine*, Florence, p. 179.

[20] Glatz, F.J.C. (1995). Biosensor for myocardial infarction diagnosis: The EUROCARDI project. *Biosensors & Bioelectronics*, **10**.

[21] Tang, H.T., Lunte, C.E., Halsall, H.B., Heineman, W.R. (1988). *p*-Aminophenyl phosphate: An improved substrate for electrochemical enzyme immunoassay. *Anal. Chim. Acta*, **214**,187–195.

[22] Schreiber, A., Feldbruegge, R., Key, G., Glatz, J.F.C., Spener, F. (1997). An immunosensor based on disposable electrodes for rapid estimation of fatty acid-binding protein, an early marker of myocardial infarction. *Biosensors & Bioelectronics*, **12**, 1131–1137.

[23] Ho, W.O., Athey, D., McNeil, C.J. (1995). Amperometric detection of alkaline phosphatase activity at a horseradish peroxidase enzyme electrode based on activated carbon: Potential application to electrochemical immunoassay. *Biosensors & Bioelectronics*, **10**, 683–691.

[24] Ivnitski, D., Rishpon, J. (1996). A one-step, separation-free amperometric enzyme immunosensor. *Biosensors & Bioelectronics*, **11**, 409–417.

[25] Kaptein, W.A., Korf, J., Cheng, S. Yang, M., Glatz, J.F.C., Renneberg, R. (1998). On-line flow displacement immunoassay for fatty acid-binding protein. *J. Immunol. Methods*, **217**, 103–111.

[26] Glatz , J.F.C., Renneberg, R., Pelsers, M.A.L., de Zwaan, C., Chu, A.E. (1998). A simple 2-min immunofiltration assay for heart fatty acid-binding protein to diagnose acute myocardial infarction, unpublished.

THE DEVELOPMENT OF THICK- AND THIN-FILM-BASED MICROBIOSENSORS

Zong-Rang Zhang, Guo-Xiong Zhang, Jian-Zhong Zhu, and Jun Hu

OUTLINE

Advances in Biosensors
Volume 4, pages 273–287.

ISBN: 0-7623-0073-6

1. INTRODUCTION

Microelectrodes are electrodes whose dimensions are a few micrometers or less. They were developed over the last two decades and display a number of desirable properties, including high current density due to nonplanar diffusional contributions to the net Faradiac current, very short time constants, and low ohmic drop. These features allow using a wider range of resistive solutions and very high scan rates. Theoretical considerations, microfabrication, and applications of microelectrode technology are attractive and valuable to electrochemists and also to experts from other fields [1,2].

Microelectronic fabrication processes are a proven technology for producing integrated circuits. This technology has also made a significant impact on related sciences and technologies. Development of microfabrication-based chemical sensors and biosensors is following the same trends in miniaturization as those seen in many other high-tech fields. The realization of potential applications of microfabrication techniques in developing microbiosensors has added new impetus to this endeavor, especially in commercializing biosensors for routine tests and *in vivo* monitoring in biological and medical research [3].

Among the microfabrication techniques, thick- and thin-film metallization processes and more recently, micromachining technology can readily be used, at least in part, to develop new generations of biosensors. It is anticipated that increasing applications of microfabrication and micromachining technologies combined with improved studies of optimal design and electrochemical characterization of microsize substrate electrodes, compatible immobilization methods of biocatalysts, and effective mediators or promotors for electron transfer processes, will encourage the development of novel microbiosensors in the forthcoming years.

The application of microfabrication technology in biosensor construction leads to miniaturization of sensor size and also provides the possibility of mass producing biosensors because of the relatively simple techniques, the consistency of different microsensors, and much lower production costs. The excellent electrochemical features of microsensors in terms of the special behavior of ultramicroelectrodes also offer a very promising feature for manufactured microsensors. This is substantiated by the recent increase in the number of conferences and publications on this subject.

In this chapter, we present a brief review of developments in thick- and thin-film microbiosensors in China. We begin with a short overview of the concepts and technologies of microfabrication processes.

2. MICROFABRICATION TECHNOLOGIES FOR MICROBIOSENSORS

In most biosensors, electrochemical elements are used as the sensing parts. The application of microfabrication technology in microbiosensor development focuses mostly on producing microworking electrodes based on metallic films, the reference electrode, and its assembly on a microchip [3].

2.1 Thick-Film Metallization Processes

Thick-film processing is similar to silk screening and uses an ink containing metallic powder and an organic binder. A photolithographic mask defines the sensor geometry or the microelectrochemical cell on the chip. A number of special pastes that have lower firing temperatures have been used and lead to more options for microbiosensor fabrication. In this process, a semiliquid medium, such as graphite and silver/silver chloride inks, are forced through a mesh screen with a specially designed pattern. Biological enzyme elements and electron transfer mediators can also be incorporated in the deposited carbon paste layer.

It is generally assumed that the electrochemical behavior of thick-film microelectrodes is similar to that of conventional planar electrodes produced from bulk metal, but this assumption is not entirely accurate. In fact, the thick film deposited on the substrate has a porous structure and a noticeably higher specific surface area than bulk metal. Therefore, it is reasonable to predict that the electrocatalytic activity of a thick-film electrode will be much higher than that of a conventional planar electrode. On the other hand, the microporous structure of a thick-film microelectrode may vary from one electrode to another. This results in minute variations of the total surface area, a factor to be considered in improving production processes. Fortunately, this defect has been much improved recently. The cross talk between the closely placed microworking electrode and the reference electrode is another problem for microsensors and has been studied carefully.

2.2 Thin-Film Metallization Processes

Thin-film processes include different kinds of chemical vapor deposition (CVD) techniques, such as hot-filament assisted chemical vapor deposition (HFCVD), microwave plasma CVD (MPCVD), ion beam enhanced deposition (IBED), low temperature silicon direct bonding (LTSDB), laser trimming (LT), and reactive sputtering (RS). These are proven microfabrication techniques used to produce well-defined geometric structures. Photoetching, plasma etching, and lift-off technology define the thin film microelctrode or sensing element.

Compared to thick-film technology, thin-film processes are much more capital- and labor-intensive. On the other hand, the thin-film electrode surface is much more uniform and reproducible than that of thick-film electrodes. The thin metal film produced by this method may have different characteristics from the bulk metal. The interface between the substrate and the metal film is smoother than that of thick films, but the adhesion between the thin film and the substrate may be problematic. Consequently, an additional metal layer may have to be deposited to enhance the adhesion between the film and substrate. This is discussed later in the review of the authors' work.

Because thin-film electrodes are relatively small, that is, in the micrometer or even in the submicrometer range, many electrode designs take advantage of this characteristic. Arrays of sensing elements, band-type microelectrodes, and other arrangements have been fabricated using thin-film techniques.

2.3 Micromachining Processes

In recent years, micromachining technology has added new impetus to the microfabrication of microchemical sensors and microbiosensors. Based on chemical anisotropical etching on a silicon wafer, this technology permits forming three-dimensional microstructures and has been applied in developing new types of microbiosensors.

Micromachining employs various microfabrication processes, including an active sacrificed layer, plasma etching, and chemical anisotropical etching (CAE), to produce three-dimensional structures, such as the chamber structure, which provides a protective and controlled environment surrounding the microsensing part. The chamber structure has the advantage of minimizing the flow effect and chemical interference and also provides a larger surface area for sensing elements. A chamber-style microsubstrate electrode may provide a good place for enzyme entrapment, especially for developing integrated systems.

3. THICK-FILM MICROBIOSENSORS

The application of thick film technology has recently been extended to sensor construction. A number of special pastes that have lower firing temperatures have appeared, leading to more options for biosensor fabrication. In addition, thick-film technology enables producing whole devices more economically than those prepared by conventional methods and will make it possible to produce disposable sensors at a low price, even on a medium mass production scale [4–6]. In this section, we outline our experiences in constructing thick-film microbiosensors from commercially available pastes, enzyme elements, and electron mediators [7–9].

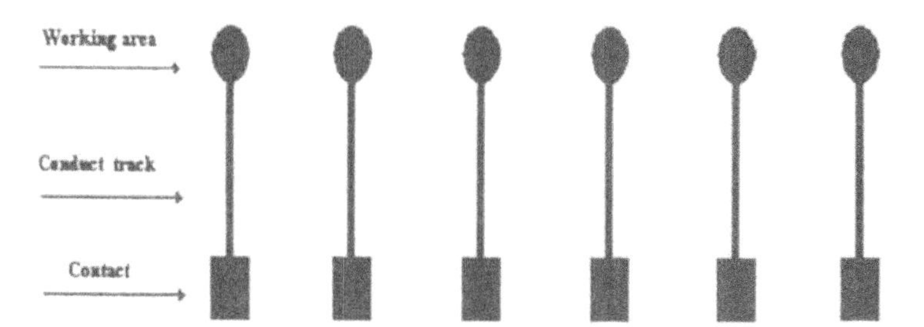

Figure 1. Schematic diagram of screen printed thick-film sensors.

3.1 Substrate Electrode Based on Thick-Film Technology

All thick-film sensors are screen printed manually or by a mechanical screen printer. In this process, a semiliquid conducting medium, such as carbon graphite ink or silver/silver chloride ink, is forced through a mesh screen onto the selected substrate, and a designed pattern for the carbon and reference electrode is produced. This process results in precision and speed of production. Figure 1 shows printed strips that have six working electrodes on a PVC substrate. Graphite ink (Acheson Colloid Co.) was used to produce the conducting track and working electrode and was dried by heating at around 65 °C. Silver/silver chloride ink (Acheson Colloid Co.) was applied to cover the conducting graphite ink track and was also allowed to dry. Then, the conducting track and reference electrode are surrounded by an insulating shroud ink. In this way, thousands of substrate electrodes are produced in a few hours.

3.2 Immobilization of Oxidoreductase on Thick-Film Substrate Sensors

Glucose oxidase sensors are being widely studied and some results have already been commercialized [4,6,9]. However, these applications require special techniques and few of the papers cited are related specifically to enzyme immobilization on a carbon layer that provides sufficient activity and shelf life for commercial purposes. Glucose oxidase is very stable under ambient laboratory conditions, but this may not be so in the commercial environment. A form of epoxypolyamine [7]

Figure 2. Cross-linking between the enzyme and the surface of the carbon electrode. (**a**) the glucose oxidase or glutamate oxidase; (**b**) the cross-linking reagent; (**c**) the surface of the carbon electrode.

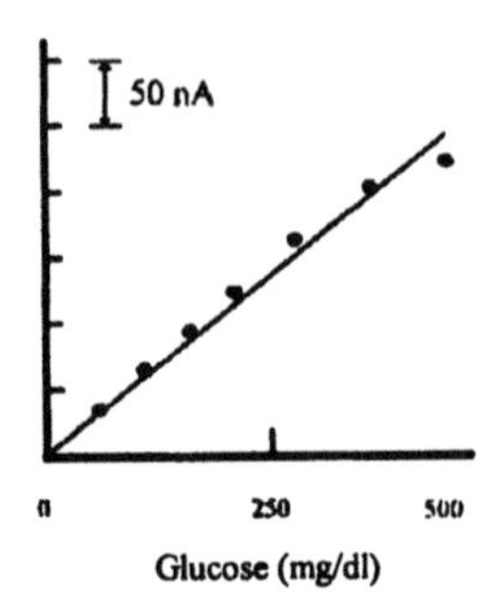

Figure 3. Calibration curve for glucose.

is used as the cross-linking reagent for immobilizing the enzyme on the surface of the substrate electrode. Figure 2 is a schematic diagram of the cross-linking reaction. The epoxypolyamine (EPA) cross-linking reagent is formed by condensation of two molecules of 1-chlor-2,3-epoxy propane and one molecule of 1,6-hexanediamine. Thus, there are two active epoxy and amino groups on the surface of the enzyme protein, and the amino groups form hydrogen bonds on the surface of the carbon electrode, which was previously modified by a hydrophilic solution. The enzyme is bound to the surface of the substrate electrode and remains stable over a long period. In addition, the C_6 chain in EPA provides a spacer residue in the cross-links and may facilitate the diffusion of substrates and products.

Glutamate oxidase is used as an example because of the wide use of monosodium glutamate in the Oriental diet. The determination of glutamic acid concentration in fermentation broth is a routine, frequent, and important measurement. Moreover, glutamate is also a neurotransmitter which plays an important role in the human brain [10]. Glutamate oxidase is expensive and not as stable as glucose oxidase. Therefore, it is treated using the process described earlier to provide good stability

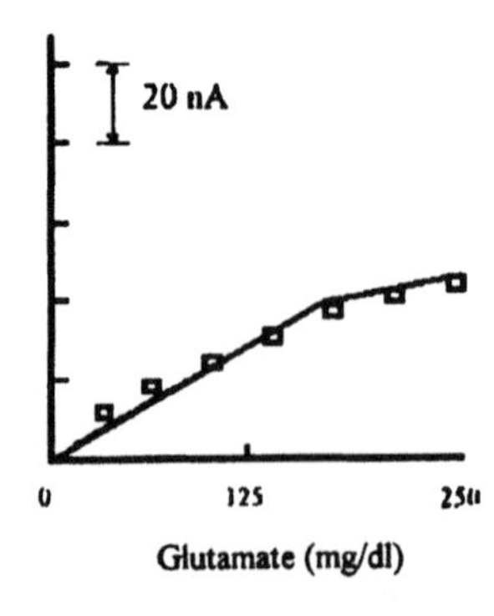

Figure 4. Calibration curve for glutamate.

and a wide linear range. Figure 3 and Figure 4 show the calibration curves for glucose and glutamate, respectively. The results demonstrate that the range of linearity of glucose is 0–500 mg/dL, which meets the requirements for routine diagnosis in diabetes. The range for glutamate is 0–180 mg/dL, indicating that this may be useful for further research in food fermentation and neuroscience.

The screen printing technique is easily controlled to print the designed pattern on a sensor strip. Mediators used for the electron shuttling can also be printed with conductive materials and/or ink-jet printed with bioactive enzyme onto the microbiosensors [9]. A pocket-sized blood-glucose analyzer that has a test strip has been commercialized. Additionally, this technology allows fabricating multifunctional enzyme sensors because the electrode area can be limited to 1 mm^2. Therefore, several working electrodes can be included on one strip together with a reference electrode.

4. BIOSENSORS BASED ON MICROARRAY ELECTRODES (MAES)

4.1 Microarray Electrodes and Their Fabrication

Microelectrodes (MEs) are electrodes whose dimensions are a few micrometers or less. In static solution, MEs display a number of desirable properties. As mentioned before, MEs have the potential to display a number of advantages over conventional electrodes and can be used for a range of electrochemical research [1–3]. The major disadvantage of a single ME is the low response current level. This shortcoming can be overcome by using microarray electrodes which greatly increase the response current level and maintain the special electrochemical characteristics of individual microelectrodes. MAEs consist of many single microelectrodes arranged in a defined pattern separated by insulating gaps but electrically connected. MEs and MAEs have both been recognized as useful and novel categories of working electrodes in modern electrochemistry, chemical sensors, and biosensors.

The clear surface of a 2-inch Si wafer that has a 1 μm thick layer of SiO_2 is patterned with a positive photoresist (AZ 1350, Shipley) so that the whole surface of the wafer is covered except for the pads, electrodes, and their paths. Ti (50 nm)/Pt (200 nm)/Ti (50 nm) sandwich layers are evaporated at 25 °C. These two thin layers of Ti are adhesion promoters for the SiO_2 and the following layer of Si_3N_4. Then, the wafer is dipped into toluene followed by ultrasonification for several minutes. The metal of the area with photoresist is lifted off leaving behind the metal on the area without photoresist (Figure 5 a). Then, a 300-nm thick silicon nitride layer is grown onto the wafer and reactive ion etching etches away the nitride in the areas of the reference electrode (RE), counter electrode (CE), working electrode (WE), and their pads (Figure 5 b). Figure 5 (c) is an enlarged picture of the MAE. The wafer is immersed in Ti etchant (H_2SO_4/H_2O = 1/1, 80 °C) to ensure that there is

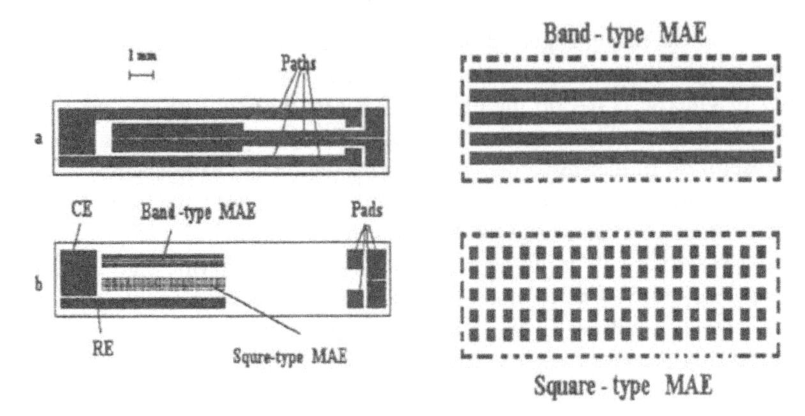

Figure 5. (**a**) Ti/Pt/Ti sandwich layers (black area) for electrodes. (**b**) After growing Si_3N_4, windows of RE, MAE WEs, CEs, and their pads are etched by reactive ion etching. (**c**) Enlargements of band- and square-type MAE (black areas). The dotted line indicates the outline of the Pt WE under the Si_3N_4 layer.

no Ti left on the surface. A silver layer is patterned on the RE area by the lift-off technique and the silver layer is partially chlorinated in a 0.25 mol/L $FeCl_3$ solution [12,20].

However, we observed that the Ti layer in the structure of SiO_2/Ti/Pt is corroded easily during the experiments, resulting in the destruction of fabricated sensors. The addition of an intermediate layer of diamond has been developed recently and is used to replace the Ti layer. The diamond film provides good adhesion between the SiO_2 substrate layer and the platinum layer. This new intermediate material has been used widely in constructing microelectronic circuits. We also tried to use the intermediate diamond layer for constructing MAEs in microbiosensor fabrication. A polycrystalline diamond film was deposited on a silicon wafer by hot-filament assisted CVD. Before deposition, the commercially polished silicon wafer is roughened with diamond powder to increase the nucleation density. The temperature of the filament and substrate are 2000 °C and 900 °C, respectively. The reactant used is a mixture of 0.5 sccm methane and 100 sccm hydrogen. After four hours of deposition, the film is about 4 μm thick. The surface of deposited polycrystalline diamond film is generally rough which is useful to enhance adhesion of the working and reference electrodes to the diamond film. The 300-nm platinum and 400-nm silver films are deposited onto the diamond film by sputtering immediately. Finally, the polyimide is used as an upper insulating layer. The pattern of the microarray electrode is formed by the lift-off technique. The silver/silver chloride reference electrode is partially chlorinated in a 0.25 mol/L $FeCl_3$ solution.

A boron-doped polycrystalline diamond thin film (BDF) electrode without external illumination or surface pretreatment was reported recently by the authors. The results of its electrochemical characterization proved that the BDF might be more favorable in electrochemical measurements. BDF was grown on the clear surface of Si(111) wafers by hot-filament CVD in a quartz low-pressure chamber. Preheated H_3BO_3 is used as a boron doping source. The growth is carried out at a filament temperature of 2000–2200 °C and a continuous film is achieved after 12 h [13,14,19].

4.2 Electrochemical Characterization of Microarray Electrodes

4.2.1 CV Behavior of MAEs

CV plots of band- and square-type MAEs were determined by using 1.0 mmol/L $K_3Fe(CN)_6$ in a 0.1 mol/L KCl electrolyte. Nonsteady-state waves for band-type MAEs were obtained even at a scan rate of 10 mV/s. The oxidation peak potential appears at about + 500 mV and the peak height increases linearly with the square root of the scan rate. Steady-state waves were observed for square-type MAEs even at a scan rate of 200 mV/s. Half-wave potentials were + 250 mV, and the steady-state current was independent of the scan rate. It follows from these findings that in spite of their very narrow width, some band-type MAEs do not always show behavior typical of microelectrodes during longer time measurements. This may be caused by overlapping of diffusion layers of individual bands and results in a wider diffusion layer that covers the whole surface area of the microarray electrode being studied.

At lower scan rates, the response curve has an S-shape but produces a normal CV plot at higher scan rates. This is somewhat different from the behavior typical of microelectrodes and might be due to the mixture diffusion controlling the behavior of linear and nonlinear diffusion of microarray electrodes.

The mass transfer intensity of MAEs is much higher than the conventional size platinum electrode. The current density of MAEs in ferricyanide solution is around 11.2 μA under the scan rate of 100 mv/s, but the latter is only 1.72 μA. Experiments were performed to compare mass transfer intensity between MAEs and a rotating electrode. The results show that the mass transfer intensity of the MAEs studied is similar to the results for RDE below 7600 rpm [15].

4.2.2 Chronoamperometric Behavior of MAEs

Chronoamperometric behavior of MAEs was detected. The observed response current of MAEs was consistent with the calculated data based on the Cottrell equation within a short timescale, but decreased later. This might be due to the edge effect of band electrodes, and it decreased noticeably later due to the overlapping of the diffusion layer of individual band electrodes. The difference between the observed and calculated data may change with the ratio of gap width and bandwidth (w_g/w_b). The optimal ratio must be 10 or more to decrease the effect of diffusion layer overlapping for MAEs [15].

4.2.3 Response Current and Current Density

Response currents and current densities were recorded for band-type and square-type MAEs in 1.0 mmol/l $K_3Fe(CN)_6$ solution at a 100 mV/s scan rate. The response currents of MAEs with 90 squares are 90/225, i.e., around two-fifths of those of MAEs that have 225 squares. Hence the response current of the MAE is the total of individual MEs. The current density of the band-type MAE decreases linearly with increasing bandwidth, whereas that of the square-type MAE is inversely proportional to the side length.

4.3 Enzyme Immobilization on MAEs

Enzymes are immobilized on MAEs as in constructing conventional biosensors. But in terms of the special requirements of MAEs-based microbiosensors, the development of a microsized compatible method of enzyme immobilization, such as precise coating on a designated place in microscale electrodes, design of mediator-assisted electron transfer, replaceability of different sensor chips, is an attractive and important step in constructing commercial microbiosensor chips. Chemical bonding, cross-linking, and coimmobilization by electropolymerization of conducting polymers (polypyrrole, polyaniline) were studied in the authors' laboratory. Nafion or gelatin film were used to protect against interference from ascorbic acid. Other substances were also studied [16]. Different mediators for promoting electron transfer (ferricyanide, ferrocene and its derivatives, tetrachloroquinone, and benzoquinone) were also studied [11].

4.4 MAE-Based Glucose Biosensors

Glucose microsensors with good performance based on MAEs were constructed in the authors' laboratory. The effects of pH, temperature, working potential, and

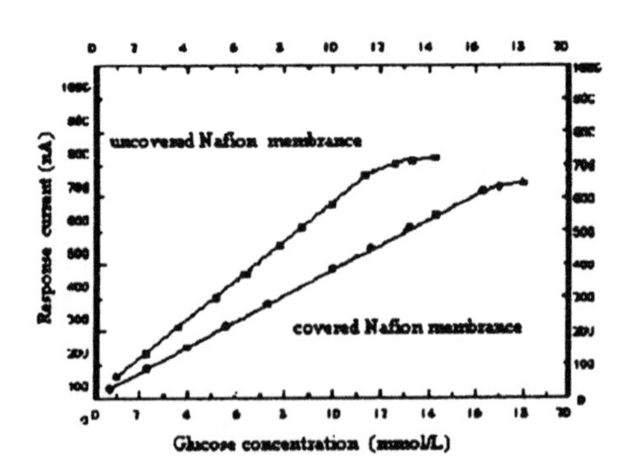

Figure 6. Calibration curve of MAE-based glucose sensor with intermediate diamond film.

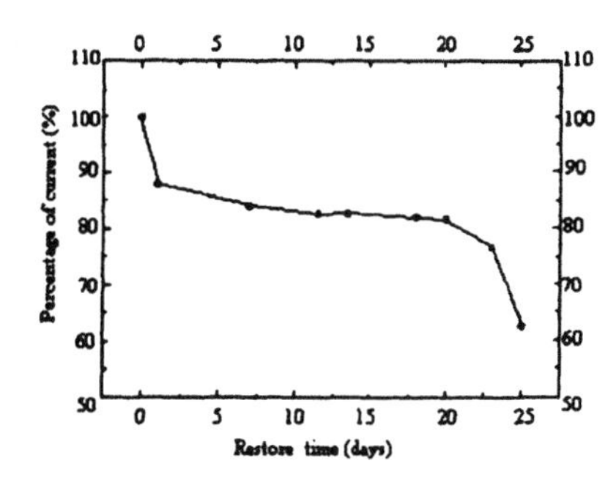

Figure 7. Lifetime of glucose microsensor.

interference were evaluated and the optimal conditions for glucose determination were identified. The results of a MAE-based glucose sensor with an intermediate diamond film are presented as an example of the microsensors fabricated. Under optimal experimental conditions, a linear response was obtained from 0.5 mmol/L to 12 mmol/L (Figure 6). The sensitivity of the designed MAE-based glucose sensor is around 58 nA/mmol/L. Such a high sensitivity may result from the edge effect which is induced by both the edge effect of the MAE and the rough surface of the intermediate diamond film. After covering it with a Nafion membrane, the linear range of the sensor increased to 15 mmol/L, and the sensitivity decreased slightly. The glucose microsensor studied was stable for more than 20 days under test conditions (Figure 7). This might be long enough for disposable strips for glucose determining. The interference of ascorbic acid and uric acid can be diminished by using double working electrodes on the same chip. We used the sensor to test a standard serum. In ten consecutive determinations on the same sample, the mean standard deviation of glucose is 0.11, and the variation coefficient is 1.6%. A comparison with spectroscopic testing resulted in a correlation coefficient of 0.99. After testing almost 100 samples of human serum (concentration of glucose is 12.0 mmol/L), the variation coefficient was 4.7% [11–13].

5. BIOSENSORS BASED ON A MINIATURE CLARK OXYGEN ELECTRODE

5.1 Introduction

The Clark electrode developed in 1962 by Clark and Lyons for glucose monitoring was historically the first transducer associated with soluble glucose oxidase, and the Clark-type oxygen electrode has become essential in biosensor development, clinical diagnosis, and fermentation monitoring. Based on the oxygen electrode, biosensors for oxalate, salicylate, urea, choline, fructose, sucrose, lactose, pyruvate, isocitrate, ascorbate, aspartame, monoamines, hypoxanthine, IMP, and

ATP/ADP have been developed by immobilizing a corresponding oxidase. There is an increasing demand in clinical diagnosis for multiple measurements with miniaturized and integrated biosensors. Micromachining and photolithographic techniques have been used to fabricate a miniature Clark oxygen electrode in our laboratory recently, and a glucose sensor based on the oxygen electrode and glucose oxidase immobilized by cross-linking has been developed.

5.2 Fabrication of a Miniature Clark Oxygen Electrode

The miniature Clark oxygen electrode consists of a lower silicon substrate, a piece of gas-permeable membrane, and an upper silicon substrate. Figure 8 is a schematic diagram of the miniature oxygen electrode [17].

Following is the fabrication process for this microelectrode:

1. Anisotropically etch grooves on the lower silicon substrate (200 μm wide, 200 μm deep except 5 μm wide and 5 μm deep on the area of working electrode) to accommodate an electrolyte solution and then form gold-working, counter, and Ag/AgCl reference electrodes after oxidizing the etched substrate.
2. Open holes (1×0.5 mm^2) on the upper silicon substrate by anisotropic etching, and then oxidize the substrate.

   It is known that when EPW (ethylene diaminepyrocatechol-water) or potassium hydroxide are used for anisotropic etching of (100) oriented silicon, undercutting on convex corners becomes a problem. To reduce or to prevent convex corner undercutting, special structures must be added to the mask at these corners.

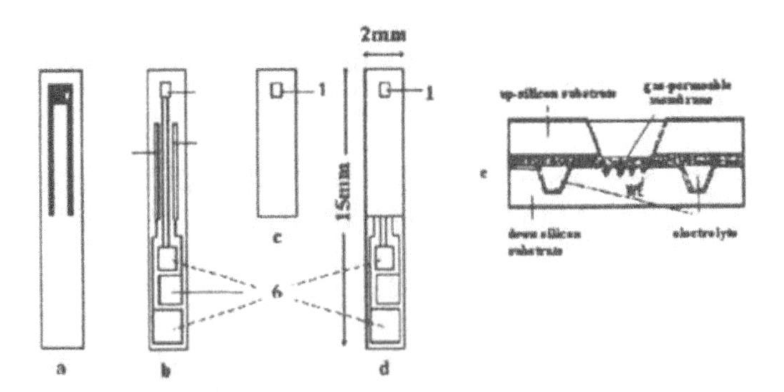

Figure 8. Schematic diagram of the miniaturized Clark oxygen electrode. (**a**) etched V-shape chamber; (**b**) electrode; (**c**) cover; (**d**) overall electrode; (**e**) intersection of 1-1′ plane (1) hole for oxygen inlet; (2) cover; (3) oxygen-permeable membrane; (4) substrate; (5) V-shaped chamber; (6) electrode path.

3. The working and counter electrodes were made from 400-nm thick gold film and the reference electrode was made from 400-nm thick silver film. An intermediate 40-nm chromium layer was used to obtain good adhesion between these films and the SiO_2.
4. Attach a piece of gas-permeable membrane (15 μm thick copolymer of tetra-and hexafluoro ethylene) to both upper and lower silicon substrates at 280 °C in a vacuum.
5. Pour the electrolyte solution, calcium alginate gel containing a 0.1 M KCl aqueous solution, into the grooves at low pressure.

5.3 Fabrication and Characteristics of the Glucose Sensor Based on the Miniature Clark Oxygen Electrode

5.3.1 Immobilization of Glucose Oxidase (GOD)

GOD (10 mg) and 14.25 mg bovine serum albumin (BSA) in 0.2 ml phosphate buffer solution (pH 6.864) are added to a 10 mL aqueous solution containing 25% glutaraldehyde, and the mixture is poured immediately into the holes of the oxygen electrode. An enzyme film forms on the gas-permeable film after drying for 24 hours. The glucose sensor has to be stored at low temperature (4 °C) and high humidity (RH > 80%) to avoid a decrease in enzyme activity [18].

5.3.2 Behavior of the Miniature Clark Oxygen Electrode-Based Glucose Sensor

The linear range and sensitivity of the miniature Clark oxygen electrode-based glucose sensor are 0.1~10.0 mmol/L and 2.2 nA/(mmol/L), respectively. The lifetime of this sensor is around 20 days.

The sensitivity can be changed by adjusting the quantity of the immobilized enzyme. Response time and stability depend on the oxygen electrode, not the enzyme-immobilized membrane.

The miniature oxygen electrode-based glucose sensor largely depends on temperature. This may be due to the nature of the oxygen electrode and the activity of GOD. During the operation of the miniature oxygen electrode-based glucose sensor, the temperature has to be strictly controlled.

Pt or Au (working electrode) and Ag (reference electrode) are susceptible to contamination and poisoning by various compounds, such as gases (H_2S, SO_2), thioorganic materials, and dissolved or solid impurities. The glucose sensor based on an oxygen electrode is more tolerant to some interference materials because the oxygen-permeable membrane is an effective barrier. If a membrane with higher permeability to oxygen but with lower permeability to other gases is used as a filter, the selectivity of the microsensor improves significantly.

6. CONCLUSION

Microfabrication technology, such as thick-film, thin-film, and micromachining techniques, has proved very useful and effective for the constructing microbiosensors that have enhanced characteristics. Furthermore, it also provides a novel method for the commercializing biosensors, especially those for producing disposable strips and for *in vivo* studies in biological and medical research and monitoring devices for environmental applications. We wish to contribute further in this research area in the future. Other laboratories in China are also involved in developing microfabricated biosensors. Information on their research is printed in this volume and in scientific journals.

ACKNOWLEDGMENTS

The authors are grateful for support from numerous organizations. In particular, the National Science Foundation of China, the Science Foundation of Shanghai Commission of Science and Technology, the Science Foundation of Shanghai Education Commission and the authors' institutions are thanked for their support. The valuable help of colleagues and students is also acknowledged.

REFERENCES

[1] Fleischmann, M., Pons, S., Rolison, D.R., Schmidt, P.R., (Eds.). (1987). *Ultramicroelectrodes*, Datatech System Inc., Morganton.

[2] Montenegro, M.I., Queiros, M.A., Daschbach, J.L. (Eds.). (1991). *Microelectrodes: Theory and Applications*, NATO ASI Series, Kluwer, London.

[3] Liu, C.C., Zhang, Z.R. (1992). Research and development of chemical sensors using microfabrication techniques. *Selective Electrode Rev.*, **14**, 147–167.

[4] Matthews, D.R. (1987). Pen-size digital 30-second blood glucose meter. *Lancet*, **4**, 778–779.

[5] Bilitewski, U. (1991). Glucose biosensors based on thick film technology. *Biosensors & Bioelectronics*, **6**, 369–373.

[6] Weetall, H.H., Hotaling, T. (1987). A simple, inexpensive, disposable electrochemical sensor for clinical and immunoassay. *Biosensors*, **3**, 57–63.

[7] Hu, J. (1986). Immobilization of cells containing glucose isomerase using a multifunctional crosslinking reagent. *Biotech. Lett.*, **8**(2), 14–16.

[8] Hu, J. (1994). Chemical modified glucose oxidase enzyme sensors. *Proceedings of Fifth International Meeting on Chemical Sensors*, p. 138.

[9] Hu, J. (1996). A soluble mediator leading to the first biosensor manufacture in China. *Proceedings of the Fifth World Conference on Biosensors*, p. 158.

[10] Bartlett, P.N., Copper, J.M. (1994). A review of the immobilization of enzymes in electropolymerised film. *J. Electroanal. Chem.*, **362**, 1–12.

[11] Zhang, Z.R., Bao, W.F., Liu, C.C. (1994). Electrochemical properties of benzoquinone and ferrocene monocarbooxylic acid at a polyaniline coated platinum electrode for glucose sensing. *Talanta*, **41**, 875–880.

[12] Zhu, J.Z., Liu, X.H., Wu, J.L., Lu, D.R., Zhang, G.X. (1994). Electrochemical characteristics of H_2O_2 microarray electrodes as base elements of biosensors. *Fresenius J. Anal. Chem.*, **348**, 277–280.

[13] Wu, J.L., Zhu, J.Z., Zhang, G.X., Lin, X.R., Cheng, N.Y. (1996). Fabrication and application of a diamond-film glucose biosensor based on a H_2O_2 microarray electrode. *Anal. Chim. Acta*, **327**, 133–137.

[14] Wu, J.L., Zhu, J.Z., Shan, L., Cheng, N.Y. (1996). Voltammetric and amperometric study of electrochemical activity of boron-doped polycrystalline diamond thin film electrodes. *Anal. Chim. Acta*, **333**, 125–130.

[15] Jia, N.Q., Tian, C.Y., Cao, D.J., Zhang, Z.R., Zhang, G.X., Zhu, J.Z. (1997). Electrochemical characterization of microarray electrodes using cyclic voltammetric and chronoamperometric techniques. *J. Shanghai Teachers Univ. (Nat. Sci. Ed.)*, (Chinese), **26**(2), 54–59.

[16] Tain, C.Y., Cao, D.J., Zhang, Z.R., Zhang, G.X., Zhu, J.Z. (1998). Planar glucose microsensors based on modified layer of polypyrrole, *Anal. Chem.* (Chinese), **26**, 854–857.

[17] Zhu, J.Z., Jiang, H.Y., Lu, D.R. (1996). Miniaturized Clark oxygen sensors fabricated by micromachining technology. *J. Transducer Tech.* (Chinese), **9**, 68–70.

[18] Zhu, J.Z., Wu, J.L., Zhang, G.X. (1997). Micromachined glucose sensor. *J. Functional Mater. Devices*, (Chinese), **3**, 109–113.

[19] Zhang, X.K., Zhu, J.Z., Wang, R., Liu, X.H., Yao, Y.F., Wu, J.L. (1994). Diamond films as substrates for glucose sensors. *Materials Lett.*, **18**, 318–319.

[20] Zhu, J.Z., Tain, C.Y., Wu, W., Wu, J.L., Zhang, H.W., Lu, D.R., Zhang, G.X. (1994). Fabrication and characterization of glucose based on a microarray H_2O_2 electrode. *Biosensors & Bioelectronics*, **9**, 295–300.

BIOSENSORS AND THEIR APPLICATION IN THE PEOPLE'S REPUBLIC OF CHINA

Feng Derong

OUTLINE

Advances in Biosensors
Volume 4, pages 289–313.

ISBN: 0-7623-0073-6

ABSTRACT

Several kinds of biosensors have been developed and used in the People's Republic of China. The lactate biosensor has been used to monitor the aerobic respiration of athletes. The glucose biosensor has been adopted as a National Standard: Glucose Analysis in Food. The glutamate biosensor together with lactate and glucose biosensors has been used in the glutamate industry to control glutamate production. In this chapter, I introduce the structure and function of these biosensors; their roles in guiding scientific training, and a method for controlling fermentation by using several kinds of biosensors simultaneously.

1. INTRODUCTION

Biosensors, which are one of the products of the biotechnology industry, are among the key projects of the Seventh, Eighth, and Ninth Five-Year Plans of China's Science & Technology Developing Programs, respectively, and they play an important role in developing and reforming traditional biotechnology. The application of biosensors will provide new methods in such fields as industrial control, clinical detection, physical fitness training, food analysis, and environmental protection.

Research on biosensors began in the early 1980s in China, and much work has subsequently concentrated on this field. In the period from 1993 to 1995, about 126 reports were published, many of which are research reports [1–60].

The first biosensor system applied in China, the Model 23-L Lactate Analyzer developed by the YSI Corporation, was introduced in the late 1980s. Because of its speed, accuracy, and convenience in detection, it attracted a lot of attention in sports and was used to determine the lactate content in the blood, thus replacing the existing complicated chemical detection method used in many organizations to guide scientific sports training. However, its high price limited its application in China. Therefore, at the request of the National Sports Committee of China and after two years of hard work, we succeeded in devising the Model SBA-30 Lactate Analyzer and developed Lactate Kits in 1989 [61]. This was the first practical biosensor analyzer in China.

Now biosensors have become generally available. The Model SBA-30 Lactate Analyzer, a monoelectrode biosensor, is widely used in physical training and its consumers number more than 140 organizations. The Model SBA-40 Analyzer, a bielectrode analyzer, simultaneously analyzes any two of the following products:

glutamate, glucose, or lactate [62]. It has been used to control fermentation, saccharification, and recovery in more than 50% of the glutamate factories in China, each of which produces more than ten thousand tons of glutamate products. Work on the newly developed on-line Model SBA-60 Analyzer, a quadric-electrode analyzer, also has a worldwide lead. On December 1, 1996, the first National Standard in Food Analysis using biosensors took effect [63]. On March 25, 1996, the Science and Technology Daily of China reported biosensors on its front page. The headline was "Breakthrough in Researching and Application of Biosensors." In the meantime, Xu Wen, the vice-chief editor and dean of the news department made a brief comment on biosensors with the title "Bring Forth New Ideas in Application." On June 20, 1997, the Science and Technology Daily of China reported the award we gained in the research on "Glutamate, Lactate and Glucose Biosensors," the China National Invention Award, which is the most honored award in the sphere of sciences and technology in China.

2. BASIC CONDITIONS FOR THE APPLICATION OF BIOSENSORS

The transition from laboratory achievement to the practical use of biosensors is sometimes a complicated process. Many people ascribe this to the instability of the recognition component, mainly proteins [64,65]. Although a problem exists with the stability of enzymes, many kinds of oxidases are comparatively stable. For instance, glucose oxidase is the preferred bioactive substance used by many people in biosensor research, and a practical glucose biosensor has been produced. Yet only a small number of imported handheld glucose biosensors for home use are sold in the Chinese market, and our SBA-serial glucose biosensors are only partly used in industry. Applications in other fields, however, are almost universal. There are more complex reasons for biosensors coming into practical use.

At present two types of biosensors can be identified by their accuracy. One is the simple biosensor that has an accuracy within 5–10% or so. The other is a highly accurate biosensor, that is accurate within 1% or so. Yet it is quick and convenient to use compared with the traditional analytical method. In this paragraph, we concentrate our discussion on the practical high-accuracy biosensors (see Table 1).

First, the manufacture and commercialization of biosensors depends on solving technical problems as stated before. Secondly, we must be clear about the demand for biosensors in the market, the status of other relevant analytical techniques, and also the acceptability of biosensors to the public.

For instance, we have predicted the application in China of the high-accuracy glucose biosensor in clinical blood-glucose analysis. Many medium- and large-scale hospitals are equipped with automatic multifunctional biochemical analyzers. Therefore, our market potential should be oriented to small-scale hospitals. Why, then, has our glucose analyzer not been wildly adopted in such hospitals? Possible

Table 1. Characteristics of High Accuracy Biosensors

	Basic Demand	*Research Issues*
Bioactive materials	Operatiinal reliability, Long lifetime, Convenience for installation of membrane	Immobilization of biological materials, Operational and storage condition
Bioactive-current exchanger	A/D conversion with no influence of circumstance	Component reliability of bioactive-current exchange, Comparison with O_2, pH, and H_2O_2 electrode
Operational systems	Major operational system consists of sample chamber, fluid system, stirrer system, sample import	Reliability of operational systems
Electronic apparatus	Control function by microcomputer	Baseline stability, temperature compensation calibration, and linearity calibration, automatically clear bubble, software structure, display, print, statistics
Operational performance	High accuracy, quickness, and convenience in detection	Manual and methods for different applications
Reagent kit	Guaranteed provision of enzyme membrane, buffer, standard solution, and other repair parts	Research for kit and its constituents, conservation, packing and mail

reasons include the following. First, keen competition exists among the available analytical techniques in this field, including the traditional spectrum analyzer and the portable simple glucose analyzer. Secondly, much work still needs to be done to generalize biosensors and to demonstrate their superiority in speed and accuracy. On the other hand, we attribute our success with the lactate biosensor to the demands of competitive sports and the superiority of this biosensor over other methods. Furthermore, we, together with the experts in sports science, have held countrywide blood-lactate analytical technique sessions for years to refine the lactate biosensor.

In China, limited funds have restricted the transition from laboratory achievements to practical application of biosensors. To turn laboratory achievements in biosensor development into production power, research work must be concerned with bioactive substances and also with the areas of mechanics and electronics. In addition, much time and human resources are required and this is why most research groups cannot afford to make this transition. Although there are several laboratory achievements in China, few of them have been transformed into practical applications.

Fortunately, with finance from the authorities, we have been able to continue to develop biosensor products and at the same time expand research in this area. We have accumulated much experience over the last decade. Our success has brought our work into a good circle of scientists, thus enabling further improvement and generalization of our biosensors.

3. DEVELOPMENT OF SERIAL SBA BIOSENSORS

3.1 SBA Enzyme Membranes, Characteristics, and Preparation

The work in serial SBA biosensors began in 1983, and now more than 10 of kinds of biosensors (see Table 2) have been developed by our group.

Among them, the glucose, the glutamate and the lactate biosensors are widely used and the number of immobilized enzyme membranes produced exceeds 10,000, that have the following characteristics:

Table 2. Performance and Application Statistics of Serial SBA Biosensors Developed by Our Group

Type of Biosensor[a]	*Enzyme*	*Stability*	*Year Developed*	*Number of Applications*[b]
Glucose (O_2)	Glucose oxidase	>1 month	1986	
Glucose (H_2O_2)	Glucose oxidase	>2 months	1989	229
L-Ascorbic acid (O_2)	Ascorbate oxidase	>15 days	1988	
L-Lactate (H_2O_2)	Lactate oxidase	>15 days	1989	335
Hypoxanthine (H_2O_2)	Xanthine oxidase (XO)	>30 days	1992	8
Inosine (H_2O_2)	Nucleoside phosphate (NP) + XO	>15 days	1992	6
Adenosine 5′-mono phosphate (H_2O_2)	Nucleotidase (EC3.1.3.5) + NP + XO	>7 days	1992	
L-Glutamate (H_2O_2)	L-Glutamate oxidase (EC 1.4.3.11)	>2 months	1992	449
Urea (H_2O_2)	L-Glutamate oxidase Urease, Glutamic dehydrogenase	>20 days	1993	12
Aspartate (H_2O_2)	L-Glutamate oxidase Glutamate-alanine transaminase		1993	
Maltose (H_2O_2)	Glucose oxidase + glucoamylase	> 1 month	1994	2
Lactose (H_2O_2)	Glucose oxidase + lactase	>15 days	1995	4
Alcohol (H_2O_2)	Alcohol oxidase	>30 days	1996	8
Choline (H_2O_2)	Choline oxidase	>20 days	1996	8

Notes: [a]See references [61], [62], and [66–73].

[b]Applications: Of the immobilized membrane pieces consumed in the Biology Institute of Shandong Academy of Sciences in 1996, 10% are used in laboratory research, and 90% are used by an SBA client.

1. Immobilized bioactive substances as catalysts; one enzyme membrane for 1000 assays or so.
2. Quick response, producing a result in one minute.
3. High precision (±1% accuracy).
4. Operating simplicity.
5. Low cost compared with enzymatic and chemical analysis; each assay costs only several fen (YMB) under continuous determination.
6. Certain biosensors reliably indicate oxygen donation and the production of by-products in microbial culture systems, and offer information to supervise processing which could originally be obtained only by using many complex physical and chemical sensors. Furthermore the information also can indicate the pathway to increasing production rate.

Most serial SBA biosensor products are based on a bioactive oxidase which helps to produce hydrogen peroxide. The key component is an immobilized enzyme-membrane ring. The production process is shown below (Figure. 1):

3.2 Compound Immobilized-Enzyme Membranes

Immobilized-enzyme membranes prepared with glucose, lactate, glutamate, and hypoxanthine oxidases are usually very stable. They can be used together with other enzymes to construct a compound immobilized-enzyme membrane. One typical example is the biosensor combining glutamate oxidase, glutamate dehydrogenase, and urease [73] which detects urea. Following is the principle of the reaction:

$$\text{Urea} + H_2O \xrightarrow{\text{Urease}} 2NH_4^+ + CO_2$$

$$\alpha\text{-Ketoglutaric acid} + \text{NADH} + NH_4^+ \xrightarrow{\text{Glutamic dehydrogenase}} \text{L-Glutamate} + \text{NAD}^+$$

$$\text{Glutamate} + O_2 \xrightarrow{\text{L-Glutamate oxidase}} \alpha\text{-Ketoglutaric acid} + H_2O_2$$

The urea biosensor determines the urea content in the blood, which indicates an athlete's fatigue levels.

Our compound immobilized-enzyme membrane products include:

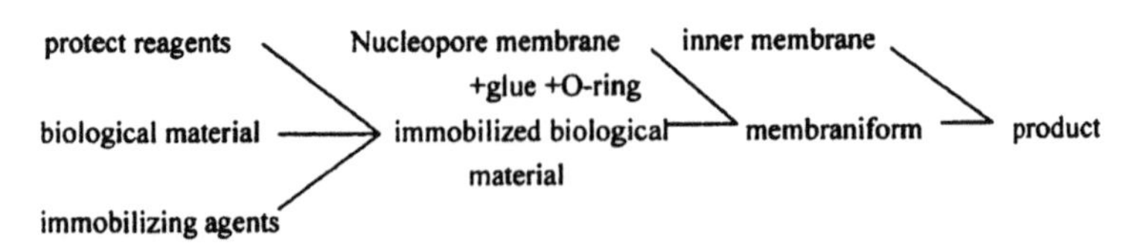

Figure 1. Preparation of the immobilized-enzyme membrane of the SBA.

1. Serial glucose biosensors:
   a. Lactose biosensor composed of glucose oxidase and lactase.
   b. Maltose biosensor composed of glucose oxidase and glucoamylase.
   c. Sucrose biosensor composed of glucose oxidase, mutarotase, and invertase.
   d. Starch biosensor that has the same composition as the maltose biosensor.

   These biosensors can be used to analyze foods, such as milk powder, honey, starch, glucoamylase, and dextrin.
2. Serial glutamate biosensors:
   a. Urea biosensor composed of glutamate oxidase, glutamate dehydrogenase, and urease.
   b. Aspartate biosensor composed of glutamate oxidase and glutamate-alanine transaminase.
   c. Alanine biosensor that has the same composition as the aspartate biosensor.
   d. $NADH^+$ and $NADPH^+$ biosensor composed of glutamate oxidase and glutamate dehydrogenase.
   e. Isocitrate biosensor composed of D-isocitrate dehydrogenase, glutamate oxidase, and glutamate dehydrogenase.
   f. Creatinine biosensor composed of creatinine deaminase (EC 3.5.4.21), glutamate dehydrogenase, and glutamate oxidase.

   The series of compound immobilized glutamate enzyme membranes are still at the laboratory stage, except that the urea biosensor has become commercial. These biosensors are important in theoretical research and also in practice. For instance, the urea/creatinine bifunctional biosensor analyzer in blood assays for kidney dialysis has become one of the highlights in biosensor research at present [74]. Another is the application of the D-isocitrate biosensor in assaying orange beverages.
3. Serial hypoxanthine biosensors:
   a. Inosine biosensor composed of hypoxanthine oxidase and nucleoside phosphorylase.
   b. Phosphate biosensor that has the same composition as the inosine biosensor except for the reagent kit, which is used with the SBA biosensor.
   c. Adenosine 5′-monophosphate biosensor composed of hypoxanthine oxidase, nucleoside phosphorylase, and 5′-nucleotidase.

   These three serial types of products are used mainly in industrial analysis.
4. Serial glycerol biosensors:
   a. Glycerol biosensor composed of glycerokinase and glycerol-3-phosphate oxidase.
   b. Creatine phosphate biosensor composed of creatine phosphokinase, glycerokinase, and glycerol-3-phosphate oxidase.

These serial products are used mainly in scientific training and industrial analysis.

Difficulties in compound immobilized-enzyme membrane in practice:

- Immobilization is complicated compared with single enzyme systems. Thus it is less easier to produce compound immobilized-enzyme membranes.
- The decreasing rates of enzymes activity are not synchronous, and this requires higher performance in linear calibration of the biosensor analyzer system.
- Oxidase is the base of compound immobilized-enzyme membrane. Yet some substrates of compound enzymes often cause interference. The solution is to mount two electrodes each equipped with a compound immobilized-enzyme membrane and an immobilized-oxidase membrane, separately. After calibration, we obtain the result by the difference between the two electrodes. This is the way our SBA-40 Analyzer is designed.

Research in compound immobilized-enzyme membranes yields far from perfect results, yet it has already shown its practical significance in expanding the application of biosensors.

4. DEVELOPMENT OF SERIAL SBA ANALYTICAL SYSTEMS

4.1 The Research on the Stability of the Energy Transformer

As a system, in addition to the biosensor we still need the auxiliary equipment listed in Table 1. The first issue is the stability of the energy transformer. Since 1988, we have used the H_2O_2 electrode as the energy transformer in the biosensor system because the result does not change with the oxygen content in the buffer system compared with the O_2 electrode. Yet the biosensor shows changes activity with a change in temperature. For instance, the lactate biosensor varies in relative activity under working conditions at different temperature (see Table 3). We must take measures to overcome biosensor instability caused by temperature variation.

4.2 Calibration of Linearity

Another feature of the immobilized-enzyme biosensor is that it gradually loses activity with time. To enable the system to work with good linearity, we must allow for periodic calibration in its design (see Table 4).

4.3 Several Kinds of SBA Biosensor Analyzers

We have developed several kinds of biosensors controlled by micro processors, Model-SBA-30, 40, 50, 60 Biosensor Analyzers, which are discussed following:

Table 3. Temperature Coefficient of Lactate Biosensor

Temperature, °C	Results	Relative Results, %	Temp. Parameter
10	5.0	100	
15	6.8	136	0.053
20	9.9	198	0.053
23	11.3	230	0.035
27	13.0	260	0.029
30	14.4	288	0.032
32.5	16.3	326	0.046
35.0	18.4	368	0.046
37.5	20.5	410	0.041

1. The Model SBA-30 Lactate Analyzer developed in 1989 became a national new product in 1990, equal to the 23L Lactate analyzer of the YSI Corporation of America in terms of quality. It is widely used in scientific training in China and has won two awards in the provincial Science and Technique Progress prize.
2. The Model SBA-50 Monoelectrode Biosensor Analyzer is a new type of SBA-30, produced in 1995. It is more automated compared with SBA-30 and is used to detect lactate and also other substances, such as glucose, by replacing the immobilized-enzyme membrane and setting multiple parameters. The SBA-50 Quick Glucoamylase Activity Analyzer which appeared in January 1996 was developed from the glucose biosensor by establishing a special sampling program. It takes only 3 minutes to determine the activity of glucoamylase (refer to Figure 2).
3. The Model SBA-40 Bielectrode Biosensor Analyzer developed in 1992 is used to analyze glutamate and glucose [62]. It became a national new product in 1993, won the Science & Technology Progress prize of Shandong Province in 1995, and was named the "eyes of glutamate fermentation." It has been widely used in glutamate factories in China and has begun to step into other industrial areas.
4. The Model SBA-60 On-Line Quadraelectrode Biosensor Analyzer System, authenticated by specialists in 1996, can be fitted out simultaneously with four kinds of biosensors. It is controlled by a computer, performs controlled

Table 4. The Effect of Linearity Calibration on Serial SBA Analyzers

Standard Solution, (u/mL)	100	120	150	200	300	375	500
C.K., u/mL	116	141	168	221	320	387	502
Test, u/mL	103	120	151	201	301	373	500

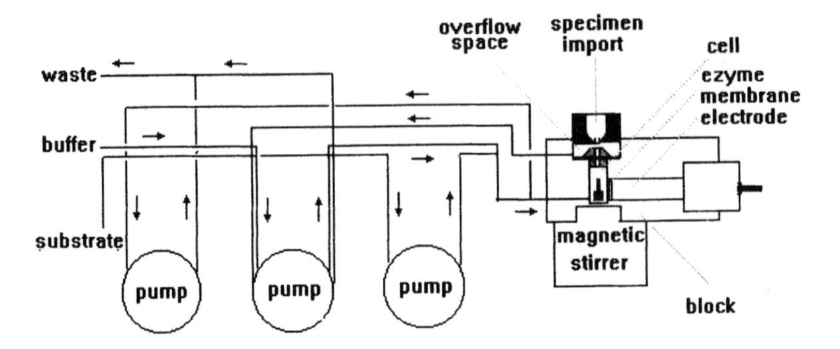

Figure 2. Schematic diagram of the Model SBA-50 biosensor analyzer.

dilution of samples, and can be connected to the fermenter or the biochemical reaction vessel to detect a substance on-line (Figure. 3).

The SBA-60 On-Line Quadraelectrode Biosensor Analytical System is composed of the following:

1. A reaction chamber connected to four electrodes and automatic disposal of foam.
2. A temperature-compensating H_2O_2 electrode that has an immobilized-enzyme membrane of glucose, lactate, glutamate, and alcohol.
3. A stirring system including electric motor and magnetic steel, stirring piece.

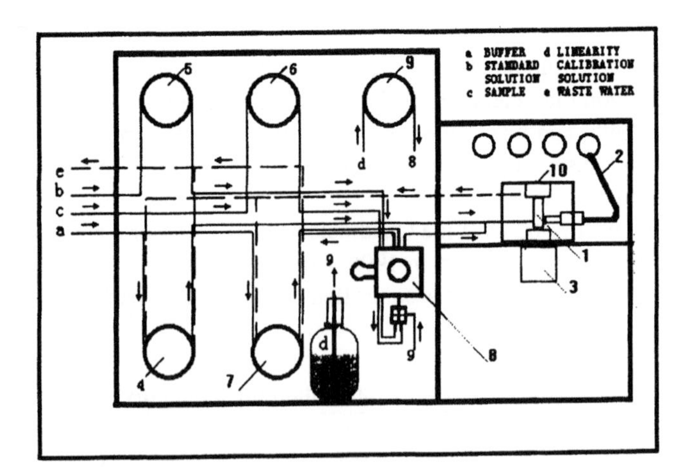

Figure 3. Schematic diagram of device for the Model SBA-60 biosensor analyzer.

4. A cleaning pump that empties the chamber, and meanwhile pumps buffer solution into the reaction chamber.
5. A calibrating pump that feeds standard solution into the measuring tube.
6. A sample pump that feeds sample solution into the measuring tube.
7. A sample or calibration pump that feeds sample or standard solution through the measuring tube (valve) into the reaction chamber. To increase the uniformity of flow during sample injection, we take advantage of a dual peristaltic pump that has 12 axles and 12 pulses per second.
8. A sample valve including a motor, a six-way valve, and a distributor.
9. A linear calibrating pump that feeds calibrating solution into the measuring quantitative tube.
10. A sample-injection opening used for injecting a sample manually.

The operating procedure is as follows. The system automatically starts up at a time set in advance. Then it cleans for a certain period of time. One of the pumps described in No. 5, No. 6, or No. 9 will work. The sample valve opens just three seconds before detection. The sample-injecting pump (No. 7) runs until a result is obtained. Next the sample valve closes, cleaning takes place for the second time, and so on until the end of the full procedure, as set in advance.

The status of the system is obtained from the computer display, including data collection, signal amplification, A/D conversion, parameter setting, and calibration of dilution. We can also change the status of the system and display and print the previous records.

For instance, Figure 4 shows the running status of system at a particular time. Following is the information you can get from it.

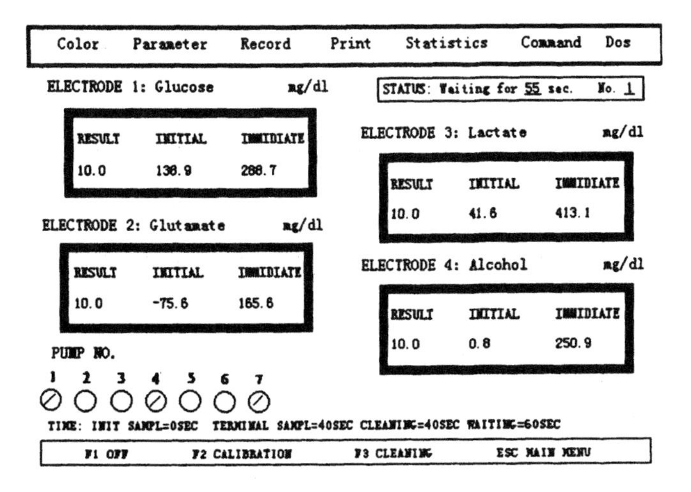

Figure 4. Screen of the Model SBA-60 biosensor analyzer.

Table 5. The Stability of Glucose, Glutamate, Lactate, and Alcohol Electrodes with the Model SBA-60 Analyzer

Samples	*Glucose*	*Glutamic Acid*	*L-Lactate*	*Alcohol*
Times	108	108	108	108
Average	10.01	10.00	10.00	9.97
CV%	0.30	0.42	0.30	0.67

1. The four biosensors shown previously are the glutamate, glucose, lactate, and alcohol biosensors.
2. The status of the system displayed is as follows. Having just finished calibration, it is ready to analyze sample No.1 55 seconds later.
3. The activity of each of the immobilized-enzyme membranes which is equal to the current value minus initial value (standard solution).
4. The initial value indicates if the system is functioning correctly.
5. The indicated pump number(s) tell us which pump(s) is (are) working.
6. The parameter setting: 60 seconds for waiting and 40 seconds for cleaning.

The SBA-60 system has excellent accuracy and stability (shown in Table 5). The SBA-60 On-Line Quadraelectrode Biosensor Analyzer System connected to a small scale fermenter has been tested in the two biggest glutamate factories on the mainland and can be used as an analog for the large-scale fermenter. Thus, it is attracting much attention from the technicians in these factories.

5. APPLICATION IN SCIENTIFIC TRAINING IN THE PEOPLE'S REPUBLIC OF CHINA

The lactate content in the blood is one of the most widely used indexes in sports science [75]. Lactate is the intermediate product of sugar metabolism in the human energy supply system. When a muscle has insufficient oxygen, glycogen generates more lactate and more energy to meet the muscle's needs during glycogen decomposition.

Urine lactate was first determined for the Chinese national swimming team in 1965. About 4500 persons were surveyed between 1979 and 1988. Many sports issues in China have made breakthroughs recently. One of the main achievements is that blood lactate analysis is now used to supervise training.

The relationship between an athlete's physical load from sports activity and blood lactate content can be divided into three main groups:

1. High endurance activity sustained over 1–2 minutes, such as swimming, middle- and long-distance running, cycling, skiing, speed skating, rowing,

and heel- and toe- walking races. The blood lactate index is quite closely related to the extent of an athlete's load.

2. Activities of fairly long duration with short rests, such as football, basketball, badminton, wrestling, judo, and boxing. The blood lactate index has a specific relationship to the load.
3. Activities of fairly short duration and long rests, such as volleyball, weight lifting, and gymnastics. The blood lactate index has no relationship to physical loading.

At present, blood lactate determination is quite popular in sports training in China because the study of biosensor applications is well developed in the country. A micromethod was developed and the blood collecting technique was standardized (20 μl) (standard treatment of blood sample, special blood solvent, sample preparation within 1 minute), and the method and check of scientific training in management were standardized.

Although lactate determination by the chemical method requires multiple test steps and is complicated, requiring 2–2.5 hr to determine 20–30 blood samples, the response time for blood lactate determination by the lactate biosensor is only 20–40 seconds. Its simplicity is welcomed by the majority of sports researchers. Over 200 lactate analyzers are now used in China's sporting arena, two-thirds of which are national serial SBA biosensors.

6. APPLICATION IN INDUSTRIAL CONTROL

6.1 The Use of Biosensors in the Production of Starch Sugar

In recent years, the bienzymatic method used in glucose production has made significant progress:

- High activity α-amylase and glucoamylase have been adopted to reduce the amount of exogenous protein in hydrolytic carbohydrate several fold.
- α-amylase that is tolerant to high temperature and glucoamylase that has low activity glucosidase are used to greatly reduce byproducts in saccharification.
- Steam-jet liquification has become widespread to resolve the difficulty in treating with insoluble starch slurry, to increase the concentration (content) of glucose solution and to completely use α-amylase that has high temperature tolerance.
- Now manufacturers know the effect of influent starch on the quality of glucose solution.

Although the bienzymatic method of producing glucose has many advantages and has developed significantly, saccharification is still controlled by the traditional method that has been used for several decades. Fehling's method is used to titrate the content of reducing sugar and its result cannot provide fully accurate information on the change of glucose content in saccharification. The determination and control of the glucose production becomes the bottleneck in saccharification technology. Therefore, we should do more to settle these problems.

The popularization and wide application of the glucose biosensor using the SBA system, combined with the implementation of National Standard GB/T 16285-1996 provides a new rapid, convenient method for the scientific analysis of glucose. Research into a new control method for producing glucose by the bienzymatic method has been advanced for consideration. The application of the glucose biosensor in glucose production is as follows:

1. *Determination of starch content.* It is necessary to study the differences between the new national standard method and the traditional method and its influences on the purchase of materials and regulation of production.
2. *Control of liquefying technology.* To select the steam-jet technique for liquefying, it is not enough to use only the iodine test as the control index. However, if it is possible to control the glucose value at the lowest level, the amount of glucoamylase substrate in saccharification will reduce, and the output and quality of bienzymatic glucose will finally increase.
3. *Detection of saccharification.* The percentage of glucose in the glucose solution on a dry substrate is defined as the DX value. The DX value is always lower than the DE value, and the specific ratio of DX to DE can be used to judge the final quality of the glucose solution, which, in fact, is equal to the ratio of the glucose value to the reducing sugar value. The concept has been utilized fairly well in some factories equipped with the glucose biosensor. The Zhoukou MSG factory has successfully used glucose biosensors to control production in the saccharification and fermentation workshop, and they have developed the technique of increasing the quality of glucose solution and the fermentation level. Table 6 is an example of the saccharification process supervised and controlled by a glucose biosensor. The various factors of saccharification, such as the amount of enzyme, temperature, pH, and dextran of different kinds, can all be examined using the same method to obtain the optional technological parameters.
4. *Rapid determination of glucoamylase activity.* When fermentation factories use the bienzymatic method to produce glucose, the amount of glucoamylase added must be correctly controlled. So the enzyme activity must be tested to properly control the saccharification process. After utilization in WuXi Xingda Bioengineering Co.—the largest enzyme-producing factory in China, the evaluation was as follows: The SBA-50 biosensor analyzer needs only three minutes to determine the glucoamylase level. The regular iodometric method needs 60 minutes for the same task. The result accords with the regular method, and the biosensor has the advantages of better repeat-

Table 6. Saccharification Process under the Control of a Glucose Biosensor

Time	0 min	1 min	5 min	10 min	30 min	1 h	2 h	3 h	5 h	7 h	12 h	24 h	30 h
Glucose	0.7	3.0	17.8	33.5	70.9	80.4	105	109	115	116	119	121.3	121.3

ability, higher precision, simple reagent requirement, and low analytical cost (see Table 4).

5. *Determination of glucosyltransferase activity.* Glucosyltransferase is the enzyme mingled in industrial glucoamylase, whose existence decreases glucose (in glucose solution) in saccharification because of polymerization. The national standard for glucose has no reference to glucosyltransferase mixed in glucoamylase, and the extent of glucosyltransferase removal from glucoamylases varies among factories. Therefore, glucosyltransferase from various factories needs to be determined. The principle of using the glucose biosensor to determine glucosyltransferase activity is as follows: The glucoamylase sample is mixed together with *p*-nitrophenyl α-D- glucoside (specific substrate of glucoamylase). Then, glucosyltransferase activity is determined in terms of the liberated glucose.

6.2 Ion Exchange Process

In ion exchange retrieval, on-line detection of the change of biochemical composition is the key to increasing the total glutamate extraction rate. In 1995, the extraction rate of glutamate in ten MSG enterprises in China ranged from 91.76% to 83.90% (according to a report in the Bulletin of Fermentational Science and Technology in February, 1996). This means that the extraction rate of glutamate in ion exchange was very low, about 43.9 ~ 55.9%. Much was wasted and lost. So we have developed a new model of a pH on-line automatic analyzer that has high precision. On this basis we have conducted simulated experiments on the ion exchange retrieval of an isoelectric liquor supernatant of glutamate. The following variables were measured: volume, concentration of glutamate (SBA-60 automatic analyzer), and pH value (SBA-pH automatic analyzer) (see results in Figure 5).

According to this simulated experiment, the retrieval rate in the test is about 90%. The previous controlling technology based on pH detection resulted in a great loss of glutamate. By using the glutamate biosensor on-line, we have the elution process under fine control we increased the retrieval rate by more than 30% and reduced the loss caused by ineffective circulation in the ion-exchange process for glutamate. The expenses of acid, alkali, and environmental pollution are also reduced.

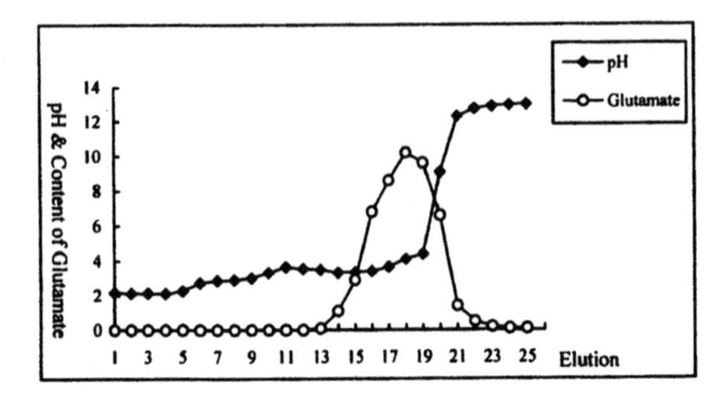

Figure 5. On-line detection of pH and glutamate content in the process of recovery elution.

6.3 Use of the Lactate Biosensor to Detect the Air-Supply Condition of Microorganisms in Fermentation

Because the lactate biosensor has a special and important use in fermentation control, many MSG factories have extensively used it. The two units of Zhoukou Lotus Enterprise Group and Jining Linghua Enterprise Group alone use more than thirty SBA-40 Model biosensor analyzers, one-third in lactate analysis.

Kou Tianping [76] in the Jining MSG factory, the first to use the lactate biosensor to determinate the lactate content in fermentation, established that the optimal lactate content in the glutamate fermentation process is 0–0.2%. At that point, the turnover of glucose-glutamate is the highest. The glutamate production rate and the turnover of glucose to glutamate decrease with increasing lactate content.

Zhang Canli, Cao Beidou, et al. [77] in the Zhoukou MSG factory also proved the relationship of lactate content to glutamate content. Lin Zhongre (general engineer) et al. in the Zhoukou MSG factory applied lactate detection in research on a "100 m^3 Air Lift Dual-Way Cyclic Flowing Fermenter," which is among the key projects of the Eighth Five-Year Plan. They regulate and select the optimal ventilation of the fermenter in terms of lactate content. This has a great effect in stabilizing the technology as fast as possible.

Various factors in fermentation such as excessive biotin, high population density of cells excessively high temperature, low ventilation, filter blockage, and a poorly corrected air flow meter increase lactate content. Analyzing the reasons for higher lactate content, which plays an important role in the workshops for obtaining high and stable yields, frequently becomes the focus of the technological analysis of fermentation.

Without proper operation, the highest content of lactate in fermentation liquid is 4.0%, and the glutamate content is only 5.0%, which cause a great loss of production. Practice indicates that limited by the present detection technique of dissolved oxygen in the fermenter to date, judging and controlling the oxygen

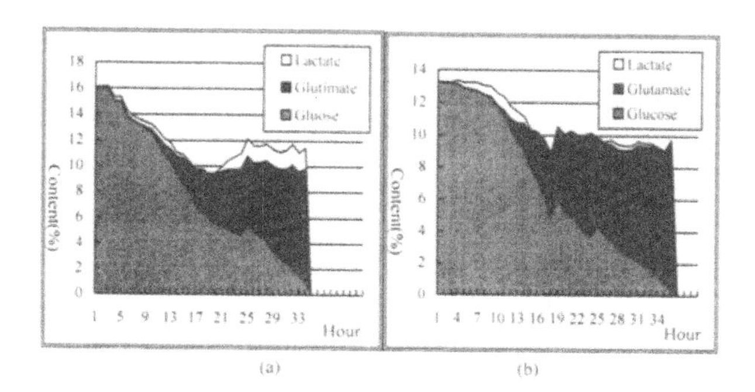

Figure 6. Influence of lactate control on fermentation: (**a**) poor lactate control; (**b**) good lactate control.

demand of microorganisms by lactate detection is the most inexpensive, effective, and reliable method for controlling fermentation. Under the control of a lactate biosensor, the production rate of glutamate increases by 0.5%, thus increasing glutamate output by 6% (see Figure 6).

6.4 Interference of Ammonium Ions in Fermentation Liquid with Enzyme Electrode

Ammonium in fermentation liquid affects to some degree the results detected by the biosensor, because ammonium ions irreversibly inhibit biosensors that contain platinum-silver. Experiment proved that data for lactate, glutamate, and glucose determination were only 94.2%, 93.0%, and 86.3% of the correct values, respectively when the ammonium ion content in fermentation liquid was 10% (see Table 7).

Liquid ammonia in glutamate fermentation is a nitrogen source and pH conditioner and is added in feeding. Generally, less is added in the initial stage, and more is added in the middle and end states. One part converts to the composition of cell protein, one part converts to glutamate, and one part is released with waste air. Therefore, quite a lot remains in the fermentation liquid as ammonium ions, and excessive ammonium ions may interfere with the determination and produce low data quality.

Adding a little NH_4Cl to the buffer may effectively resist interference by ammonium ions. Table 7 indicates that determination deviation of three electrodes is reduced below 1% by using buffer containing added NH_4Cl. This method can satisfy the determination in the normal fermentation liquid. Buffer with added NH_4Cl inhibits the enzyme membrane, but the inhibition is reversible, and once changed back to the previous buffer, the enzyme membrane activity recovers. Another method of reducing deviation often used in factories is that of adding an

Table 7. Interference of NH_4^+ with Biosensors for Lactate, Glutamate, and Glucose

	Buffer			*Buffer + NH_4Cl*		
	Lactate	*Glutamic Acid*	*Glucose*	*Lactate*	*Glutamic Acid*	*Glucose*
CK	100.3	99.9	99.4	99.4	99.5	100.4
2% NH_4Cl	97.3	98.0	94.1	99.0	99.3	100.0
4% NH_4Cl	95.5	96.0	90.6	99.6	99.1	99.7
6% NH_4Cl	95.0	94.7	88.5	99.8	99.0	98.4
8% NH_4Cl	94.3	93.8	87.5	100.3	98.7	99.0
10% NH_4Cl	94.2	93.0	86.3	99.6	98.2	98.6

experience coefficient during the setup calibration. The coefficient varies with the technology, but is generally in the range of 1.06–1.15.

At one time, the interference of ammonium ions in the fermentation liquid (especially in glutamate fermentation) limited the use of biosensors in fermentation factories. However, because the biosensor has fairly good stability in experiments, and the data detected by biosensors plays a significant role in guiding fermentation control, biosensors are gradually spreading in the Chinese fermentation industry. Now many MSG factories, including those in which money from Japan, Korea, France, and Taiwan is invested, are using glucose, glutamate, and lactate biosensors. Their application in other industries is also expanding.

6.5 What Have Biosensors Brought to Glutamate Fermentation Control?

1. Provided new fermentation control parameters [71].
   The main parameters of the technological process used continuously in the glutamate industry for thirty years are O.D. value, reducing sugar, and glutamate. Limited by the speed of determination and the degree of accuracy, control information cannot be collected in terms of practical requirements. But biosensors provide a rapid and accurate method for obtaining a wide range of useful information about glutamate fermentation (see Table 8).
2. Early stage predictions about abnormal phenomena.
   a. Infection: Glucose consumption rate appears unusual; glutamate does not appear.
   b. Seed overgrowth: glutamate appears in early stage of fermentation.
   c. Quality of glucose material does not pass: difference between reducing sugar and glucose too high; saccharification not complete.
   d. Lactate content too high: ventilation insufficient, broth not stirred sufficiently, excessive cell density, air system block, or inadequate correction to air-flow meter.

Table 8. New Parameters for Glutamate Fermentation Control

Parameter	*Traditional Method*	*New Method*
Cell density	OD determination every 2 hours	Indicating population density during fermentation on-line
Substrate concentration	Reducing sugar determination every 2–6 hours	1. To determine the growth rate of microbe according to the consumption of glucose; 2. To determine the transformational rate together with the production rate of glutamate 3. To determine the end of fermentation
Product concentration	An assay every six hours during the initial 12–28 hours; then every 2 hours.	1. To find out if the amount of biotin added is suitable during the initial 6–12 hours (fermentation with no addition of penicillin); To determine the addition time for biotin (fermentation with addition of penicillin) 2. To determine the transformational rate together with the consumption of glucose 3. Seed quality control; glutamate existence shows seeds aging
By-product concentration	To determine the production of lactate during fermentation	To control the lactate below a concentration of 0.2%

   e. Biotin not enough or excessive: time when glutamate appears accelerated or retarded.

3. Biosensors provide a new way to use a computer to control fermentation.

From the 1980s onward, the application of computers in biotechnology has increased. They first were used in kinetics and optimization research on the off-line fermentation process. This means setting up a mathematical model to simulate fermentation. Secondly, they were used to control the fermentation process. Third, they were used for data storage and manipulation. Now, more advanced commercial fermenters use computers to control parameters, such as temperature, fermenter pressure, agitation speed, air flow rate, pH, dissolved oxygen, foam, and oxidation-reduction potential, but some essential parameters, such as substrate, product, intermediate product, cell concentration, viscosity of fermentation liquid, air-liquid interface area, still cannot be successfully determined and controlled. The main reason is lack of proper sensors which use the potential advantages of the computer and can automatically supervise and control some parameters in fermentation. The development of the biosensor automatic analyzing method, will enlarge the application of the computer in biotechnology and more highly automatic fermentation.

In recent years, because of application of biosensors in the glutamate industry, it is estimated that on-line fermentation control will be realized step-by-step during the Ninth Five-Year Plan period. The first step is to realize off-line detection with the biosensor and gather relative information for on-line supervision and control. The second step is to adapt the biosensor analyzer to a continuous on-line working system and realize on-line track analysis of the drainage method. The third step is to realize a pH system and control system for automatic addition of glucose, to research dealing with a large amount of data in the computer, and finally to reach the goal of total automatic control in glutamate fermentation. Now there are three favorable factors. The first is that biosensors used in some MSG factories for more than five years have gathered much important firsthand data by off-line detection. The second is that more scientific control parameters fit for the current glutamate fermentation industry in our country have been selected. These parameters include physicochemical parameters, such as temperature, ventilation, fermenter pressure, amount of liquid ammonia, and biological parameters, such as the glucose, lactate, and glutamate concentration, cell density, and pH. The third is that the Model SBA-60 four-electrode biosensor analyzing system and computer-controlled automatic dilution system have been developed.

The four-electrode biosensor on-line analyzing system has two uses in production, one to simulate production with a small-scale fermenter parallel to a large-scale fermenter, and the other to do on-line analysis directly connected to a large fermenter. The former can resolve production problems and explore new technological conditions. The latter can keep production in an optimal condition. But the connection of a large-scale fermenter to biosensors still needs to be studied. This is one of the Ninth Five-Year Plans that we are undertaking in China. The main aim of the project is to settle the problem caused by a biosensor that cannot be sterilized. We intend to drain the sample from the fermenter and dilute it to the concentration range that can be detected by the biosensor. The equipment must have reliable long-term reliability. Therefore, before settling on the connection technique for connecting a large-scale fermenter to a biosensor, a lot of experimentation must be done on the small-scale fermenter.

In 1996, we developed an automatic dilution system with peristaltic pumps. The degree of dilution can be controlled by a 486 computer in this system. We used a small-scale fermenter together with the model SBA-60 four-electrode on-line analyzing system and pH automatic analyzer to simulate a fermentation test. The total initial 6000 mL of fermentation liquor was put into the small-scale fermenter, and 1075 mL of 30% glucose was added in the middle of the fermentation. At the end of the fermentation, 4190 mL remained. The concentrations of glutamate, glucose, and lactate were 9.26%, 0.38%, and 0.085%, respectively. The fermentation data are listed in Figure 7.

According to the data in Figure 7, the target product in fermentation is glutamate and the by-product is lactate. In the fermentation process, the amount of glucose decreases continuously. From 0 to 8 h, the rate of glucose consumption is 1%. In

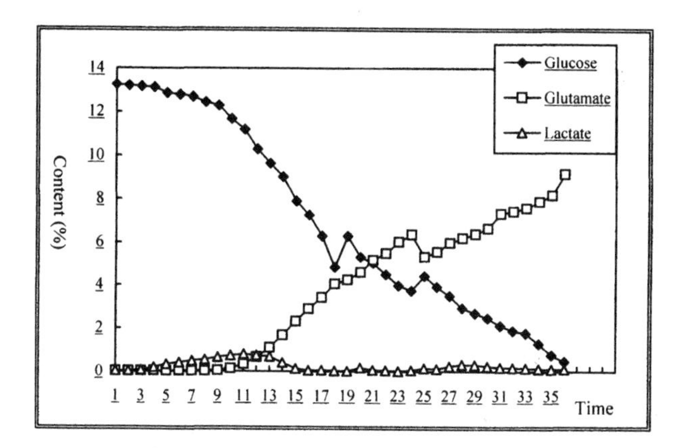

Figure 7. Fermentation controlled by the SBA60 on-line analytic system in a small-scale Fermenter analogous to that in a large-scale fermenter.

the period of rapid production of glutamate (11 to 18 h), the rate of glucose consumption is 0.91% per hour and the inversion rate is 58.24%. The highest rate of glucose consumption is 1.13% per hour. From 19 to 24 h, the rate of glucose consumption is 0.512% per hour, and the inversion rate is 83.2%. In the later stage of fermentation, the glucose consumption rate is 0.2993%. From 25 to 33 h the inversion rate of glucose reaches 85.67%. In the middle and later stages, the inversion rate is higher than the theoretical value. This may be the result of self-cytolysis.

It is worth noticing that the appearance of lactate is related to cell growth. There are three lactate peak values in this test (11 h, 20 h, and 29 h). From this, we can conclude that the increase of lactate is related to the amount of biotin in the glucose added in fermentation when ventilating insufficiently, or with excessive biotin, excessive inoculation, too high a temperature, or adding glucose too late. These frequently cause the lactate to go out of control and cause glutamate output to decrease.

7. FOOD ANALYSIS

Biosensors used for food analysis have been approved by the National Technique Regulating Bureau. We determined the content of glucose with the glucose biosensor in comparison with the method using liquid enzyme from the North Food Inspection Center for China's Farming. The results accord with the demands. Using a biosensor to detect the glucose content in food (two kinds of fruit, two kinds of vegetables, three kinds of drink, and two kinds of wine) has the advantages of rapid response and high precision (Table 9).

Table 9. Analysis of Glucose Content in Food by Biosensor

Food Samples		Times n	Average Glucose Content, %	SD	CV, %
Fruits	Hawthorn	10	1.85	0.012	0.67
	Pear	10	1.89	0.009	0.5
Vegetables	Chinese cabbage	10	0.65	0.008	1.26
	Cucumber	10	0.71	0.005	0.68
	Tomato		0.53	0.005	0.9
Wines	Red wine	10	5.6	0.5	0.9
	Yellow wine	10	9.44	0.117	1.24

8. PERSPECTIVE

In the past few years, we have succeeded in developing the lactate biosensor first applied in physical training. Afterwards we made the breakthrough into the industrial fermentation control and formulated a national standard by using the glucose biosensor.

In the People's Republic of China, compared with other countries, the application of biosensors in industrial control and scientific physical training has reached quite an extent, and every year more than 1,000,000 samples are detected by taking advantage of glutamate, lactate, and glucose biosensors. Yet there is still much work to be done in applying biosensors. Due to the difference in pricing systems between China and Western countries, it is difficult to assess the comparative output value. Based on the rough number of 240 sets of biosensor analyzers used, the output value is about 2,000,000 U.S. dollars.

It is predicted that in several years Chinese biosensors will be generally adopted in food analysis, biosensors, on-line in industrial control will come into practice, and the application of biosensors in environmental protection will be greatly expanded.

REFERENCES

[1] Sun, C.Q., Guo, L.P. (1993). Application of ion associate of $(C_6H_5)_4B^- \cdot N^+(CH_3)_4$ to amperometric biosensor for glucose based on carbon paste modified by 1,1′-bimethylferrocene. *Chin. J. Anal. Chem.*, **21**(8), 882–886.

[2] Tian, C.Y., Zhang, G.X. (1993). Study of micro amperometric type biosensor. *Chin. J. Chem. Sensor.*, **13**(1), 16–17.

[3] Yu, S.G., Zhuo, D. (1992). Study on membrane electrode biosensor. *Elec. Tech. Heilongjiang.*, **17**(1), 1–4.

[4] Wang, Q., Deng, J.Q. (1993). Development of sensor for L-ascorbic acid based on ginger tissue and its application in determination of pill. *Chin. J. Anal. Chem.*, **21**(9), 1018–1021.

[5] Sun, C.Q., Song, W.B. et al. (1993). Studies on application of ion associate to amperometric biosensors for glucose based on paste modified by $FeCp_2$. *Chem. J. Chin. Univ.*, **14**(8), 1073–1075.

[6] Jin, L.T., Jie, J.S. et al. (1993). Pyruvate oxidase biosensor for the determination of GPT based on the glassy carbon electrode modified with Nafion and methyl viologen. *Chem. J. Chin. Univ.*, **14**(9), 1210–1213.

[7] Zhang, Z.J., Ma, W.B. (1993). A fiber optic biosensor for the determination of cholesterol. *Chem. J. Chin. Univ.*, **14**(10), 1366–1369.

[8] Ji, X.F., Zhang, Y.H. (1993). Glucose sensor based on glassy carbon electrode modified with platinum sinked in AQ membrane. *Chin. J. Biochem. Biophys.*, **20**(5), 395–397.

[9] Sun, C.Q., Song, W.B. et al. (1994). Study on amperometric glucose biosensor based on TTF. *Chin. J. Chem. Sensor.*, **14**(1), 21–27.

[10] Yie, B.C., Li, Y.R. (1994). The response kinetic model of amperometric enzyme electrode. *Chin. J. Chem. Sensor.*, **14**(1), 44–48.

[11] Zhuang, Y.L., Qi, D.Y. (1994). Study on L-alanine biosensor based on cactus tissue. *Chin. J. Biomed. Eng.*, **13**(2), 164–169.

[12] Zhuang, Y.L., Qi, D.Y. (1994). Biosensor based on apple tissue. *Chin. J. Anal. Instrum.*, **4**, 9–13.

[13] Jin, J.T., Mao, Y.P. et al. (1994). Study on glucose biosensor based on tetathiafulvalene. *Chin. J. Anal. Lab.*, **13**(3), 13–15.

[14] Shen, H.M., Xu, C.X.(1993). Study on amperometric urea sensor. *J. Transducer Tech. Addition*, 12–14.

[15] Yie, B.C., Li, Y.R.(1993). Study on the construction and character of platinum electrode with enzyme linked to platinum. *Chin. J. Ind. Microbiol.*, **23**(6), 11–15.

[16] Guo, D.L., Liu, S.H. (1993). Study of biosensor. V. Study on urea electrode based on tissue. *Chin. J. Chem. Res. Appl.*, **5**(1), 32–37.

[17] Cao, D.J., Mao, D.Y. (1993). Electric aggregational immobilization of aniline-glucose oxidase. *Chin. J. Chem. Sensor.*, **13**(4), 30–33.

[18] Zhuang, L., Zhang, Y.X. (1993). Study on the compatibility of blood using glucose enzyme biosensor. *Chin. J. Chem. Sensor.*, **13**(4), 49–53.

[19] Yie, B.C., Li, Q.X. (1994). The glutamate biosensor and its application to flow injection analysis system. *Chin. J. Biotechnol.* **10**(2), 109–113.

[20] Zhuang, Y.L. (1994). Study on biosensor based on jade plant tissue. *Chin. J. Biotechnol.*, **10**(2), 151–156.

[21] Wang, S.G., Ji, X.S. et al. (1995). Development of glucose biosensor based on acetate film. *Chin. J. Biotechnol.*, **11**(3), 260–265.

[22] Sun, S.Q., Wang, N. (1994). Study on the determination of concentration of glucose in glucose injectional solution by using enzyme electrode. *Chin. Pharmacol. J.*, **29**(9), 546–548.

[23] Deng, J.Q., Wang, Q. (1994). Current progress of bioelectrode. *Chin. Acta Anal. Sci.*, **10**(3), 73–84.

[24] Hu, W.P., Zhang, X.E. (1993). A multi-function enzyme sensor for sucrose and glucose determination. *Chin. Biochem. J.*, **9**(6), 741–747.

[25] Qiao, Y.G., Yin, T.L. (1994). Glucose biosensor based on carbon paste electrode modified with ubiquinone. *Chin. J. Anal. Chem.*, **22**(7), 709–711.

[26] Wang, Q., Deng, J.Q. (1994). Study on improving the selectivity of microbial enzyme sensor. *Chin. J. Chem. Sensors*, **14**(4), 243–244.

[27] Wang, L.G., Wang, X. (1995). Study on SBLM biosensitive apparatus. *J. Nankai Univ. (Nat. ed.)*, **28** (4), 51–53, 63.

[28] Zuang, Y.L. (1995). Study on the kinetic response mechanism of tissue biosensor for L-arginine determination. *Chin. J. Biotechnol.*, **11** (1), 93–95.

[29] Li, W., Chen, J. (1995). A fiber optic biosensor for immunoanalysis. *Chin. J. Chem. Sensors*, (1), 22–26.

[30] Dong, Z.H., Li, Y.J. (1994). Biosensors based on completeable chemical receptor. *Chin. J. Sensor Tech.*, **94** (1), 14–19.

[31] Guo, L.P., Sun, C.Q. (1994). Study on glucose sensor based on platinum electrode modified with tetrathiafulvalene. *J. Northeast China Teacher's Univ.*, **94** (2), 57–59.

[32] Liu, H.Y., Deng, J.Q. (1995). A glucose sensor based on ferrocene and cross-linker modified. *Chin. J. Biochem. Biophys.*, **22** (1), 68–71.

[33] Huang, H.C., Fu, M.G. (1994). Study on glucose biosensor. *J. Nanchang Univ. (Sci. ed.)*, **18** (3), 303–308.

[34] Liu, H.Y., Deng, J.Q. (1994). Hypoxanthine biosensor modified by Nafion-ferrocene. *Chin. Acta Anal. Sci.*, **10** (4), 7–10.

[35] Yu, P., Xu, C.X. (1994). Development and comparison of the two kinds of uric acid biosensors. *Chin. Acta Sensor Sci.*, **7** (4), 28–33.

[36] Jin, C., Shun, G. (1995). Fiber optic biosensor of immobilized firefly luciferase. *Chin. J. Biochem. Biophys.*, **27** (1), 91–94.

[37] Liu, B.H., Kong, J.L. (1994). Study on the method of immobilizing microorganism membrane of sensor for rapid determination on BOD. *Chin. J. Environ. Sci.*, **15** (6), 8–11, 27.

[38] Zhang, Y.H., Yang, S.P. (1995). Ferrocene derivative modified initiative hydrogen peroxide biosensor. *Chin. J. Anal. Chem.*, **23** (1), 9–13.

[39] Liu, H.Y., Deng, P. (1995). Glucose sensor modified with Nafion-ferrocene-bienzyme. *Chin. J. Anal. Chem.*, **23** (2), 154–158.

[40] Hu, X.Y., Leng, Z.Z. (1995). Study on tyrosinase amperometric biosensor for the determination of low concentration of inhibitor of cyanidum. *Chem. J. Chin. Univ.*, **16** (3), 359–362.

[41] Zhang, H., Xu, C.X. (1995). Study of cholesterol biosensor. *Chin. J. Transducer Tech.*, **95** (2), 20–25.

[42] Na, X.L., Zhang, H.G. (1995). Development of flowing cell sensor. *Chin. J. Transducer Tech.*, **95** (6), 24–26.

[43] Xie, P.H., Xu, C.X. (1995). Study on voltammetric-amperometric biosensor for distinguishing and calculating microbial cell. *Chin. Acta Sensor Sci.*, **8**(3), 24–29.

[44] Zhang, H.W., Xu, C.X. (1994). Development of bienzyme membrane electrodes for determination of the activities of GOT and GPT in blood serum. *Chin. J. Transducer Tech.*, **94**(6), 10–14.

[45] An, L.C., Zeng, H.J. (1994). Study of glucose biosensor. *Nangjing Univ. Tech.*, **94**(6), 92–96.

[46] Xie, H.F., Song, G. et al. (1995). The application of biosensor membrane electrode to environmental protect. *Chin. J. Microbio.*, **15**(2), 66–69.

[47] Hu, X.Y., Leng, Z.Z. (1995). Response character of catechol on polyphenol oxidase amperometric sensor with preactivated polyamide. *Chin. J. Anal. Chem.*, **23**(4), 416–418.

[48] Cai, C.X., Chen, H.Y. (1995). Immobilized, character and application of glucose oxidase on electrode with nano-degree amount of glod. *Acta Chim. Sin.*, **53**(4), 336–339.

[49] Na, X. L., Guo, J.W. (1995). Development of biosensor based on white blood cell membrane and the determination of the activity of hydrogen peroxidase. *Chin. J. Anal. Chem.*, **23**(3), 446–448.

[50] Zhang, X.H., Wang, Q. (1995). Study on hydrogen peroxide sensor based on ginger powder tissue. *Chin. J. Anal. Chem.*, **23**(3), 336–339.

[51] Liu, H.Y., Deng, J.Q. (195). Amperometric glucose sensor based on Eastman-AQ-tetrathiafulvalene modified glassy carbon electrode. *J. Fudan Univ, Nat. Sci. ed.*, **34**(1), 112–114.

[52] Wu, H.H., Chen, W.B. et al. (1994). Biofunctional electrode. *Chin. J. Xiamen Univ.*, **22**(4), 111–115.

[53] Liu, H.Y., Qian, J.H. et al. (1995). 1,1′-Dimethylferrocene-Nafion modified glucose sensor. *J. Fudan Univ, Nat. Sci. ed.*, **34**(3), 333–338.

[54] Shi, Q.Z., Wu, J.M. (1995). Development and application of lactate oxidase biosensor based on mediator. *Chin. J. Anal. Chem.*, **23**(8), 926–929.

[55] Fan, Q.C., Yan, J.H. (1994). Application of hydrogenic L-ascorbic acid electrode based on plant tissue membrane. *Chin. J. Hyg. Insp.* **4**(6), 350–352.
[56] Liao, H., Yin, I.Z. (1995). Development of hydrogen peroxide biosensor. *Chin. J. Anal. Chem.* **23**(6), 734.
[57] Hu, W.P., Zhang, X.M. (1995). A GOD electrode free from the influence of alcohol. *Chin. J. Biotech.*, **11**(1), 45–50.
[58] Lin, Z.H., Shen, G.L. (1995). A new kind of immuno-electrode for thyroid hormone based on carrier membrane. *Chin. J. Chem. Sensors*, **15**(1), 2–3.
[59] Liao, H., Yin, L.Z. (1995). Development of hydrogen peroxidase electrode for flow injection. *Chin. J. Chem. Sensor* **15**(1), 4–6.
[60] Guo, M.D., Hu, S.S. (1995). Study on lactate oxidase electrode based on tetrathiafulvalene. *Chin. J. Chem. Sensor*, **15**(1), 7–8.
[61] Feng, D.R., Feng, Y.S. (1990). The research of immobilized lactate oxidase enzyme-membrane-ring. *Shandong Sci.*, **3**(3), 1–6.
[62] Feng, D.R., Shang, X.Q. (1993). Study on the model SBA-40 L-glutamate and glucose biofunctional analyser used in the fermentation process. *Chin. J. Food. Fermentation Ind.*, **93**(2), 33–37.
[63] National Standard GB/T 16285–1996, *Method for determination of glucose in food—Enzyme - colorimetric method and enzyme - electrode method.*
[64] Wang, Y.X. (1987). Perspective of biosensors. *Chin. J. Biotech.*, **3**(1), 22–23.
[65] Fan, C.Y. (1995). Developments and applications of biosensors. *Chin. J. Transducer Tech.*, **2**, 1–5.
[66] Feng, D.R., Feng, Y.S. et al. (1992). A multi-functional biosensor system for assaying fish meat freshness. *The Second China-Japan Joint Symposium on Enzyme Engineering*, Beijing, China.
[67] Shi, J.G., Zhou, F.Z. et al. (1996). Study on the determination of glucoamylase with glucose enzyme electrode. *Chin. J. Biotech. Addition*, **12**, 226–231.
[68] Feng, D.R., Qiu, W.Z. (1990). Rapid analysis of Vitamin C by enzyme electrode. *Chin. J. Pharm.* **21**(6), 261–263.
[69] Feng, D.R., Sun, S.Q. (1995). Study on biosensor for the determination of aminotransferase. *Shangdong Sci.*, **8**(1), 45–49.
[70] Feng, D.R., Sun, S.Q. (1995). The application of the series of SBA biosensors. *Chin. Fermentation Sci. Tech. Commum.*, **24**(1), 24–25.
[71] Feng, D.R. (1996). Biosensor and glutamate fermentational production. *Chin. Fermentation Sci. Tech. Commun.*, **25**(3), 10–13.
[72] Feng, D.R. (1996). The development and applicational progress of biosensors. *Chin. Proc. Biotechnol.*, **16**(3), 13–15.
[73] Feng, D.R., Li, Z.H. (1996). Study on bifunctional analyser of urea-nitrogen and glucose. *Chin. J. Biotechnol.*, **12**(2), 189–193.
[74] 10400694 BIOSIS Number 9600694 *Multifunctional flow-injection biosensor for simultaneous measurement of creatinine, glucose and urea.*
[75] Feng, W.Q., Wong, Q.Z. (1990). *Blood Lactate and Sports Training—Applied Handbook.* Beijing.
[76] Kou, T.P. (1994). Studies on product rate of glutamate and the concentration of lactate. *Chinese Fermentation Sci. Tech. Commun.*, **23**(2), 38.
[77] Zhang, C.L., Cao, B.D. et al. The application of lactate biosensor in glutamate fermentation. *Chinese Fermentation Sci. Tech. Commun.*, **24**(2), 38–39.

SenTest™ FROM SINGLE CARBON ENZYME ELECTRODE TO MASS PRODUCTION OF BIOSTRIPS

Jun Hu and Bixia Ge

OUTLINE

Advances in Biosensors
Volume 4, pages 315–325.

ISBN: 0-7623-0073-6

ABSTRACT

A single graphite carbon rod electrode was used in research leading to the mass production of biosensors in China. Two soluble mediators, benzoquinone and ferricyanide, that have different kinds of glucose oxidase are discussed. The properties of the electrodes indicate that screen and ink-jet printing are feasible techniques for machine production of glucose sensors. An overview of the development of a successful commercialized instrument, the SenTest^{TM} glucose sensor system, is presented. It will reach a total sales figure of a half million dollars in 1997. Future research on multifunctional probes for neonatal screening programs is also discussed.

1. INTRODUCTION

Biosensors have attracted considerable attention in recent years because of their potential to provide solutions to a wide range of analytical problems. Amperometric enzyme electrodes normally consist of a biological sensing element in intimate contact with an electrochemical transducer. The electrode is generally poised at a specific potential and the current generated by redox processes at the surface of the electrode is recorded. Some approaches realizing electron transfer from oxidoreductase enzymes to amperometric electrodes have been demonstrated [1].

Initial work exploited soluble mediators, such as ferricyanide [2] and benzoquinone [3] in solution, but these failed to provide commercially marketable instruments. This review sums up our studies of biocatalysis and bioelectrochemistry on both soluble mediators using glucose oxidase in the Yicheng company. These have proved particularly useful to date and are the basis of a highly successful commercial instrument, the SenTest^{TM} handheld glucose sensor.

2. FUNDAMENTAL RESEARCH ON A SINGLE CARBON ENZYME ELECTRODE

2.1 Chemically Modified Carbon Electrode

Because its porous, conductive properties and compatibility with biological materials, carbon is normally used as a base electrode in experimental bioelectrochemistry [4]. A graphite carbon rod is easily fixed in a suitable glass tube by using epoxy resin. Lengths of insulated wire are attached to the reverse end of the carbon rod with a silver-filled epoxy adhesive. The other opening of the tube can be used repeatedly by polishing it with fine aluminum oxide powder after each experiment. Figure 1 shows the simple construction of a carbon rod electrode. Benzoquinone [5–7] and ferricyanide [8,9] adsorbed on graphite to produce chemically modified electrodes are efficient mediators for electrochemical oxidation by an oxidoreductase. The electrons are shuttled to the electrode surface by the redoxidation of the bound mediators at the electrode.

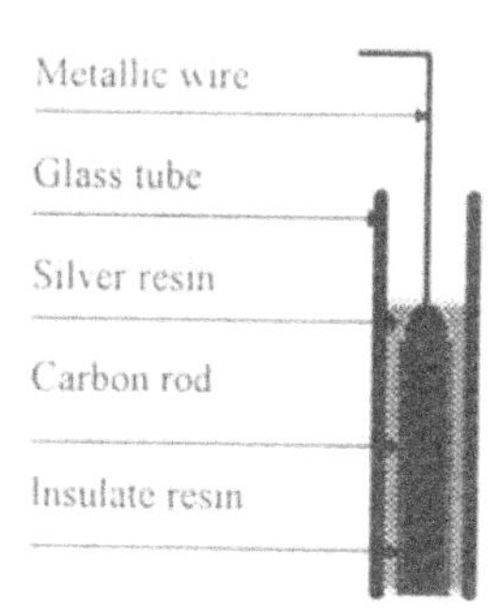

Figure 1. Schematic diagram of the simple type graphite carbod rod electrode showing the relationship of the parts.

In glucose determination, benzoquinone or ferricyanide adsorbed on a carbon rod are used to replace oxygen as the electron acceptors for glucose oxidase. Because the $E^{0'}$ of the mediator mainly dictates the potential at which electron transfer occurs, oxidoreductase enzymes can be oxidized with an applied potential by using a properly substituted mediator [10].

$$\text{Glucose} + \text{Quinone} + H_2O \rightarrow \text{Gluconic acid} + \text{Hydroquinone}$$

$$\text{Hydroquinone} \rightarrow \text{Quinone} + 2H + 2e \tag{1}$$

$$\text{Glucose} + \text{Ferricyanide} + H_2O \rightarrow \text{Gluconic acid} + \text{Ferrocyanide} + H_2O_2$$

$$H_2O_2 \rightarrow 2H + 2e \tag{2}$$

Initial cyclic voltammetric investigations of benzoquinone and ferricyanide adsorbed on a solid graphite rod electrode showed promising properties for electrocatalytically oxidizing glucose and glucose oxidase. In our research, the electrochemical experiments gave an $E^{0'}$ value of +160–250 mV versus SCE at pH 7.5 in contact with the graphite carbon rod surface. This indicated that both soluble mediators were adsorbed, compared with +10–30 mV when the electrode is bare. This provided a good basis for further research.

2.2 Glucose Oxidase, Its Structure, and Electron Mediation

Glucose oxidase is used to produce a large number of commercial biosensors [11]. The enzyme derived from different sources differs little in molecular weight and sedimentation coefficient [12]. There are very few detailed reports on the effect of sources on biosensor fabrication and production because mediation depends on molecular weight and also on the enzyme structure [13]. Thus, enzyme sources and structure have been considered in the research and applications of commercial glucose sensors.

Our knowledge of the glucose oxidase enzyme structure shows that there is a large gap between the enzyme active site and the surface of the electrode, which is difficult for electrons to cross. However, transfer of electrons between the active site and the electrode should be facilitated if a small electron acceptor is used as mediator [14].

Enzyme reoxidation is even achieved by using electron relays attached to the enzyme structure [13]. Information about the glucose oxidase enzyme structure also indicates that the enzyme active site is accessible through a deep pocket [15]. The pocket is funnel-shaped and has an opening of only 100 Å at the enzyme surface, but it is obviously sufficient to put the substrate molecule at the bottom. This also suggests that only low molecular weight mediators can move all the way down to the active site where they can exchange electrons [16]. Mediation of the enzyme is also affected by the pH. At low pH the enzyme active site becomes hydrophilic and rises closer to the surface of the enzyme, becoming more amenable to the entry of soluble mediators, such as ferricyanide [17] and/or benzoquinone. It is also suggested that FAD of glucose oxidase enzyme dissociates under these conditions and mediates the transfer of electrons from the active site to ferricyanide and similar electron acceptors.

We examined enzymes from different commercial sources. The experimental results with different enzymes from *Penicillium amagasakiense* (Yicheng Co., Ltd.) and *A. niger* (Sigma, Biozyme, and Toyobe) are shown in Figure 2.

The pure enzyme preparation without any stabilizer deposits on the carbon electrodes to give the best linear range for glucose. Enzymes that contain stabilizers showed poor linearity, even at high enzyme concentrations. It is possible that the results for the enzyme without any stabilizer confirm our ideas proposed previously and enzyme works very well with a low molecular weight mediator. In this case, electron transfer can be facilitated by the low molecular weight mediators ferricyanide and benzoquinone, even in high concentrations of substrate. The enzymes

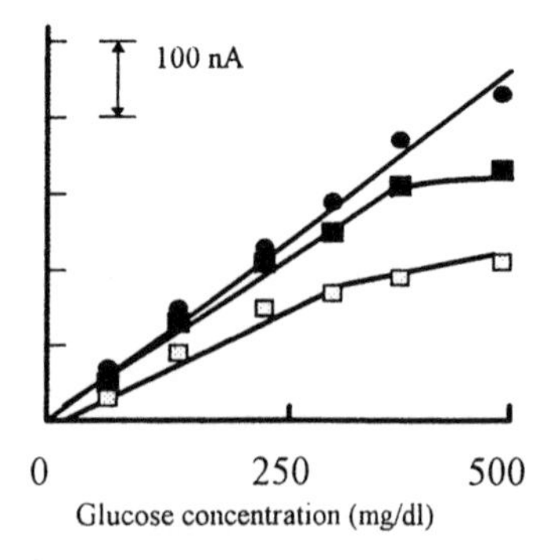

Figure 2. The effect of enzyme sources on the biosensors making ●---●, enzyme from *Penicillium amagasakiense*, Yicheng Co., Ltd; ■---■, enzyme from *A. niger*, Bio-zyme; □_□, enzyme from *A. niger*, Sigma.

that have stabilizers showed poor linearity. Arguably this limitation was caused by using a high molecular weight polysaccharide stabilizer. Such a stabilizer would attach closely to the enzyme's active site and would block the path of the mediator, therefore preventing it from reaching the pocket where the enzyme active site is located. Furthermore, it would inhibit the mediation of electron transfer. Separation of the stabilizers from the commercial enzymes failed in the laboratory because it was difficult to identify the stabilizer used.

3. MASS PRODUCTION OF BIOSENSORS

3.1 Screen Printing Technology

Research papers reported that thousands of biosensors have been produced and that most are handmade. Construction of biosensors by hand is appropriate to demonstrate a principle, but it is difficult to achieve the precision and quantity of production required for devices to be used outside centralized institutes. However screen printing can be used to mass produce inexpensive, reproducible, and sensitive disposable electrochemical sensors for determining trace levels of compounds imported into biological fluids. Figure 3 shows the construction of the screen printing stencil that was used to make the carbon electrodes. It consists of photographic positives of the desired electrode shapes placed directly on photosensitive emulsion layer, which is deposited on top of a 220-mesh monofilament polyester screen, pretensioned over a 250 mm × 250 mm aluminum frame.

Then, the emulsion is exposed to ultraviolet light to stabilize and fixes the emulsion permanently to the screen, thereby blocking the mesh pores. The areas

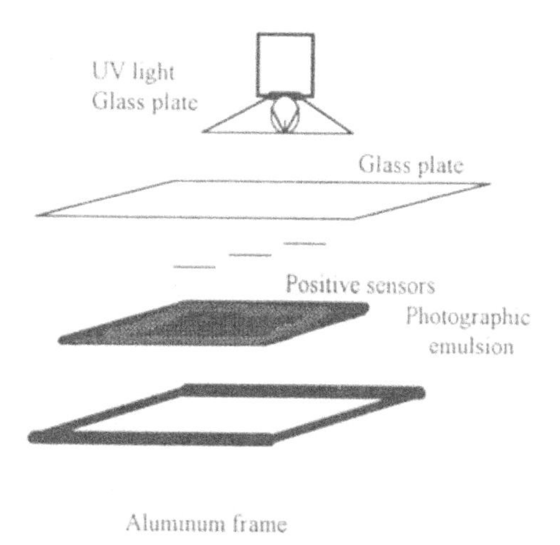

Figure 3. Design and components used for producing the screen stencil.

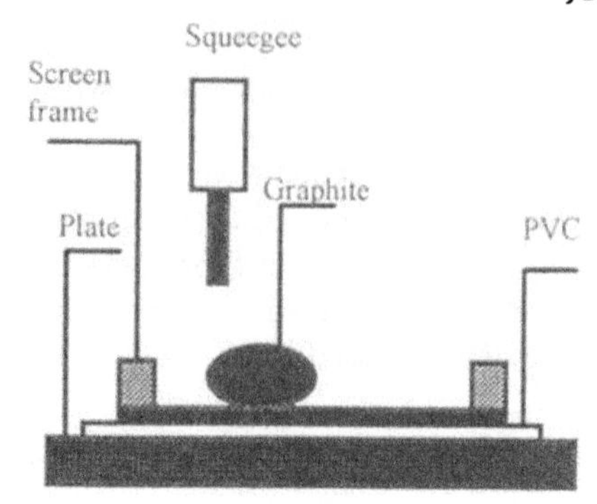

Figure 4. The screen printer and the screen printing process.

under the opaque sections of the positives remain unchanged and are subsequently washed out with a jet of water. Then, the stencil is force-dried at a temperature of around 40 °C.

3.1.1 Screen Printing Process

A TY30 printer (Tung Yuan Ind. Corp. Taiwan) was used to produce both working and reference electrodes. First, a sized piece of PVC support material was cleaned with solvent and attached to the solid base of the screen printing equipment. Graphite ink was applied to print the tracks. The screen had a mesh size of 220 counts per inch and an emulsion thickness of 20 μ.

The tracks were allowed to dry at a temperature of 70 °C for 1 h. The screen used to print the silver/silver chloride reference electrode had a mesh of 350 counts per inch and a emulsion thickness of 18 μ. Figure 4 is a schematic diagram of the screen printer and the screen printing process. Figure 5 presents a prototype of the screen printed electrode.

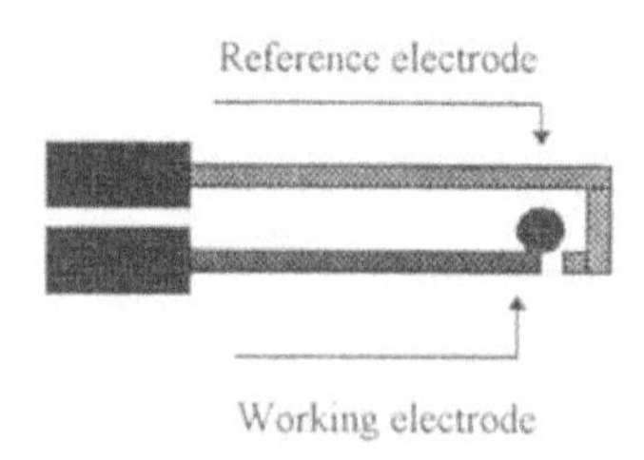

Figure 5. Schematic diagram of a screen printed glucose electrode.

3.2 Ink-Jet Printing Technology

Ink-jet printing has also been demonstrated as a manufacturing technique that facilitates fast, precise, and economical production of disposable glucose biosensors. Deposition of a biological reagent can be achieved with an ink-jet nozzle [18–20]. Such a nozzle allows placing fluid accurately at volumes as low as 1 nL on any surface at high speed. The ink-jet printer (Biodot) has been used to produce a large number of glucose sensors. Glucose oxidase and the soluble mediator were both printed on the surface of carbon electrodes using the same print pattern. Figure 6 shows the processing of ink-jet printed enzyme electrodes.

3.3 Marketing the Glucose Sensor

To test the practical application of the strip enzyme electrode, it was compared in a clinical study with the routine method for detecting diabetes. More than 500 volunteers who had diabetes were tested in both assays. Figure 7 shows the results of a clinical study comparing the COBAS MIRA Auto-Biochemical Instrument with the conventional method. The results demonstrate good agreement between the two methods. The following linear regression was calculated from a study of 153 patients: $y = 18.01 + 0.992x$, whose correlation coefficient is $\gamma = 0.9746$ [21].

Before commercializing the product, the long-term stability of the glucose sensor should be seriously considered for product shelf life and the activity of the enzyme electrode itself. Fortunately, glucose oxidase is a stable enzyme when it is in a dry base [11].

According to a survey by the Chinese Health and Medical Ministry in December 1996 [22], there are more then 20 million diabetics on the mainland and this figure will probably increase to 50 million by the end of the century. Because of this huge number of patients, various kinds of glucose meters are now imported into the local market. Most of them are optical devices that require complicated manual procedures. They are expensive, and there is concern about serious cross-contamination

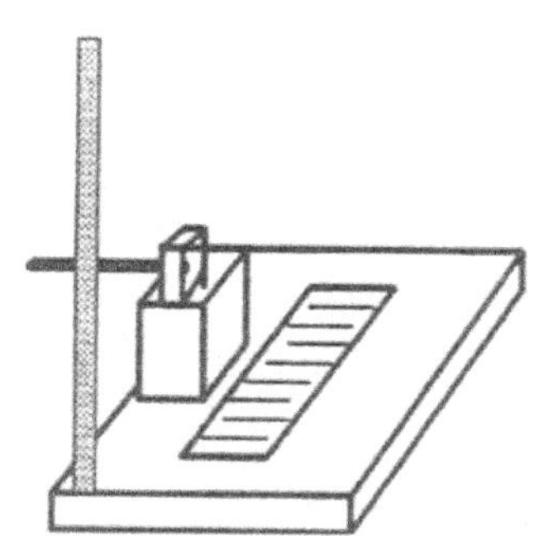

Figure 6. Schematic diagram of the ink-jet printer and the process of printing.

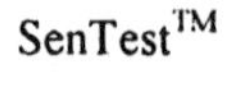

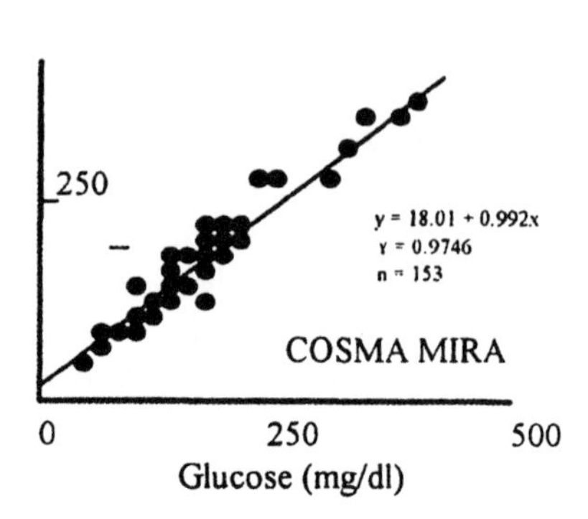

Figure 7. The correlation regression curves for SenTest™ with COBAS MIRA.

when such devices are used in hospital and clinics. The hepatitis virus is of particular concern for Chinese people. In response to these problems, SenTest™, the first electrochemical device for whole blood glucose self-monitoring, was launched in 1995.

Like other successful amperometric devices, SenTest™ has a large LC display and a one-button on/off switch. It is small, easy to manage, and importantly, it is competitively priced at US$ 50 for the meter and US$ 0.25 for the strip. This compares with more than $100 for an imported test meter and at least US$ 0.50 for an imported test strip. In 1996, the total sales of the SenTest™ were about US$ 150,000, and 200,000 strips were produced. This year it is expected that the total sales of the SenTest™ system will be US$ 500,000.

3.4 Future Research Program

There is a cautious but distinct move from centralized clinical testing strategies to local schemes, which is expected in several health-care areas during the next decade. In keeping with many of the emerging medical technologies, "drop-of-care" blood analysis promises to raise a number of contentious issues about the reliability of the test procedures.

The quality of the neonate, which has now reached several tens of million a year is very important for the future of the country. Therefore, discussions of neonatal health diagnosis and genetic disease screening programs were scheduled at a recent meeting on National Health Care held by the Chinese Health and Medical Ministry.

Central clinical testing in neonatology is well established in the Western world where reliable blood testing for metabolic diseases in newborn babies is used in early diagnosis and later in therapy. Most of these central laboratories are organized at a national level and are equipped with the most advanced automated instrumen-

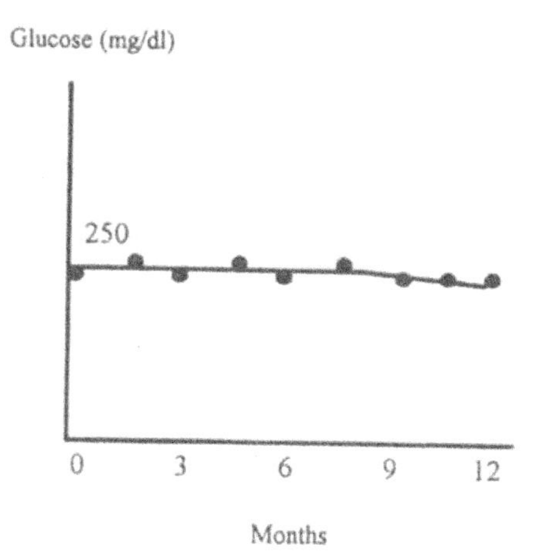

Figure 8. The long-term stability of a disposable strip enzyme electrode.

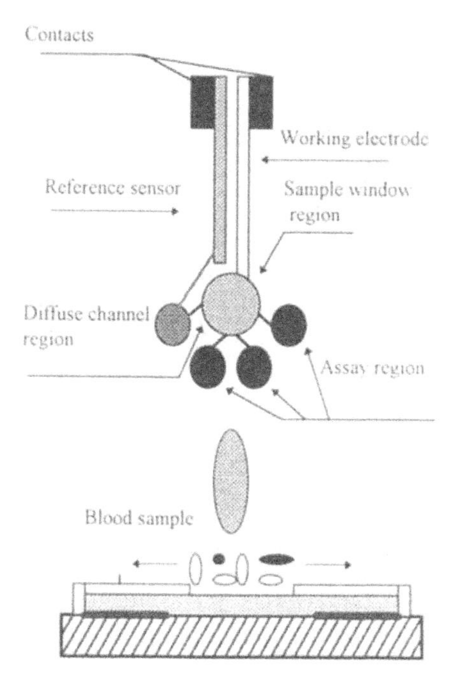

Figure 9. Membrane separator and amperometric sensor systems with enzyme hydrolysis, immunoreaction, capillary flow, plasma fraction, and electrochemical reaction.

tation and accredited tests for neonatal programs. Normally, ten tests are made routinely in screening programs worldwide, depending on the policies of the health-care systems in question. During the first few days after birth, several blood tests are performed on babies to achieve early diagnosis of genetic disorders and metabolic diseases. A sample of blood usually measuring 100 μL is taken from a heel puncture and applied to a paper disc. Then the paper is carefully air-dried for a day, labeled, packaged and sent to a hospital center accredited for the range of tests needed. Positive or negative results of the tests are made available three or four days after dispatching the samples.

In China, few hospitals are doing the tests, and those that do conduct single tests, such as galactose, phenylalanine and thyroid stimulating hormone compared with ten tests by their Western counterparts. A lack of instrumentation and trained technicians means that it is difficult to do such tests all over mainland China. However, if we can provide a simple method similar to the glucose test strip, a neonatal screening program will be possible in China.

We propose to design and fabricate a multianalyte electrochemical sensor, which can simultaneously test for phenylalanine, galactose, and thyroid stimulating hormone from a single drop of blood.

The research program is focused on developing a new methodology for an advanced testing system for newborn clinics. A combination of a research facility that has expertise in printed disposable sensor technology forms an excellent basis for future research and development of decentralized analyzers in this clinical field. The research group is looking to the study to provide fundamental data on the proposed assay system to give a valid assessment of the technology for clinical diagnostic testing. The research group's interests cover the scope of test chemistry design, advanced sensor fabrication, and instrument development, and the group is open to cooperate with biosensor researchers, both domestic and international.

4. CONCLUSIONS

It is now universally recognized that biosensors play an increasingly vital role in most rapid and simple analytical methodologies that are required by skilled analysts and technicians and also by untrained personnel, who want to use them in hospitals, doctors' clinics, and other workplaces in remote areas. This is a matter of great concern in China, especially for self-monitoring blood in illnesses, such as diabetes.

We have presented an overview of the development of mass produced disposable enzyme strips by a research group based in the MICROsense. Undoubtedly, the future development of increasingly selective, sensitive, multifunctional, and more stable sensors will present many challenges. It is hoped that the information given in this paper will provide an opportunity for researchers working in this important and exciting area to become familiar with and understand the issues that are being investigated by the group at MICROsense, Shanghai.

REFERENCES

[1] Cardosi, M.F., Turner, A.P.F. (1989). The realization of electron transfer from biological molecule to electrodes. In *Biosensors: Fundamentals and Applications*, Turner, A.P.F., Karube, I., Wilson, G.S., eds., Oxford University Press, Oxford, pp. 257–265.

[2] Racine, P., Engelhardt, R., Higelin, J.G., Mindt, W. (1975). An instrument for the rapid determination of L-lactate in biological fluids. *Med. Instrum.*, **9**, 11–16.

[3] Williams, D.L., Doig, A.R., Jr., Korosi, A. (1970). Electrochemical-enzymatic analysis of blood glucose and lactate. *Anal. Chem.*, **42**, 118–128.

[4] Inaiello, R.M., Yacynch, A.M. (1981). Immobilized enzyme chemically modified electrodes as an amperometric sensor. *Anal. Chem.*, **53**, 2090–2098.

[5] Ikeda, T., Hamada, H., Senda, M. (1986). Electrocatalytic oxidation of glucose at a glucose oxidase-immobilized benzoquinone-mixed carbon paste electrode. *Agric. Biol. Chem.*, **50**, 883–890.

[6] Hu, J. (1989). Benzoquinone modified glucose oxidase enzyme electrode. *Chin. J. Biotechnol.*, **6**, 328–331.

[7] Hu, J., Turner, A.P.F. (1991). An enzyme electrode for glucose consisting of glucose oxidase immobilized at a benzoquinone-modified carbon electrode. *Anal. Lett.*, **24**, 15–24.

[8] Richter, T., Rayner, M.H., Bilitewski, U. (1996). Application of screen-printed fructose and glucose sensors to food analysis. *4th World Congress on Biosensors*, May 29–31, Bangkok, Thailand. Elsevier Applied Science, Oxford, UK, p. 168.

[9] Selkirk, J.V., Turner, A.P.F., Saini, S. (1996). A hexacyanoferrate film L-lactate biosensor. *4th World Congress on Biosensors*, May 29–31, Bangkok, Thailand, Elsevier Applied Science, Oxford, UK, p. 198.

[10] Persson, B., Lan, H.L., Gonton, L., Kamoto, Y., Hale, P.D., Boguslavsky, L.I., Skotheim, T. (1993). Amperometric biosensors based on electrocatalytic regeneration of NAD^+ at redox polymer-modified electrodes. *Biosensors & Bioelectronics*, **8**, 81–88.

[11] Wilson, R., Turner, A.P.F. (1992). Glucose oxidase: An ideal enzyme. *Biosensors & Bioelectronics*, **7**, 165–185.

[12] Kusai, K. (1960). Crystalline glucose oxidase. *Annu. Rep. Sci. Works. Fac. Sci. Osaka Univ.* 43–74.

[13] Degani, Y., Heller, A. (1987). Direct electrical communication between chemical modified enzymes and metal electrodes *via* electron relays, bound covalently to the enzyme. *J. Phys. Chem.*, **91**, 1285–1280.

[14] Turner, A.P.F. (1988). Amperometric biosensors based on mediator modified electrodes. *Methods Enzymol.*, **137**, 90–105.

[15] Hecht, H.T., Kalisz, H.M., Hendle, J., Schimid, R.D., Schemburg, D. (1993). Crystal structure of glucose oxidase from *Aspergillus niger* refined at 2.3 Å resolution. *J. Mol. Biol.*, **229**, 151–172.

[16] Alvarez-Icaza, M., Kalisz, H.M., Hecht, H.J., Aumann, K.-D., Schomburg, D., Schmid, R.D. (1995). The design of enzyme sensors based on enzyme structure. *Biosensors & Bioelectronics*, **10**, 735–742.

[17] Kulys, J.J., Cenas, N.K. (1983). Oxidation of glucose oxidase from *Penicillium vitale* by one- and two- electron acceptors. *Biochim. Biophys. Acta*, **744**, 57–63.

[18] Kimura, J., Kawana, Y., Kuriyama, T. (1988). An immobilized enzyme membrane fabrication method using an ink-jet nozzle. *Biosensors*, **3/4**, 41–52.

[19] Newman, J.D., Turner, A.P.F. (1992). Ink-jet printing for the fabrication of amperometric glucose biosensors. *Anal. Chim. Acta*, **262**, 13–17.

[20] Hart, A.L., Turner, A.P.F., Hopcroft, D. (1996). On the use of screen- and ink-jet printing to produce amperometric enzyme electrodes for lactate. *Biosensors & Bioelectronics*, **11**, 263–270.

[21] Hu, J. (1996). Research and development of biosensors in China. *Sensors Mater.*, **8**, 477–484.

[22] Zhang, L. (1996). Medical reform program. *People's Daily*, Dec. 20, 1996.

INDEX

Printed and bound by CPI Group (UK) Ltd, Croydon, CR0 4YY
08/05/2025
01864830-0001